Materials Horizons: From Nature to Nanomaterials

Series Editor

Vijay Kumar Thakur, School of Aerospace, Transport and Manufacturing, Cranfield University, Cranfield, UK

Materials are an indispensable part of human civilization since the inception of life on earth. With the passage of time, innumerable new materials have been explored as well as developed and the search for new innovative materials continues briskly. Keeping in mind the immense perspectives of various classes of materials, this series aims at providing a comprehensive collection of works across the breadth of materials research at cutting-edge interface of materials science with physics, chemistry, biology and engineering.

This series covers a galaxy of materials ranging from natural materials to nanomaterials. Some of the topics include but not limited to: biological materials, biomimetic materials, ceramics, composites, coatings, functional materials, glasses, inorganic materials, inorganic-organic hybrids, metals, membranes, magnetic materials, manufacturing of materials, nanomaterials, organic materials and pigments to name a few. The series provides most timely and comprehensive information on advanced synthesis, processing, characterization, manufacturing and applications in a broad range of interdisciplinary fields in science, engineering and technology.

This series accepts both authored and edited works, including textbooks, monographs, reference works, and professional books. The books in this series will provide a deep insight into the state-of-art of Materials Horizons and serve students, academic, government and industrial scientists involved in all aspects of materials research.

Review Process

The proposal for each volume is reviewed by the following:

1. Responsible (in-house) editor
2. One external subject expert
3. One of the editorial board members.

The chapters in each volume are individually reviewed single blind by expert reviewers and the volume editor.

Satyam Suwas · David P. Field
Editors

Advances in Texture, Microtexture, and Allied Techniques

Measurements, Analyses, and Application

Springer

Editors
Satyam Suwas
Department of Materials Engineering
Indian Institute of Science
Bengaluru, Karnataka, India

David P. Field
Institute of Materials Research, School
of Mechanical and Materials Engineering
Washington State University
Pullman, WA, USA

ISSN 2524-5384 ISSN 2524-5392 (electronic)
Materials Horizons: From Nature to Nanomaterials
ISBN 978-981-96-5345-4 ISBN 978-981-96-5346-1 (eBook)
https://doi.org/10.1007/978-981-96-5346-1

This Springer imprint is published by the registered company Springer Nature Singapore Pte Ltd.
The registered company address is: 152 Beach Road, #21-01/04 Gateway East, Singapore 189721, Singapore

Preface

The intricate relationship between microstructure and material properties forms the source of advancements in materials science and engineering. Microstructure is true sense is a combination of geometrical morphology and crystallographic texture. For polycrystalline materials, understanding and analyzing bulk texture and local texture, also known as microtexture, is essential to unravel the mechanisms that govern their performance and durability. This reference text provides a comprehensive and state-of-the-art exploration of the techniques required to measure and analyze local texture or microtexture in polycrystalline materials.

A distinctive feature of this work is its focus on microstructural evolution under various conditions. Detailed observations are provided into the transformations that occur during processes such as deformation, annealing, fatigue, and corrosion. The book discusses both in-situ and ex-situ observation strategies, offering a holistic view of how microstructures adapt and evolve in response to environmental and operational stresses. While metals remain the primary focus, the text also extends its coverage to include geological materials, emphasizing the universality of the discussed techniques and concepts.

This work is a result of collective efforts by leading academicians and researchers from across the globe, with significant contributions from Asia, North America, and Europe. Their expertise ensures that each chapter is meticulously detailed, highly informative, and practically oriented. This collaborative effort makes the book an invaluable resource for professionals, researchers, and students aspiring to gain an understanding of microtexture analysis and its applications.

Beyond the technical depth, this book underscores the pivotal role of microtexture analysis in advancing materials science and engineering. It not only elaborates on the measurement and quantification of microtexture but also highlights its applications in optimizing material properties. Furthermore, it explores the use of microtexture as a forensic tool to decode histories of deformation and annealing processes, providing unique insights into the behavior and life cycles of materials.

This book is organized into three distinct parts, each addressing a key area of study to provide a comprehensive understanding of crystallographic texture and its applications in materials science. Part I *Introduction* comprises five chapters, starting with Chap. 1, which offers a historical overview and modern advancements in the study of crystallographic texture. This chapter traces the evolution of the field, highlighting foundational discoveries and contemporary breakthroughs that have shaped current knowledge. Chapter 2 lays the groundwork for understanding Electron Backscatter Diffraction (EBSD). It delves into the principles, operational requirements, and critical factors influencing EBSD performance, equipping the knowledge necessary for crystallographic and microstructural analyses in materials research. In Chap. 3, the focus shifts to strain analysis using EBSD within the Scanning Electron Microscope (SEM). The chapter also explores the integration of EBSD with complementary techniques, presenting a holistic approach to strain analysis. Chapter 4 showcases the capabilities of in-situ EBSD studies, emphasizing their contributions to the Integrated Computational Materials Engineering (ICME) framework. Through case studies, it demonstrates how this technique aids in unraveling complex microstructural phenomena. The part concludes with Chap. 5, which introduces Electron Channeling Contrast Imaging (ECCI). This technique, based on Scanning Electron Microscopy (SEM), is highlighted as a powerful tool for analyzing crystalline defects in bulk samples, eliminating the need for the thin foils required by Transmission Electron Microscopy (TEM).

Part II *Application of Advances in Electron Backscattered Diffraction* includes four chapters. Chapter 6 underscores the importance of micro and mesotextures in understanding microstructural development. It highlights the integration of these textures into computational models for simulating microstructural evolution within the Grain Boundary Engineering (GBE) framework. Chapter 7 explores the application of EBSD for determining grain boundary characteristics, with a particular focus on twinning-induced GBE in stainless steels and nickel-based alloys, illustrating practical implications for material design. In Chap. 8, the reconstruction of the parent phase following displacive phase transformations is discussed. This chapter explains reconstruction algorithms, addresses challenges, and provides resources to facilitate such studies. Chapter 9 examines the formation of Microtextured Regions (MTRs) in two-phase ($\alpha + \beta$) titanium alloys. The mechanisms driving these formations and their impact on fatigue performance are critically analyzed.

Part III *Application of EBSD to Unravel the Critical Problems in Materials Investigation* contains four chapters. Chapter 10 investigates microstructural and crystallographic texture developments during deformation and recrystallization in Duplex Stainless Steels (DSSs) and Dual Phase Steels (DPSs), detailing their distinct features. Chapter 11 surveys the role of microtexture in cyclic plastic deformation across metallic materials. It also explores the uses of EBSD and ECCI to study the evolution of microstructure, microtexture, and dislocation structures in 304LN austenitic stainless steel under cyclic deformation. Chapter 12 provides a detailed review of how microtexture and grain boundary character distribution (GBCD) influence the electrochemical degradation behavior of materials. This chapter highlights their roles in corrosion resistance, offering insights into material design and performance enhancement in corrosive environments. Finally, Chap. 13 explores the application of EBSD in analyzing the kinematics of tectonically deformed rocks. It details advancements in identifying kinematic reference frames and preparing geological samples for high-quality analysis.

Bengaluru, India Satyam Suwas
Pullman, USA David P. Field

Contents

Editors and Contributors

About the Editors

Satyam Suwas is a professor and chair of Materials Engineering at Indian Institute of Science, Bengaluru, India. He obtained his Ph.D. in Materials and Metallurgical Engineering from IIT Kanpur and worked at Defence Metallurgical Research Laboratory, Hyderabad; University of Lorraine, France and RWTH Aachen, Germany before joining Indian Institute of Science. His specialization includes materials processing, crystallographic texture, and mechanical behavior of materials. He is a Humboldt fellow and has also been conferred with the Friedrich Wilhelm Bessel Research Award by the Humboldt foundation, Germany. He serves as editorial board member of many journals, and is the associate editor for Materials Characterization, and Current Science.

David P. Field is a professor in the School of Mechanical and Materials Engineering, and the director of the Institute of Materials Research at Washington State University. He received his Ph.D. in Mechanical Engineering in 1991 from Yale University. From 1990 to 2000 he was employed as a Sr. engineer at Alcoa Technical Center and then as the director of Technology at TexSEM Laboratories. He joined Washington State University in 2000. He has served in many editorial capacities and is currently the Editor-in-Chief of Materials Characterization. His research interest focusses upon the statistical description of microstructures with emphasis on polycrystalline materials including metals, ceramics, and electronic materials.

Contributors

Qin Bai School of Materials Science and Engineering, Shanghai University, Shanghai, PR China;
Key Laboratory for Microstructures, Shanghai University, Shanghai, PR China

Darothi Bairagi Department of Metallurgical and Materials Engineering, Indian Institute of Technology Kharagpur, Kharagpur, India

Pallabi Bhuyan Department of Metallurgical and Materials Engineering, Indian Institute of Technology Kharagpur, Kharagpur, India

Pravash Chandra Chakraborti Metallurgical and Material Engineering Department, Jadavpur University, Kolkata, India

Martin A. Crimp Chemical Engineering and Materials Science, Michigan State University, East Lansing, MI, USA

David P. Field Institute of Materials Research, School of Mechanical and Materials Engineering, Washington State University, Pullman, WA, USA

Sushil Kumar Giri R&D, Tata Steel Limited, Jamshedpur, India

Saumya Gupta Materials Science Centre, Indian Institute of Technology, Kharagpur, India

N. P. Gurao Department of Materials Science and Engineering, Indian Institute of Technology Kanpur, Kanpur, India

R. Jeyaraam NAND Technology Integration, Micron Semiconductor Asia Pte Ltd, Singapore, Singapore

Josh Kacher School of Materials Science and Engineering, Georgia Institute of Technology, Atlanta, USA

Rajesh K. Khatirkar Department of Metallurgical and Materials Engineering, Visvesvaraya National Institute of Technology (VNIT), Nagpur, Maharashtra, India

B. Sravan Kumar Department of Metallurgical Engineering and Materials Science, IIT Bombay, Mumbai, India

Amrita Kundu Metallurgical and Material Engineering Department, Jadavpur University, Kolkata, India

Arjun Mahato School of Nano Science and Technology, Indian Institute of Technology, Kharagpur, India

Manish A. Mamtani Department of Geology and Geophysics, Indian Institute of Technology, Kharagpur, India

Sumantra Mandal Department of Metallurgical and Materials Engineering, Indian Institute of Technology Kharagpur, Kharagpur, India

Rahul Rakshit Department of Metallurgical and Materials Engineering, Indian Institute of Technology Kharagpur, Kharagpur, India

Ranjit Kumar Ray AICTE-INAE Distinguished Visiting Professor, SRMIST Chennai and CMRIT, Bengaluru, India

Shibayan Roy Materials Science Center, Indian Institute of Technology, Kharagpur, India

Tim Ruggles Sandia National Laboratories, Albuquerque, USA

Souvik Sahoo Materials Science Centre, Indian Institute of Technology, Kharagpur, India

I. Samajdar Department of Metallurgical Engineering and Materials Science, IIT Bombay, Mumbai, India

V. Subramanya Sarma Department of Metallurgical and Materials Engineering, Indian Institute of Technology Madras, Chennai, India

Subhasis Sinha Department of Metallurgical Engineering, Indian Institute of Technology (BHU) Varanasi, Varanasi, India

Reshma Sonkusare Institute of Materials and Process Design, Helmholtz Zentrum Hereon, Geesthacht, Germany

Satyam Suwas Department of Materials Engineering, Indian Institute of Science, Bangalore, India

Srikanth Vedantam Department of Engineering Design, Indian Institute of Technology Madras, Chennai, India

Stuart I. Wright EDAX, Ametek Materials Analysis Division, Mahwah, NJ, USA;
Gatan, Pleasanton, USA

Shuang Xia School of Materials Science and Engineering, Shanghai University, Shanghai, PR China;
State Key Laboratory of Materials for Advanced Nuclear Energy, Shanghai University, Shanghai, PR China

Part I
Introduction

Chapter 1
Crystallographic Texture—Historical Perspective and Modern Developments

Ranjit Kumar Ray and Satyam Suwas

1 Introduction

The term "crystallographic texture" or simply "texture" is used to denote the statistical distribution of crystal (grain) orientations in a polycrystalline material, which may be metallic, ceramic, polymeric, or a bio-material. Although the subject of crystallography or science of crystals evolved over a period of last few centuries, the term "texture" was first used by d'Halloy and d'Omalius [1] to describe the directional characteristics in polycrystalline rocks, as a result of alignments of minerals therein. The primary concern of using texture as a characterization tool by the geologists and earth scientists originated from the idea that it could help understanding the deformation mechanisms in rock minerals present along the faults in the earth's crust. The textural information on these rocks, which deform by continuous aseismic creep and micro-seismicity, could contribute to the understanding of the micro-mechanisms of their deformation, leading to a better prediction of future earthquakes in the region. Later, mathematicians and physicists came into the picture, and they contributed immensely to develop the theoretical basis of texture measurements. In the twentieth century, mainly due to the advent of X-ray diffraction, texture research gained momentum among the materials scientists, who now are the biggest beneficiaries of this characterization technique.

A polycrystalline material usually comprises many grains (or single crystals), separated from one another by distinct boundaries. If it so happens that, in a piece of material, the crystallographic orientation of each grain is different from those of all other grains, then the material is termed "random", possessing isotropic properties.

R. K. Ray (✉)
AICTE-INAE Distinguished Visiting Professor, SRMIST Chennai and CMRIT, Bengaluru, India
e-mail: rkray@iitk.ac.in

S. Suwas
Department of Materials Engineering, Indian Institute of Science, Bangalore, India

S. Suwas and D. P. Field (eds.), *Advances in Texture, Microtexture, and Allied Techniques*, Materials Horizons: From Nature to Nanomaterials,
https://doi.org/10.1007/978-981-96-5346-1_1

By contrast, if one or more than one orientation(s) is preferred by a reasonably large number of the grains, then the material is termed "textured", having anisotropic properties. There is hardly any material under the sun, which is not textured, with varying degrees of sharpness. During processing, all materials assume textures which are unique for the applied processing method and are also characteristic of the material itself. As already defined, texture of a material refers to the statistical distribution of the grain orientations, irrespective of the spatial location of any grain or grain aggregate. Therefore, this definition of texture corresponds to the average texture of the material, or the "macrotexture" (or global texture). Sometimes it is necessary to measure texture from a much smaller population of grains in a polycrystalline material, which involves measurement of the individual grains in the ensemble, in a highly localized region. The resultant texture pertains to only the grains in the chosen ensemble, and therefore the spatial locations of the grains can be clearly specified. This texture is known as "microtexture" (or local texture) from a grain aggregate, and as such may not be representative of the global texture of the material.

The mainstream techniques for both "macrotexture" and "microtexture" measurement are indicated in Fig. 1. While X-ray diffraction is the most popular method for macrotexture measurement, it is the Scanning Electron Microscope (SEM), coupled with the Electron Backscattered Diffraction (EBSD) attachment, which is generally used for microtexture measurement. In the former method, a sheet sample of the material to be investigated is placed on the specimen holder for diffraction to occur from a specific {h k l} group of planes in the sample. Keeping the diffraction conditions fixed, the sample is given a number of rotations, in such a way as to allow as many grains or crystals as possible, in the irradiated volume of the sample, to contribute to the total diffracted radiation. The intensity of this diffracted radiation, which is proportional to the density of the {h k l} group of planes, is plotted in the form of a diagram, known as a "pole figure". This figure essentially represents a two-dimensional distribution of poles or perpendiculars (normals) to the {h k l} planes in the different grains of the sample. In fact, it is nothing but a stereographic projection of the {h k l} poles, on which the sample parameters are superimposed. For example, for a sheet material, the relevant sample parameters will be the Rolling Direction (RD), the Transverse Direction (TD), and the Normal Direction (ND). For having a better resolution, the X-ray data used for drawing pole figures can be re-plotted in a three-dimensional imaginary "orientation space", defined by the three Euler angles ϕ_1, ϕ, and ϕ_2. From the plots of a series of two-dimensional sections of the pole density data, given in the three-dimensional orientation space, which are also known as ODFs (orientation distribution functions), it becomes much easier to read out the specific texture components, possessed by the sample.

It will be pertinent at this point to ask the question why the study of texture of materials is at all necessary. Suffice it to say that crystallographic texture or preferred orientation of grains in a polycrystalline material affects all material properties, be it physical, mechanical, chemical, electrical, magnetic, and what not. For illustration purpose, the variations of Young's modulus E and that of the Lankford parameter r with texture have been illustrated in Figs. 2 [2] and 3 [2]. Many more such examples could be given here, however, limitation on space will not allow that. The dependence

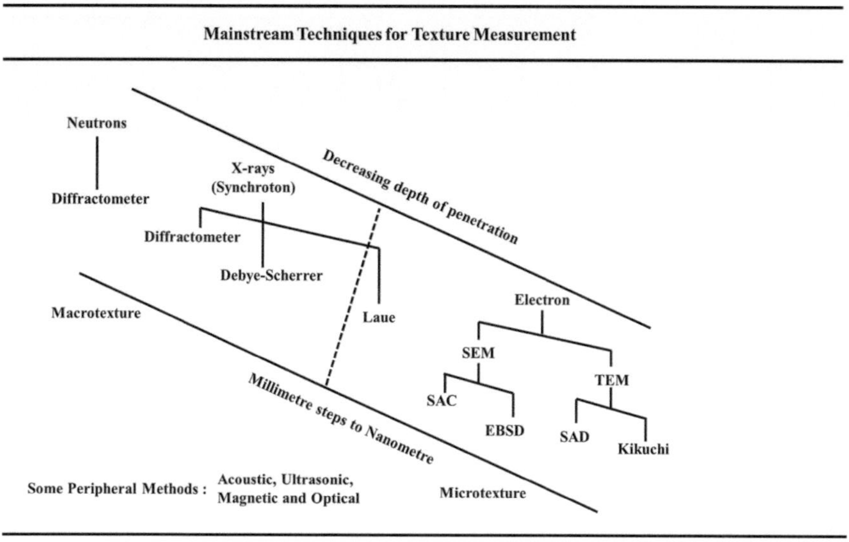

Fig. 1 Texture measurement techniques

of properties on texture has paved the way for tailoring the texture of a material to derive the desired properties in a way very similar to the control of microstructure of a material to impart specific properties.

2 Historical Perspective

The historical development of the science of crystallographic texture, like any other science, involves three main stages, such as (1) experimentation and gathering of empirical knowledge; (2) providing plausible explanation, theorizing, and modeling; and (3) deciding future research strategy. The whole exercise comprises two major activities, such as (1) systematic recording of the results of texture research and their skilled interpretation and (2) continuous development of relevant experimental and analytical techniques. The major developments in texture research in material science right from the beginning are shown in a tabular form in Fig. 4. In this description, an attempt has been made to put the research highlights in a chronological order, as far as practicable. The reader will obviously find the absence of many notable researchers' names in this table. However, this has been done simply to economize space, and, in no way, reflects on the merit or importance of their contribution in the field. The discerning reader will observe that, surprisingly, all these research milestones refer to a particular problem or "puzzle" in material science, a complete solution of which has been eluding the researchers for well over six decades. This puzzle relates to the formation of the "copper"-type and "brass"-type rolling textures in *fcc* metals

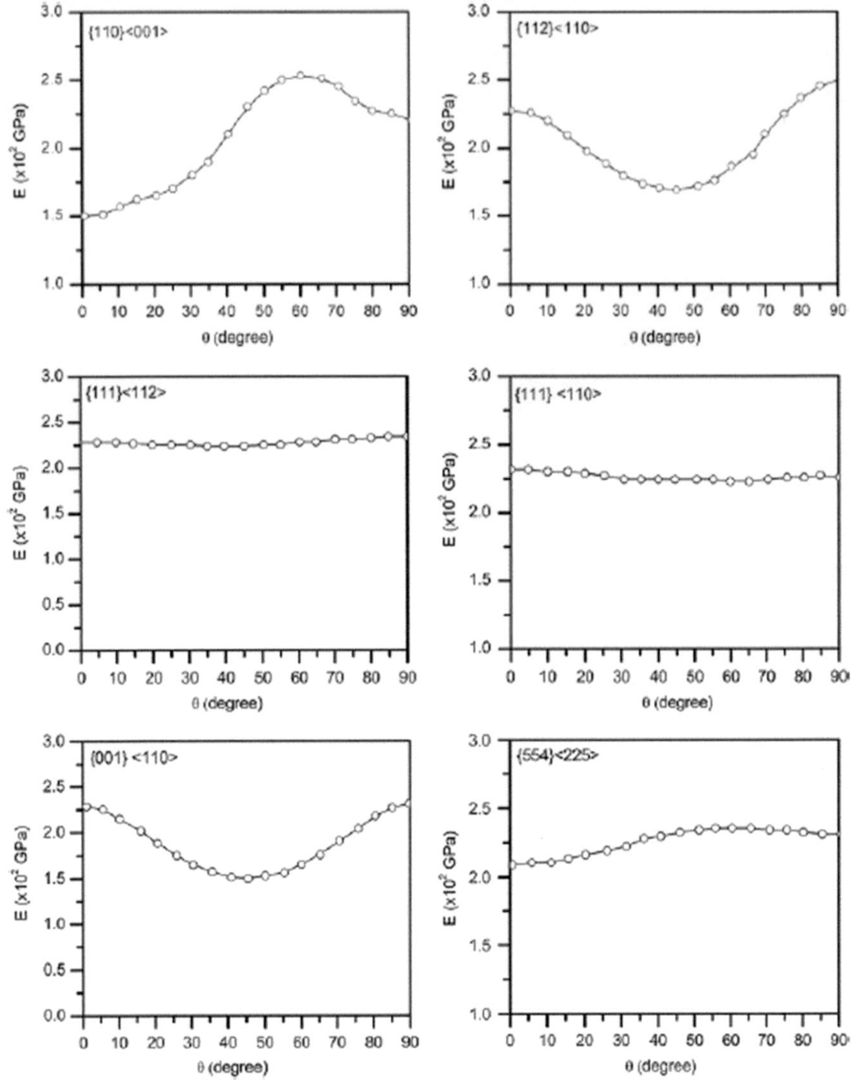

Fig. 2 Variation of E versus θ (angle from the rolling direction) for various orientations [2]

and alloys. It is in the fitness of things that this puzzle and various attempts to solve it from time to time are mentioned here, in some detail, to give a proper historical perspective. This will also illustrate how researchers throughout the world have come together to understand a very important problem in texture research, thereby making enormous contribution in furthering its domain. It is rather interesting to note that even many decades after the first investigation of the above phenomenon, the interest of the texture community has not waned, and every now and then new publications

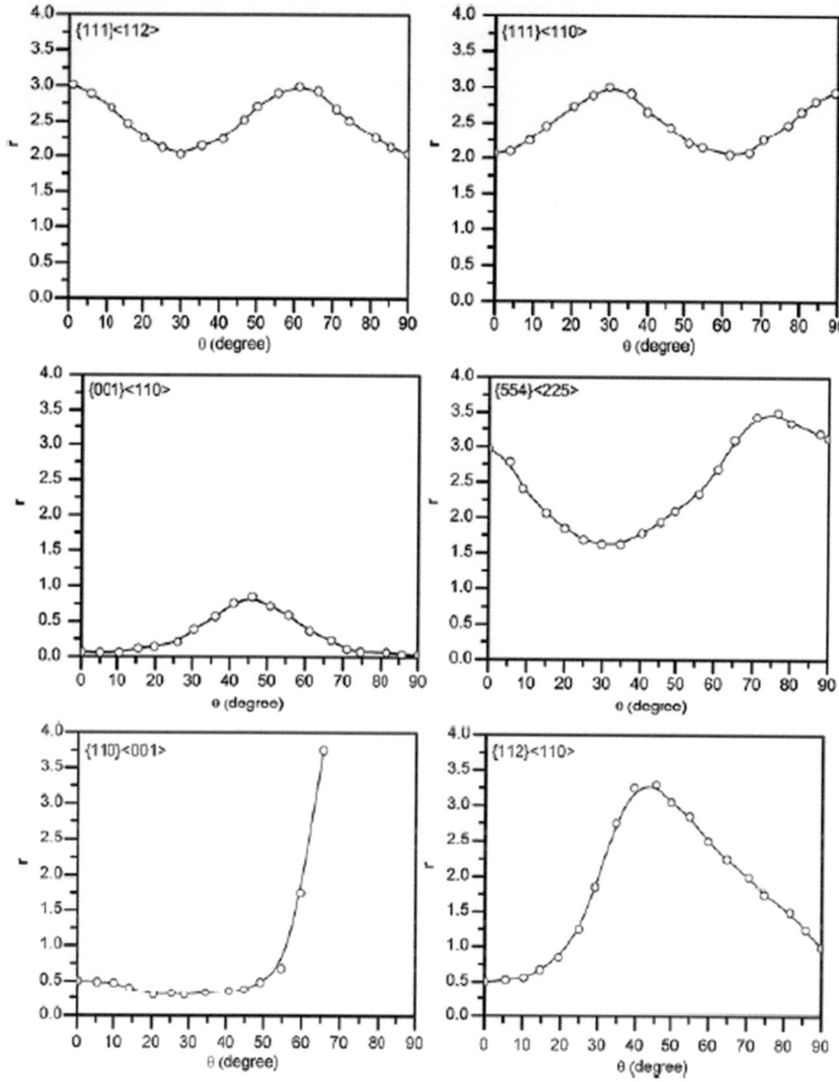

Fig. 3 Variation of r versus θ (angle from the rolling direction) for various orientations [3]

on the subject keep on appearing in learned journals. Thus, in a sense, this topic is very much alive even today.

Research Highlights

Hu et al. (1952) - Qualitative description of Cu- & Bs-type textures

Wassermann (1963) - Twinning theory for texture transition

Smallman & Green (1964)

Dillamore & Roberts (1964) - Cross slip phenomenon

Hu et al. (1966) - Slip by partials

Leffers & Grum-Jensen (1968) - Countered twinning theory based on volume effect

Duggan et al. (1978) - Twins Overshooting {111} || ND components

Hirsch et al. (1988) - Cu-Zn alloys - Bs-type → Twinning → Shear banding

Leffers et al. (1990) - Modified Sachs' model - single slip phenomenon

Ball & Gottstein (1993) - Ni_3Al - Strong Bs - Shear banding

G. Chowdhury, Bhattacharya and Ray (1998, 2000) Ni_3Al (B), Ni_3Al (B, Zr) → Strong Bs → Reordering & Twinning

Engler (2000) - Cu-Mn alloys - Bs-texture -Short range ordering

Fig. 4 Research highlights

3 Copper" and "Brass" Deformation Textures in FCC Materials

It has been generally observed that the rolling textures in heavily deformed Cu and Al are quite different from the rolling textures of Cu–Zn, Cu-Sn, Ag, and a few other metallic alloys, although all the above materials belong to the *fcc* system. Figure 5 a–d shows possibly the earliest pole figures from 95% cold rolled pure Cu and Cu-30% Zn alloy (α-brass), taken by Hu and his co-workers [3], which qualitatively depict the differences between the two types of deformation texture, namely, the "copper" type (for pure Cu) and the "brass" type (for the Cu–Zn alloy). A quantitative description of the differences between these two types of *fcc* deformation textures was later provided by Hirsch and Lucke [4] from the ODFs of the two materials, under similar rolling conditions (Fig. 6a, b). In general, qualitatively, the "copper"-type texture is primarily represented by the ideal orientation, $\{112\} < 111 >$ (also called the "Cu" component), while the "brass"-type texture is characterized by the ideal orientation $\{110\} < 112 >$ (also called the "Bs component"). In each of the above cases, however, the entire texture is made up of the four major components, Cu, Bs, S ($\{123\} < 634 >$), and G or Goss ($\{110\} < 001 >$). The relative proportions of these texture components are different for the "copper"- and the "brass"-type textures. It is generally believed that material characteristics and rolling conditions decide the deformation textures of fcc materials. It was in the 1960s that the empirical condition that the "copper"-type texture is favored by high stacking fault energy (SFE) and that the "brass"-type texture is favored by low SFE was indicated by Haessner [5, 6]. Hu et al. [7–9] suggested that the "copper" type is promoted by high rolling temperature, whereas "brass" type is promoted by low rolling temperature. It was Leffers [10] who suggested that low strain rate favors the formation of "copper"-type texture, while high strain rate promotes the formation of "brass"-type texture.

The above empirical relationships and experimental findings led to the advancement of a number of theories right from the 1960s to explain the differences between the "copper"-type and the "brass"-type textures, the so-called "rolling texture transition". The first and foremost among these theories was the one given by Wassermann [11], who suggested that the "brass"-type texture is formed due to the formation of deformation twinning, while the "copper"-type texture arises due to normal slip, without any twinning taking place. Smallman and Green [12] and Dillamore and

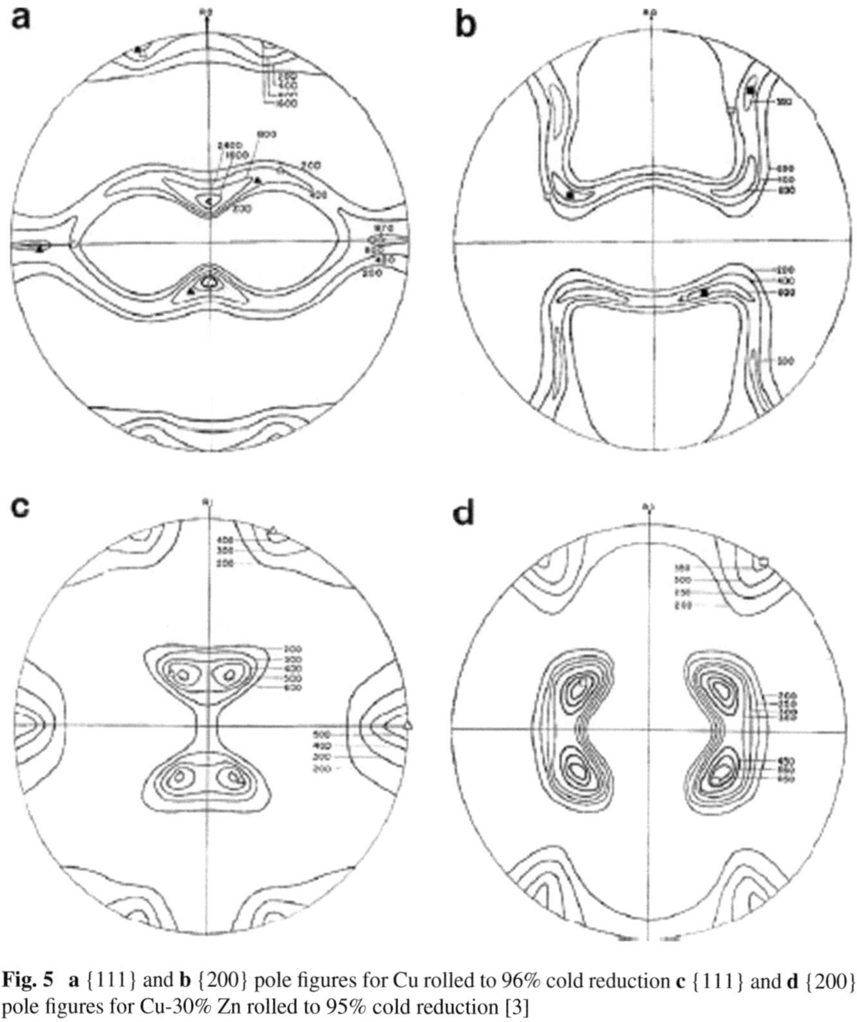

Fig. 5 **a** {111} and **b** {200} pole figures for Cu rolled to 96% cold reduction **c** {111} and **d** {200} pole figures for Cu-30% Zn rolled to 95% cold reduction [3]

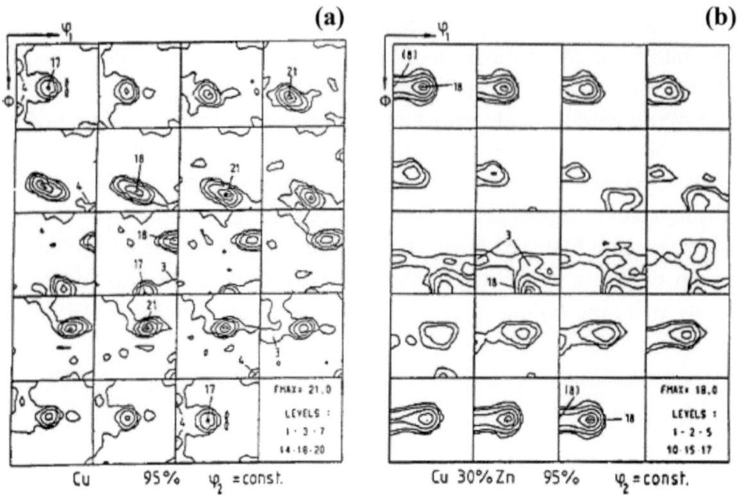

Fig. 6 **a** ODF representation for texture of Cu, 95% cold rolled, **b** ODF representation for texture of Cu-30% Zn, 95% cold rolled [4]

Roberts [13] suggested that "copper"-type texture formation involves extensive cross-slip, while "brass"-type texture forms when cross-slip frequency is low or zero. Hu [14] opined that "copper"-type texture forms when slip occurs by perfect dislocations only, whereas slip by both partial and perfect dislocations is necessary for the development of "brass"-type texture. There were other theories also, a reference to which can be found in ref. [15]. All the above theories are based on mostly pole figure data, without any detailed microstructural investigation. Further, all these theories refer to the final rolling textures only, without any reference to the intermediate textures that form during the various processing steps.

Out of all these theories, the idea of Wassermann caught the imagination of the texture community; however, at a later stage, from microstructural observations, Leffers and Grum-Jensen [16] found that the volume fraction of twins implied in Wassermann's theory is far greater than what is observed experimentally. Later work, involving ODFs and detailed microstructural evaluation, has thrown much new light on the subject. A rather widely accepted scheme for "brass" texture formation at that time was given by Duggan et al. [17], and this was based on detailed textural and microstructural observations. According to this scheme, the texture development, in both "copper" type and "brass" type, is rather similar till about 50% deformation, and caused by octahedral slip only. Beyond this level of deformation, in case of the "brass"-type texture, profuse twinning occurs, notably in the {112} < 111 > oriented grains. Since further slip, on slip planes, other than those parallel to the twin boundaries becomes difficult, on further deformation, an overshooting phenomenon and twin plane rotation toward the rolling plane occur, due to which orientations of the type {111} < uvw > develop. Within the range of deformation reduction 60–95%, shear bands appear to play a major role, since the twin structure inhibits homogeneous deformation. Formation of newer shear bands continue till the twinned volume is fully eliminated, at about 95% rolling reduction. Beyond this level of reduction, deformation again occurs by the resumption of homogeneous slip, leading to the development of the characteristic "brass" texture.

Although, in the beginning, it was believed that the development of both the "copper"-type and the "brass"-type textures is rather identical till about 50% rolling reduction [17–19], citing several earlier references, Leffers [20] claimed to show that a distinct difference exists in the two textures, at substantially lower reductions. A detailed study of deformation textures of a series of *fcc* Ni-Co alloys, with varying amounts of Co, after 95% cold rolling reduction, was carried out by Ray [21], using both pole figures and ODFs. The results show that while pure Ni and Ni–Co alloys with up to 30% Co possess the "copper"-type and the Ni–60% Co alloy the "brass"-type deformation textures, the Ni–40% Co alloy has a texture which can best be described as somewhat intermediate between the two types. Carstensen et al. [22] chose these very alloy compositions and recorded systematically, at low strain intervals, their textures up to 35% rolling reduction, using an EBSD, coupled with an SEM. They could observe a very clear difference in texture development between the Ni–30% Co and the Ni–60% Co alloys. While, in the former, the density of the "Cu" orientation was found to increase linearly with strain, with the "Bs" orientation density remaining at a constant low level, the reverse was found to be true in the latter alloy. These observed early differences between the two types of textures give credence to the earlier ideas of Leffers [20].

More recently, Madhavan et al. [23], using extensive Orientation Imaging Microscopy (OIM™) and ODF measurement, made a very detailed investigation on the above Ni–Co series of alloys. They observe that pure Ni and Ni–20 wt% Co, which have high SFE, develop a characteristic "copper"-type texture, while in the medium SFE Ni–40 wt% Co the texture is "copper" type up to 95% reduction, with the texture shifting toward the "brass" type after 98% cold reduction. Microstructural investigations suggest that this texture transition has been caused by Cu-type shear bands, forming preferentially in the "Cu"-oriented $\{112\} < 111 >$ grains. Although deformation twins are found to form after 70% cold reduction, their volume fraction is low and confined to limited number of grains only. Tendency of twin formation diminishes completely at large strain and extensive shear banding takes over beyond 90% reduction. In case of the low SFE Ni–60 wt% Co alloy, the texture is "brass" type from the early stages of rolling. Deformation twins form in this alloy very early, after about 10% reduction. Extensive twinning occurs after 40% straining, suggesting that twinning is the major deformation mode in the initial stages. Beyond this level of reduction, twinning almost ceases and extensive Bs-type shear banding, preferentially found confined to $< 111 > \parallel$ND grains, takes over as the primary mode of deformation. Therefore, the final rolling texture in this alloy appears to form by the joint action of twinning and shear banding. These results clearly support the earlier ideas of Leffers [20] and Leffers and Ray [15] that the "brass"-type texture deviates from the "copper"-type texture right from the very early stages of development. In addition, these results also modify and supplement Wassermann's [11] hypothesis of twinning as solely responsible for the texture transition in *fcc* materials. The very important role of shear bands, in addition to twinning, has been clearly highlighted in the above results.

A twist to the above results has been provided by the findings of a comprehensive study by Engler et al. [24] and Engler [25] on the formation of rolling texture in a series of *fcc* Cu-Mn alloys. Steffens et al. [26] showed that unlike many other elements, the addition of Mn up to 11.6 at. %, in pure Cu, does not lead to any significant change in the SFE. Despite that, the texture results clearly indicate a transition from the "copper"-type to the "brass"-type texture for the 90% cold rolled materials, when the Mn content of the alloys exceeds 8 at. %. This texture transition has been attributed to the increase in yield strength and formation of short-range ordering in the alloys, which lead to shear banding. In a further twist, in a series of 95% cold rolled Ni–Fe alloys with up to 40 wt.% Fe, the rolling texture is found to be always of the "copper" type, for all compositions [27], which is expected since addition of Fe does not cause any significant change in the SFE of the alloys [28]. Curiously enough, the addition of Fe to Ni also leads to extensive shear banding, caused by short-range ordering [27], as in the Cu–Mn alloys; however, the final texture results are totally different in these two *fcc* systems.

A literal "texture transition", implying a change from "copper" type to "brass" type, during rolling sequence, has been observed in case of a Ni–40% Co alloy, as mentioned above [23]. A similar phenomenon has been reported [29] in case of the *fcc* boron-doped intermetallic Ni_{76} Al_{24} alloy, where the rolling texture is of "copper" type till 65% reduction, and this changes into "brass" type after 85% rolling reduction. This change-over has been attributed as due to a transformation from the $L1_2$ to DO_{22} structure [30]. In an earlier work, Ball and Gottstein [31], however, suggested that the "brass" texture in boron-doped Ni_3 Al is due to shear banding.

Based on the results, enumerated above, it is now generally accepted that: (1) the "brass"-type texture deviates from the "copper" type from the very early stages of development and (2) the "brass"-type texture is accompanied by the formation of deformation twins in the beginning, however, this mode of deformation changes into shear band formation in the very high strain regime. The agreement seems to stop here, and several relevant questions have started bothering the texture community. For example, how to reconcile the texture formation in Cu–Mn alloys vis-à-vis the fundamental premise of the role of SFE in the development of deformation textures in *fcc* materials, and how to explain the phenomenon of texture transition as a function of the severity of rolling deformation. These are precisely the reasons why the "copper" to "brass" texture transition has been a puzzle for the metallurgical community to solve for the last more than six decades, and although substantial progress in the understanding of this age-old problem has been made over the years, the entire phenomenon, including the details, has not been clearly understood yet.

During the course of many studies, as enumerated above, it was increasingly realized that the advances made on the subject, over the years, have been very much dependent on two important factors: (1) continuous and rapid development in the experimental and analytical techniques of texture measurement with time and (2) liberal use of characterization methods, other than texture, such as microstructure measurement, which helped immensely in arriving at a much better appreciation of the problem and its solution.

4 Development of the Experimental and Analytical Techniques

A glimpse of the general experimental techniques of texture measurement that are available today has already been provided in Fig. 1. To give a historical perspective, the timelines of the various developments, over the years, in both macrotexture and microtexture evaluation are illustrated in Figs. 7 and 8, respectively. As mentioned earlier, while X-ray diffraction has been the most popular method for macrotexture measurement, right from the beginning, the SEM-EBSD method is now universally used for microtexture determination.

Before proceeding further, a few words must be said about the relative effectiveness of the different macrotexture measurement techniques. Due to very low penetration depth (only a few μm) of X-rays in most materials, texture measurement by X-rays relates practically to the surfaces of materials. Due to the "defocusing effect" and because X-ray diffraction from other than very thin samples, in the transmission mode, is almost impossible, complete pole figures cannot be obtained from routine

Pole figures and ODFs	
1924:	**Wever (Berlin) generated 1st pole figure from evaluating the inhomogeneous intensity distribution along the Debye-Scherrer rings.**
1948:	**Introduction of texture goniometer and Geiger counters (Decker et al., Norton).**
1949:	**Schulz initiated modern quantitative X-ray texture analysis.**
Since 1960s:	**Pole figure determination by neutron diffraction.**
1966: R.J. Roe-	**Introduced the Generalised Spherical Harmonic methods for ODF plots.**
1968: R.O. Williams-	**Introduced the Iterative Least Square method to produce biaxial pole figures**
1969: H.J. Bunge-	**Independently developed The Generalised Spherical Harmonic methods for ODF generation.**
Today,	**Automated computer-controlled systems are available and the corresponding measuring techniques and the necessary correction methods are well established.**

Fig. 7 Timeline of macrotexture measurement (pole figure and ODF) techniques

EBSD Development

1. **1928: Observation of Kikuchi lines in TEM by Kikuchi.**
2. **1954: Back reflected Kikuchi lines obtained in TEM by Alam**
3. **1969 – 1979: 3 Diffraction technique in SEM**
    - **i. SACP: 10 μm in practical resolution [Joy et al. Oxford Univ.]**
    - **ii. Kossel diffraction: 20 μm practical resolution [Dingley et al. at Bristol Univ.]**
    - **iii. EBSD: Less than 1 μm practical resolution [Venables et al. at Sussex Univ.]**
4. **1982 – 1984: Computer assisted indexing of EBSD, subtraction of background value**
5. **1990: Fully automated indexing of EBSD patterns by Hough transformation [Yale, Clausthal, Risφ]**
6. **2000: Very Fast Automated Pattern Analysis**
    - **i. High resolution EBSD**
    - **ii. In-situ Experiment EBSD**
    - **iii. High current EBSD**

Fig. 8 Timeline for microtexture measurement (EBSD) techniques

XRD measurements of texture. The very small beam size of X-ray synchrotron radiation, however, makes it better suited to mapping of orientations on the grain scale. As a result, even though the resolution is much higher, such measurements are far from statistically representative. Compared to the above two radiations, the main advantage of the neutron diffraction method is the low absorption of neutron by most elements. Thus, measurements can be carried out in the transmission mode, even from massive specimens, and complete pole figures can be obtained from a single experiment. Sample volumes to the tune of a cm^3 can be used and, therefore, the measured data are statistically more representative, with significantly low error margins. Despite these advantages, by virtue of the fact that it requires a nuclear reactor or a reliable thermal neutron source, this technique is not so commonly used.

Coming to the microtexture measurement techniques, the SEM-EBSD method, which is now universally used, effectively links microstructure and orientation together, resulting in what may be termed as orientation mapping or orientation microscopy or more commonly as Orientation Imaging Microscopy (OIM™). OIM™ presents a complete quantification (and visualization) of the orientational aspects of microstructure, which gives a spatial connotation of crystallographic orientation vis-à-vis microstructure. Although some orientation mapping techniques, using a Transmission Electron Microscope (TEM), are also available, these are hardly used, keeping in view certain limitations with the TEM approach.

The methods of analysis of macrotexture data (see Fig. 7), obtained by using the X-ray diffraction texture measurement technique, developed over a period of the last more than hundred years. Historically speaking, the first pole figure was generated as far back as in 1924 by Wever [32] in Berlin, from evaluation of the inhomogeneous intensity distribution along Debye–Scherrer rings in X-ray powder photographs. Decker et al., in 1948, introduced the first texture goniometer that helped in the fast acquisition of texture data. In 1949, Schulz [33] initiated the modern quantitative texture analysis technique, using the X-ray method. Pole figure determination by the neutron diffraction method started in the 1960s. During the period 1966–1969, the mathematical basis of ODF determination and the related analytical techniques were independently introduced by Williams, Roe and Bunge [34–36]. Due to rapid scientific developments, advanced highly automated computer-controlled data acquisition systems and corresponding measuring techniques are now very much well established.

Considering the microtexture analysis techniques (see Fig. 8), it was Kikuchi [37] who first observed in 1928, the "Kikuchi lines" from thick single crystals during diffraction in a TEM. The Kikuchi line pattern, produced by diffraction from a particular grain, gives an idea about the exact orientation of that grain. In fact, the SEM-EBSD method is essentially based on collecting all the Kikuchi line patterns from the grains in a selected area of a specimen, from which the texture can be determined. In 1954, Alam et al. [38] reported observation of "back-reflected Kikuchi lines" in a TEM. In the 10 years, spanning between 1969 and 1979, three different SEM-related techniques, namely, Selected Area Channeling Pattern (SACP), Kossel diffraction, and Electron Backscattered Diffraction (EBSD) were developed. The first computer-assisted indexing of EBSD started between 1982 and 1984, and fully automated indexing of EBSD patterns using Hough transformation was introduced in 1990. From the year 2000, very fast fully automated modern pattern analysis systems have become available.

Side by side with the modern developments in the analytical techniques, a lot of effort was directed toward activities related to theoretical texture modeling, starting from 1928. As a result of concerted efforts by many researchers, several satisfactory models are now available to explain the development of texture during deformation of materials, although models for simulating the phase transformation and recrystallization processes are still not that successful. A comprehensive list of the proposed models is given in Fig. 9.

5 Modern Developments

Out of the numerous experimental methods, the SEM-EBSD has become the most popular, for obvious and simple reasons. The earliest method for texture measurement, the X-ray diffraction (XRD) technique, provides a statistical knowledge of the total orientational aspects of many grains in a polycrystalline sample, without giving any clue about which grain in the sample has what orientation. This lack of connectivity between the spatial location of a grain and its orientation is the biggest lacuna of the measured bulk texture of a material sample, as obtained by XRD. The

Texture Modeling

- ❑ **First attempt: German Physicists Sachs, Cox and Sopwith (1928), based on single slip**
- ❑ **G.I. Taylor, English physicist, mid-20th Century ; first to provide a quantitative explanation of texture development.**
- ❑ **Later many modifications were proposed to improve the predictions from Taylor's model, the notable ones are due to**
  - ❑ **Bishop and Hill**
  - ❑ **Chin and Mammel**
  - ❑ **Van Houtte – Lamel and Alamel models**
  - ❑ **Gottstein – GIA model**
  - ❑ **Kalidindi and many others – FEM based models**
- ❑ **An altogether different approach was parallely developed to model texture based on Eshelby's inclusion model. The main contributors of this type of modeling are**
  - ❑ **Alain Molinari and Gills Canova**
  - ❑ **Carlos Tome and Ricardo Lebensohn**
  - ❑ **Laszlo Toth and many others**
  - ❑ **Jonas and Co-workers**
- ❑ **Modeling activities are pursued in India at IITs (Bombay, Madras and Kanpur) and IISc**
- ❑ **Recrystallization texture modeling: A few attempts (Szpunar, Gottstein, Raabe, …….)**
- ❑ **Transformation texture modeling: Primarily by Humbert and co-workers, a few others also are involved.**

Fig. 9 Efforts in texture modeling

SEM-EBSD technique has bridged this knowledge gap. It is precisely for this reason that there has been an enormous amount of work to develop a host of analytical methods based on the SEM-EBSD system, resulting in several innovations. These developments have yielded a complete knowledge about the microstructure of materials, encompassing almost every aspect, starting from the minute internal details of grains and grain boundaries, and leading to information about crystallography and topological parameters. This exercise provides knowledge about "orientation connectivity", which means the way in which grains of multiple orientations are connected in a three-dimensional sample.

6 Analytical Tools Associated with EBSD

Figure 10 [39] shows a schematic representation of a SEM-EBSD orientation imaging system. The schematics of the EBSD measurement set-up is displayed in Fig. 11. Modern EBSD packages evaluate the acquired diffraction patterns from crystals automatically and extremely fast. The sampling points on the specimen surface are selected either manually or by automatically moving the electron beam or the specimen stage. The goodness of fit between an experimental diffraction pattern and its indexed solution is often expressed as the "Confidence Index" (CI), which is a numerical assessment of the "mean angular deviation" between the above two. Once a population of orientations, from a selected area, has been indexed from the EBSD patterns, the information is generally provided, by all standard commercially available EBSD packages, in the form of displays, depicting details regarding microstructure, texture, grain boundaries, and internal strain. In fact, the EBSD datasets allow the researchers to construct several maps, each map depicting some specific kind of information. Some of the most important such maps are as follows.

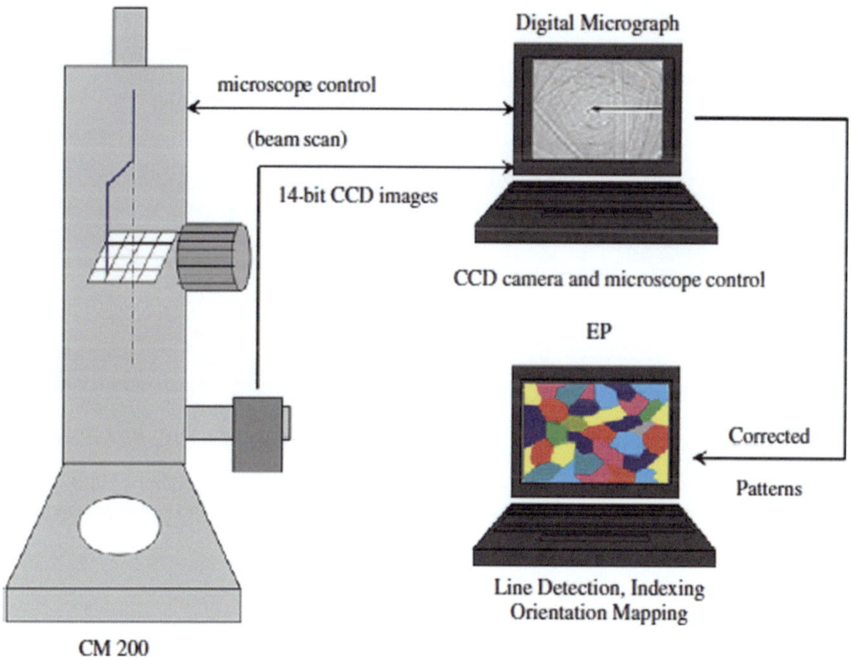

Fig. 10 Schematic representation of a SEM-EBSD orientation imaging system [39]

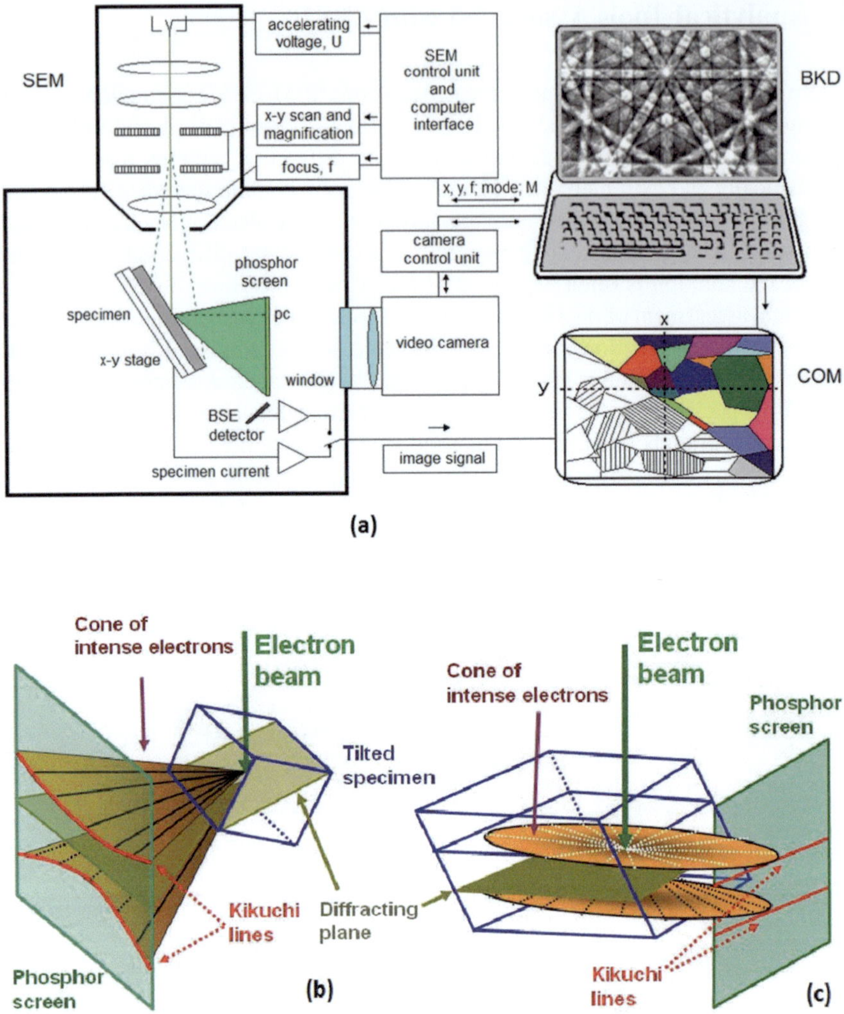

Fig. 11 **a** Schematic of EBSD measurement set-up (courtesy, http://ebsd.info/basics.html); **b** and **c** schematic of Kikuchi pattern formation (courtesy, https://www.globalsino.com/EM/page2342.html)

6.1 Image Quality" (IQ) Maps

IQ is a measure of the quality of an EBSD pattern. An IQ map is constructed by mapping the IQ values measured from all the diffraction patterns in the area of interest within the sample. In a way, IQ maps provide a useful visualization of a microstructure. Generally, several different metrics are used for quantifying image quality. For example, the conventional IQ metric provides superior grain boundary and strain contrast, while the IQ metric based on the average intensity of the diffraction pattern gives better topological and phase contrast.

6.2 *Inverse Pole Figure" (IPF) Maps*

IPF maps depict the crystallographic texture of the sample being investigated. This map is essentially a projection of the sample coordinates (such as RD, TD, and ND, for a rectangular parallelepiped sample) on the crystal coordinate system (such as 100, 010, and 001, for the cubic system). Thus, IPF representation is just the reverse of that of Pole Figure (PF), which is a stereographic projection of crystal orientations on the specimen coordinate system.

6.3 *Grain Boundary Character Distribution" (GBCD) Maps*

GBCD maps display the relative proportions of different types of grain boundaries, such as High-Angle Grain Boundary (HAGB), Low-Angle Grain Boundary (LAGB), and Coincidence Site Lattice (CSL) boundaries.

6.4 *Kernel Average Misorientation" (KAM) Maps*

KAM is a measure of local misorientation within a grain. It quantifies the average misorientation around a point of interest, with respect to a pre-decided set of nearest neighbor points. Mathematically, KAM can be expressed as

$$\mathrm{KAM}_{i,j} = \frac{1}{|N(i,j)|} \sum_{(k,l)\in N(i,j)} \omega\left(o_{i,j}, o_{k,l}\right)$$

where $|N(i,j)|$ stands for the total number of neighboring pixels; $\omega(o_{ij}, o_{kl})$, the angle of misorientation between the orientation o_{ij} at the center; and o_{kl}, the neighboring orientation. It is obvious that for the computation of KAM, the choice of the set $N(i,j)$ of the neighboring pixels is of utmost importance. Usually, the value of KAM is higher in deformed grains, due to the higher dislocation density. Thus, KAM measurements provide a method to distinguish between deformed and recrystallized grains. Further, local lattice distortion, density of geometrically necessary dislocations (GND), and an estimation of the stored strain energy, all these parameters can be calculated from KAM analysis.

6.5 Grain Orientation Spread" (GOS) Maps

GOS for a grain is the mean value of misorientations between all pixels in the grain of interest and the average orientation of the grain. GOS for a set of grains is the mean value of GOS values of all the grains in the set. The value of GOS is quite sensitive to the physical state of the material. Lattice distortion of grains due to deformation results in a high value of GOS, while strain-free recrystallized grains have low GOS values. The GOS values, therefore, provide a method to distinguish recrystallized grains from unrecrystallized ones, the threshold value of GOS for effecting this separation, usually varies between 1 and 2° in many materials. Mathematically, GOS can be expressed as

$$GOS(i) = \frac{1}{J(i)} \sum_{j} \omega_{ij}$$

where *GOS (i)* is the GOS of grain i, *J(i)* is the number of pixels in grain i, and ω_{ij} is the misorientation angle between the orientation of pixel j and the mean orientation of grain i.

6.6 "Grain Reference Orientation Deviation" (GROD) Maps

GROD maps display the orientation heterogeneities that arise inside any grain, due to deformation. In this method of measurement, each pixel within a grain is given one color based on the misorientation of the point, relative to a reference orientation for the grain in question, to which the pixel belongs. Thus, these maps essentially show the fragmentation, suffered by a grain, in terms of lattice rotation, which is necessary for maintaining compatibility with respect to the neighboring grains. There are several variations in GROD, which is based on the selection criteria for the reference orientation, such as

(a) $GROD_{CI}$: to denote orientation of point in grain with the highest CI;
(b) $GROD_{FIT}$: to denote orientation of point in grain with the smallest fit;
(c) $GROD_{MAX}$: to denote orientation of point in grain producing the maximum GROD;
(d) $GROD_{MIN}$: to denote orientation of point in grain producing the minimum GROD;
(e) $GROD_{IQ}$: to denote orientation of point in grain with the highest IQ;
(f) $GROD_{XY}$: to denote orientation of point in grain nearest to the centroid;
(g) $GROD_{KAM}$: to denote orientation of point in grain producing the minimum KAM;
(h) $GROD_{AVG}$: to denote average orientation of all points in a grain.

In addition to the above, the following maps may also be constructed from the EBSD datasets, such as the "phase maps" to display the distribution of different phases in the sample, the "grain size maps" to provide the average grain size, as well as the grain size distribution, and the "Taylor and Schmid factor maps" to give the yield strength and plastic strain at grain level, etc. All these different methods will be discussed in detail in the subsequent chapters.

7 Recent Developments

Over a period of the last two decades, there emerged a new area of materials research, known as "Grain Boundary Engineering" (GBE). The object of GBE is to achieve better properties by altering and manipulating the crystallography of grain boundaries, by using suitable processing regimes. EBSD and orientation mapping have played very important roles here, by allowing the routine acquisition of statistically significant data on grain misorientations. Grain boundaries are essential surfaces between two adjacent grains, the complete elucidation of the crystallographic orientations of which needs five independent variables. The traditional EBSD technique is concerned with collecting orientation and misorientation data from a polished surface of a material, where a grain boundary appears as a line trace between two grains. The crystallographic direction of this boundary trace, along with the three misorientation parameters between the neighboring grains, provides four out of the five essential parameters to characterize the grain boundary completely. The determination of this last parameter requires a strategy of sectioning of the three-dimensional material, parallel to the surface of observation, and then collecting the EBSD data from all those sections, followed by combining the information elicited from the different sectioned surfaces. This is the basis of three-dimensional or 3-D EBSD. The 3-D EBSD is still not an essential part of most of the commercial EBSD packages, and plenty of new developments are taking place in this technique. In addition to grain boundary surfaces, the 3-D EBSD method can be used to study any other surfaces also, such as microcracks, phase boundary surfaces, and fracture surfaces or facets. It may be argued that the TEM may be another tool for examination of surfaces from a series of parallel thin foils made from a three-dimensional material. However, the major obstacle to such an exercise is the enormous amount of difficulty that could be faced during specimen preparation. Thus, for all practical purposes, TEM-based techniques can be precluded for detailed surface examination, and the SEM-EBSD has emerged as by far the most convenient and useful technique for the crystallographic analysis of surfaces. Moreover, the advantages of an SEM in imaging capabilities, such as large depth of field, stereo-imaging, and simple specimen tilt facilities, have made the SEM-based EBSD as a formidable tool for such an exercise.

The successful application of the above technique depends on the use of a robust and stable sectioning method. It may be possible to utilize advanced mechanical sectioning techniques to expose new surfaces at different depths, along the thickness direction of the sample, followed by EBSD analyses on those surfaces. The other

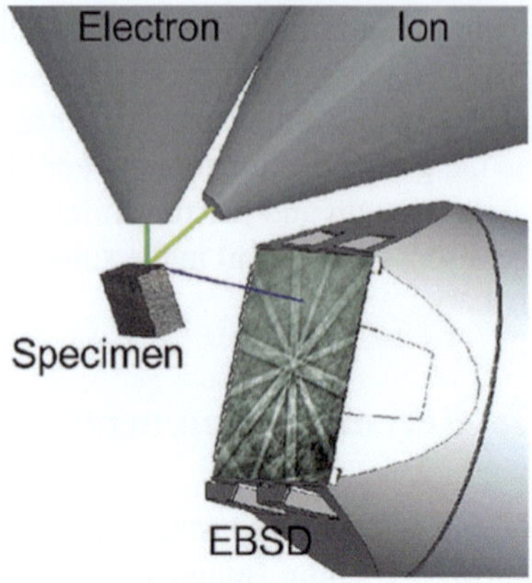

Fig. 12 Set-up geometry of the electron and ion beam in a FIB-SEM-EBSD system (courtesy, Oxford Instruments: https://www.ebsd.com/ebsd-techniques/3d-ebsd)

more advanced approach will be to use a carefully controlled ion beam milling facility for the sectioning purpose. There are automated versions available for both the techniques. For very precision work, it may not be practical to take the specimen out of the SEM chamber, section it outside, and then reposition it for further collection of new data. A solution to this has been achieved by combining the SEM-EBSD system with a Focused Ion Beam (FIB). Here, the ion beam is used to mill away the surface in situ, after collection of EBSD data from one surface. The whole process is then continued to compile data from all the relevant sections, followed by combining them. The entire process has been automated nowadays. A schematic of this "dual-beam" technique is presented in Fig. 12 [courtesy, Oxford Instruments]. Recent developments in FIB technology have been in the use of plasma FIB and the incorporation of femtosecond lasers into the FIB-SEM system. This exercise has speeded up the milling process considerably. Side by side, the use of CMOS (complementary metal oxide semiconductor)-based EBSD detectors, in place of the older CCD (charge coupled device)-based ones, has led to much higher speed acquisition of EBSD data.

8 The Way Forward and Conclusions

The question that now comes naturally to the minds of every member of the texture community is: "what next"? "Quo vadis" texture? or where is texture going from here? From the days when texture research began in a very modest way, we have now come to a stage when considerable progress has been made in understanding the origin of crystallographic texture, how processing leads to the formation of different types of texture in materials, and how texture affects all material properties, be it mechanical, chemical, electrical, and the like. Side by side, enormous advances have been made in equipment dedicated to texture measurement and texture research, the

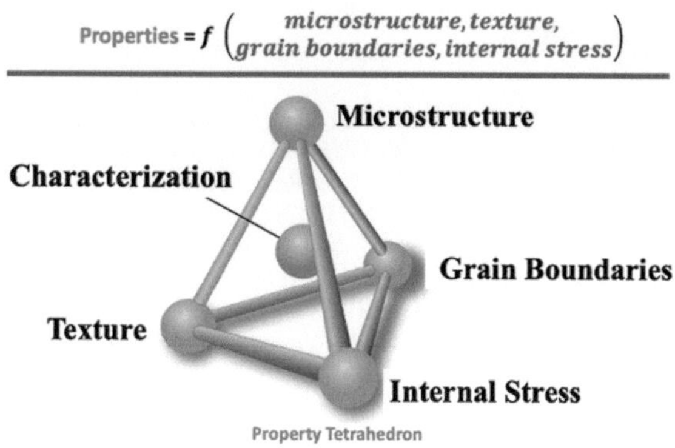

Fig. 13 The property tetrahedron

latest being the FIB-SEM-EBSD system. As has been explained above, the FIB-SEM-EBSD system is a unique and versatile one, using which a researcher can evaluate and measure all the four characterization parameters for any material, namely, the microstructure, texture, grain boundary configuration, as well as internal energy. Over and above this, this equipment can also be used for dynamic in situ measurements, for example, when the sample is mechanically tested or is kept in a special environment, within the specimen chamber of the SEM. Thus, the FIB-SEM-EBSD is possibly the ultimate equipment which the materials scientist could wish for, and which came primarily out of the quest and effort of the texture community, over the years, with the aim of arriving at a holistic understanding of the texture of materials. Such an equipment, therefore, is bound to enormously benefit the materials scientists and engineers, engaged in the task of enhancing and modifying the existing properties of conventional materials, as well as in developing better and newer materials possessing novel and unique properties. It is no wonder, therefore, that the scope and usage of such a "complete" characterization system is fast enlarging and spreading over diverse materials, ranging from metals and alloys, metal-matrix composites, nanocomposites, and high entropy alloys to such diverse areas, such as recovery and recrystallization (including dynamic recrystallization), welding and joining, and additive manufacturing. Future activities in texture research will primarily center around further improvement in the FIB-SEM-EBSD system, specifically in the areas of data collection and analysis, surface and interface characterization methods, strain assessment capability, and, finally, further improvement in the three-dimensional EBSD technique. All these will result in much faster acquisition of more reliable and more accurate data, paving the way for better understanding and prediction of material properties, based on the characterization of the four parameters, microstructure, texture, grain boundaries, and internal strain, as shown pictorially, by way of the "Property Tetrahedron", in Fig. 13.

References

1. D'Omalius JJ, D'Halloy JJ (1833) Introduction a la geologie. Levrault, Paris
2. Ray RK, Jonas JJ, Hook RE (1994) Cold rolling and annealing textures in low carbon and extra low carbon steels. Inter Mater Rev 39:129–172
3. Hu H, Sperry PR, Beck PA (1952) Rolling textures in face-centered cubic metals. J Met 4:76–81
4. Hirsch J, Lucke K (1988) Mechanism of deformation and development of rolling textures in polycrystalline f.c.c. metals— I. Description of rolling texture development in homogeneous CuZn alloys. Acta Metall 36:2863–2882
5. Haessner F (1963) Habilitation dissertation. Technical University, Stuttgart
6. Haessner F (1963) The conditions for the occurrence of the copper-and brass-type texture in rolled face-centered cubic metals. Z Metallkd 54:79–85
7. Hu H, Cline RS, Goodmann SR (1961) Texture transition in high-purity silver and its correlation with stacking fault frequency. J Appl Phys 32:1392–1399
8. Hu H, Cline RS (1961) Temperature dependence of rolling textures in high-purity silver. J Appl Phys 32:760–763
9. Hu H, Goodmann SR (1963) Texture transition in copper. Trans Met Soc AIME 227:627–639
10. Leffers T (1968) Deformation rate dependence of rolling texture in brass containing 5% zinc. Scr Metall 2:447–452
11. Wassermann G (1963) The effect of mechanical twinning, on the formation of rolling textures in face-centered cubic metals. Z Metallkd 54:61–65
12. Smallman RE, Green D (1964) The dependence of rolling texture on stacking fault energy. Acta Metall 12:145–154
13. Dillamore IL, Roberts WT (1964) Rolling textures in f.c.c. and b.c.c. metals. Acta Metall 12:281–293
14. Hu H, Cline RS, Goodmann SR (1966) Recrystallization, grain growth and texture. ASM 295–367
15. Leffers T, Ray RK (2009) The brass-type texture and its deviation from the copper-type texture. Prog Mater Sci 54:351–396
16. Leffers T, Jensen AG (1968) Development of rolling texture in copper and brass. Trans Met Soc AIME 242:314–319
17. Duggan BJ, Hatherly M, Hutchinson WB, Wakefield PT (1978) Deformation structures and textures in cold-rolled 70:30 brass. Metal Sci 12:343–351
18. Kallend JS, Davies GJ (1972) The development of texture in copper and copper-zinc alloys. Texture 1:51–69
19. Hirsch J (1984) Ph. D. Dissertation. RWTH Aachen, Germany
20. Leffers T (1996) Proc. ICOTOM 11. International Academic Publishers, Beijing, pp 299–306
21. Ray RK (1995) Rolling textures of pure nickel, nickel-iron and nickel-cobalt alloys. Acta Metall Mater 43:3861–3872
22. Carstensen JV, Ray RK, Leffers T (2002) Texture development in Ni-Co alloys rolled to moderate reductions. Mater Sci For 408–412:607–612
23. Madhavan R, Ray RK, Suwas S (2016) Micro-mechanical aspects of texture evolution in nickel and nickel–cobalt alloys: role of stacking fault energy. Philos Magaz 96:3177–3199
24. Engler O, Pithan C, Lucke K (1994) Rolling texture development in Cu-Mn alloys. Mater Sci Forum 157–162:679–683
25. Engler O (2000) Deformation and texture of copper–manganese alloys. Acta Mater 48:4827–4840
26. Steffens T, Schwink C, Korner A, Karnthaler HP (1987) Transmission electron microscopy study of stacking-fault energy and dislocation structure in CuMn alloys. Philos Magaz A 56:161–173
27. Shankar G, Madhavan R, Kumar R, Sahoo B, Ray RK, Suwas S (2020) Micro-mechanism of evolution of microstructure and texture in Ni-Fe alloys. Materia 13:100811
28. Gallagher PCJ (1970) The influence of alloying, temperature and related effects on the stacking fault energy. Metall Trans 1:2429

29. Chowdhury SG, Ray RK, Jena AK (1998) Rolling texture in the intermetallic compound $Ni_{76}Al_{24}$(B). Mater Sci Engg A A246:289–301
30. Chowdhury SG, Ray RK, Jena AK (1995) Structural transformation in Ni_3Al(B) due to cold rolling. Scr Metall 32:1501–1506
31. Ball J, Gottstein G (1993) Large strain deformation of $Ni_3Al + B$: Part I. Microstruct Text Evolut During Roll Inter 1:171–185
32. Wever F (1924) Über die Walzstruktur kubisch kristallisierender metalle. Z Phys 28:69–90
33. Schulz LG (1949) A direct method of determining preferred orientation of a flat reflection sample using geiger counter x-ray spectrometer. J Appl Phys 20:1030–1033
34. Williams RO (1968) Analytical methods for representing complex textures by biaxial pole figures. J Appl Phys 39:4329–4335
35. Roe RJ (1965) Description of crystallite orientation in polycrystalline materials. III. general solution to pole figure inversion. J Appl Phys 36:2024–2031
36. Bunge HJ (1965) Zur Darstellung allgemeiner Texturen. Z Metallkde 56:872–874
37. Kikuchi S (1928) Electron diffraction in single crystals. J Appl Phys 5:83
38. Alam MN, Blackman M, Pashley DW (1954) High-angle kikuchi patterns. Proc R Soc 221A:224
39. Fundenberger JJ, Morawiec A, Bouzy E, Lecomte JS (2003) Polycrystal orientation maps from TEM. Ultramicro 96:127

Chapter 2
Basics of SEM-Based Techniques for Electron Backscatter Diffraction

David P. Field, Satyam Suwas, and Stuart I. Wright

1 Introduction

Since electron backscatter diffraction (EBSD) is used extensively throughout the book, we have included this chapter to offer a brief description of the technique. EBSD is one of various SEM-based techniques that can be used to obtain local information on the microstructures of bulk crystalline or polycrystalline materials. It specifically lends itself to mapping of crystal-dependent features in the microstructure such as orientation mapping and defect mapping. EBSD can be implemented on virtually any instrument that produces a coherent beam of electrons, but most typically on scanning electron microscopes (SEMs). The spatial and angular resolution of the technique is dependent upon the electron beam system and the specimen. It can be claimed that the first EBSD patterns were obtained by Nishikawa and Kikuchi [1] who placed photographic film above the specimen in an early electron beam instrument and recorded patterns now known as Kikuchi patterns. In fact, a more correct term for EBSD is probably backscatter Kikuchi diffraction (BKD) since it is directly analogous to Kikuchi diffraction in the transmission electron microscope (TEM), but this term has mostly gone out of the literature in favor of EBSD. Another early observation of EBSD patterns was that of Alam, Blackman, and Pashley in

---

D. P. Field (✉)
Institute of Materials Research, School of Mechanical and Materials Engineering, Washington State University, Pullman, WA, USA
e-mail: dfield@wsu.edu

S. Suwas
Department of Materials Engineering, Indian Institute of Science, Bangalore, India
e-mail: satyamsuwas@iisc.ac.in

S. I. Wright
EDAX, Ametek Materials Analysis Division, Mahwah, NJ, USA
e-mail: stuart.wright@ametek.com

S. Suwas and D. P. Field (eds.), *Advances in Texture, Microtexture, and Allied Techniques*, Materials Horizons: From Nature to Nanomaterials,
https://doi.org/10.1007/978-981-96-5346-1_2

1954 who called the measurement wide-angle Kikuchi diffraction [2]. A couple of decades later, Venables and co-workers investigated the technique in detail [3, 4], describing attributes like spatial resolution as a function of electron beam source and accelerating potential. The values obtained by Venables are remarkably accurate for even modern-day instruments.

To obtain EBSD patterns, the specimen is typically tilted to relatively high angles so that the angle of the beam relative to the specimen surface normal direction is on the order of 70 degrees. The areal detector (a phosphor screen or other position-sensitive detector) is positioned near the specimen to detect the diffraction pattern as shown in Fig. 1. The required hardware consists of the detector system, and the requisite mechanical interface that allows proper mounting of the detector in the SEM chamber at the appropriate geometry. Conventional applications include a phosphor screen and a low-light camera to detect the diffraction pattern from the back side of the screen. Position-sensitive electron detectors are particularly useful for applications of low accelerating potentials used during EBSD. Digital interfacing with the SEM for control of the electron beam and/or specimen stage is required for automated and unattended operation. Specimen preparation is generally minimal for patterns to be obtained, but specialized techniques are useful in ensuring high-quality EBSD mapping for complex structures. For thin films, and many other smooth and flat surfaces, there are often no specimen preparation requirements other than ensuring that the bare polycrystalline surface is exposed. For bulk specimens, conventional metallurgical techniques with a final mechanical polishing step using 0.05 μm colloidal silica suspension on a vibration polishing system are typically adequate. In rare instances or for special requirements, such as determination of stresses or geometrically necessary dislocations, the final step is often chemical or electrochemical polishing to remove residual surface deformation caused by mechanical surface preparation techniques. Broad-beam ion etching has also been shown to be effective in preparing both metal and ceramic specimen surfaces for high-quality EBSD imaging [5, 6].

Spatial resolution for EBSD is dependent upon many factors and is typically direction dependent with lateral resolution being about a factor of 3 better than the spatial resolution in the direction of the specimen tilt. As with all techniques that depend upon the specimen/beam interaction volume for signal, the resolution improves with

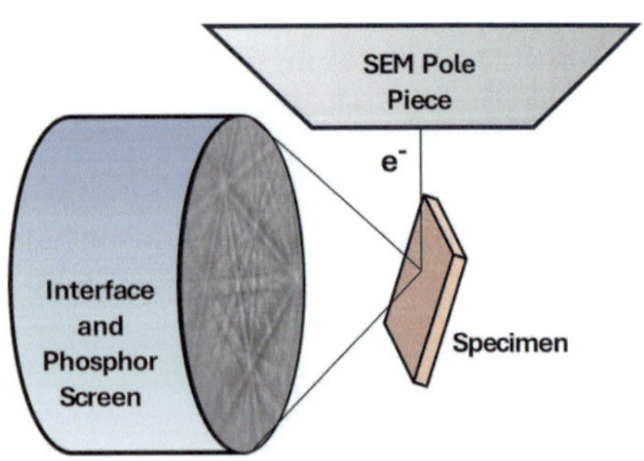

Fig. 1 Schematic showing the relative positions of the SEM pole piece, the specimen (tilted at a high angle from horizontal), and the interface and phosphor screen (or other EBSD pattern detector)

higher atomic number and lower accelerating potential of the electron beam. Spatial resolution in the lateral direction is generally accepted to be approximately 10 nm while using a field-emission source electron beam [7, 8]. This high spatial resolution is a result of the EBSD patterns being formed by electrons that are elastically scattered and retain a high fraction of the incident beam energy. Those that contribute to the signal in the EBSD pattern are scattered from the top 10 nm of the surface (for high Z materials at relatively lower beam energies) [3, 7].

Improved resolution can be obtained by thinning the specimens to the point where they are electron transparent, such as it is done for TEM imaging. The EBSD patterns can then be obtained from volumes of closer to 2 nm in diameter for ideal conditions [9]. This technique has been dubbed transmission Kikuchi diffraction (TKD) or sometimes transmission EBSD (t-EBSD). TKD can be performed using standard EBSD systems with the electron transparent specimen being tilted slightly away from the detector so that transmitted electrons are viewed instead of those that are backscattered. In addition, specialized "on-axis" TKD detectors have been developed that yield a stronger signal-to-noise ratio that results in better performance. TKD will not be further discussed in this chapter, but it is understood generally that improved spatial resolution can be obtained using TKD systems as opposed to more conventional EBSD measurements because the thin specimens do not result in the same sub-surface beam spreading as that which occurs with bulk specimens. Virtually every option is available for TKD as it exists with EBSD as far as data processing and interpretation.

Orientation and phase information can be obtained from EBSD patterns using various strategies. Early in automated EBSD development, the Hough (or Radon) transform was successfully implemented to determine the Kikuchi band positions within EBSD patterns [10–14]. This type of EBSD analysis has been commercially available since 1994. More recently, dictionary indexing of EBSD patterns has been adopted wherein a set (or "dictionary") of candidate EBSD patterns is simulated on a regular array of orientations. Experimental patterns are compared against the dictionary patterns using a similarity metric to identify the simulated pattern that is closest to that obtained by measurement. A newer technique termed "spherical indexing" has also been introduced to improve the computational efficiency and the large memory requirements inherent to standard dictionary indexing. Spherical indexing is gaining traction with EBSD users and the technique will likely be adopted universally in the near future. Automated EBSD data collection includes identification of individual orientations with respect to a previously identified sample coordinate frame, the phase, and a measure of local lattice integrity sometimes called image quality (IQ) or band contrast (BC). These values are stored digitally for given x–y coordinates (and sometimes z) and are used to generate images with desired features highlighted (crystallographic poles, grain boundaries, etc.). Inherent in the EBSD imaging technique are microstructural statistics of various measures such as grain size, crystallographic texture, grain boundary structure, and phase morphology.

There have been several reviews and complete descriptions of the details of EBSD systems including indexing algorithms, necessary hardware, and applications [cf.

15–17]. This information will not be repeated here in detail, but we offer a cursory overview to assist the reader who may not be familiar with the technique.

2 EBSD System Overview

Since the inception of automated EBSD systems in the early 1990s, the technique has evolved considerably. Thousands of measurements per second can be reliably obtained and indexed by various analytical approaches. Orientation imaging requires rapid identification of the patterns in a systematic approach, such as scanning over a regular array of points. Figure 2 shows an orientation image of a Cu thin film with colors inside the grains representing crystallographic poles aligned with the film normal orientation, as indicated by colored stereographic unit triangle key. Grain boundaries are shows as black lines with twin boundaries indicated by yellow lines. Grain boundaries were defined as having five degrees or more misorientation between neighboring measurement points and twin boundaries as being within five degrees of the exact twin boundary relationship (60 degrees rotation about a <111> axis). A discretely plotted inverse pole figure is also shown, indicating the distribution of poles aligned with the film normal direction. There is also a grain size chart plotted on a histogram showing the area fraction of grains with a given grain diameter. These are typical types of plots (out of dozens of possibilities) that are used in representing data collected using automated EBSD techniques. High-quality imaging requires that a high fraction of data points is properly indexed. This typically requires high-quality EBSD patterns although Herculean efforts are always underway to extract more information from patterns of marginal or poor quality (including "dictionary indexing" and "spherical indexing" that are discussed briefly in this chapter). EBSD pattern quality depends partially upon specimen preparation and partially upon the hardware and electronics that are used to capture the patterns.

EBSD imaging systems include the following:

- An electron source (typically an SEM),
- Mechanical interface/electron detector, and
- Software/indexing algorithms.

2.1 SEM/Electron Source Considerations

While a scanning electron microscope is the instrument of choice for EBSD systems, other types of instruments such as ultra-high vacuum chambers are also regularly used. The important aspect for EBSD is that a coherent electron beam is produced. The beam typically has an accelerating potential of 10–30 kV, but is sometimes higher. Coherent EBSD patterns have been observed at voltages as low as 2.5 kV [18], and a concerted effort is currently underway to obtain patterns at low potentials to avoid charging and to improve spatial resolution [19–21]. If a reflective aluminum coating

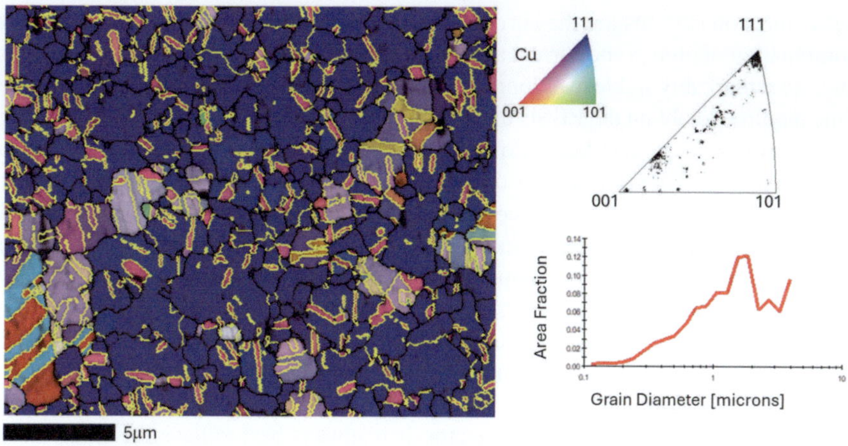

Fig. 2 Typical map, plot, and chart from data obtained using automated EBSD. Orientation map with grains colored according to poles that align with the specimen normal direction. Also shown is a discretely plotted inverse pole figure showing orientations aligned with the specimen normal direction, and a grain size histogram

is applied to a phosphor screen, as has been typical for conventional detector systems, the coating tends to absorb lower energy electrons and it is likely that no patterns will be seen below 5 kV of incident beam voltage. The advantages of lower beam potential include fewer problems with issues of charging on non-conductive specimens and improved spatial resolution, since the beam spread is less pronounced within the specimen. The current in the electron beam is also important as higher currents produce patterns with improved signal-to-noise ratios. The minimum probe current for high-quality imaging is dependent upon the EBSD pattern detector system. In general, several nA of probe current is the minimum required for good EBSD pattern quality, and several μA are ideal. Detectors that can integrate the received electron signal over time may be able to operate reasonably at lower probe currents, but the pattern acquisition time will increase to unreasonably slow levels if automated EBSD imaging is to be performed. Typical current imaging systems for EBSD are slow-scan charge-coupled device (CCD) cameras or complementary metal-oxide semiconductors (CMOS) sensors. Direct electron detectors are also being introduced commercially, particularly for low-voltage and low-current operation.

Slow-scan CCD cameras along with current PC technology (as of early 2024) are capable of obtaining and reliably indexing several hundred to a thousand or so images per second. CMOS imaging systems can perform EBSD data collection at rates over 5000 patterns per second. The direct electron detectors are currently much slower but offer opportunities for high-quality EBSD analysis at conditions where the standard phosphor screen and CCD or CMOS camera systems tend to falter. These systems will invariably become faster in the coming months.

Additional considerations for the SEM (or electron beam) system include beam diameter at given probe currents, and long-term stability of the beam. Schottky-source field-emission systems are the current electron source of choice that offers the best combination of high probe current and small effective beam diameter. These systems are also remarkably stable over long periods of time. If the beam is not stable over the time required to obtain the EBSD data, the patterns may be reduced in quality during the scan or unacceptable beam drift could occur. In current operation, it is rare to perform a single EBSD scan over times longer than a couple of hours, so the longer term stability is of less importance than previously when imaging required several hours. In addition, the general beam stability of systems has increased. One case where long-term stability is of importance in current practice is when in situ EBSD imaging is performed on a heating stage. Observation on a heating stage may capture processes that occur very slowly and constant EBSD imaging must be performed to capture the desired kinetic phenomena.

Finally, it is important to recognize the fact that the specimens are generally tilted to high angles toward the EBSD detector. It is always best to have the stage tilted toward the detector and allow x–y stage motion within the plane of the specimen surface. Alternatively, a pre-tilted specimen holder could be used to face the detector position, and the stage could either be stationary during data collection or calibrated to move in x–y–z positions to maintain a given working distance (and focus). Tilting toward the detector assumes that ports on the SEM chamber are available at the proper positions to allow detectors to be inserted to a location near the specimen, as indicated in Fig. 1. If simultaneous energy-dispersive X-ray spectroscopy (EDS) and EBSD are to be performed simultaneously, it is necessary that the ports, specimen tilting, and specimen detector allow for proper positioning. The introduction of specialized stages for in situ experiments such as heating or tensile stages places additional geometric restrictions on the geometrical configuration within the SEM chamber.

2.2 *Mechanical Interface*

EBSD data collection generally requires that the electron detector be placed in a position near the highly tilted specimen. As such, there is a high probability that the detector, specimen, stage, or even the pole piece of the SEM could be damaged unless extreme care is taken to optimize the interface design and the data collection operating procedures. The mechanical interface generally slides in and out of the SEM chamber along an axis aligned with the detector normal orientation. There are mechanical stops to place the detector at the ideal position for EBSD data collection and a touch sensor is often included to automatically retract the interface in the event that any part of the system comes too close to the pole piece, specimen, or stage.

Sometimes a forward scatter detector is included at the bottom of the EBSD detector. This is because the strongest signal of high energy electrons is in the forward scatter position when the specimen is at such a high tilt angle. The ideal position would be to have the detectors at the same position as the phosphor screen, but this would interrupt EBSD detection. Alternatively, the EBSD detector itself can be divided into sections and used as an electron detector by analyzing the average intensity over that region. It has become common to use the EBSD detector as an electron detector for imaging using portions of the detector (phosphor screen) or the entire detector as a standard backscatter electron detector [22–24]. This is readily done over the same grid used for EBSD data collection but is not as straightforward if one attempts to use the EBSD detector in standard imaging scan modes.

2.3 EBSD Detector Systems

As with Kikuchi diffraction in the TEM, the earliest EBSD patterns were recorded on photograph film that was placed inside the chamber and near the specimens [3, 25]. The first detectors used for computer-aided EBSD pattern indexing systems phosphor screens placed at a strategic position near the specimen with a low-light camera focused on the back side of the phosphor [26]. Charge-coupled device (CCD) or silicon-intensified target (SIT) cameras were used because of their ability to image in low-light conditions. The slow scanning rates for these cameras (generally television rates or approximately 30 frames per second) were not a problem because only about one pattern per second was obtained and indexed. As computers and algorithms became more efficient, it was necessary to increase the speed of these systems so slow-scan CCD cameras and complementary metal-oxide semiconductor (CMOS) detectors were employed. This enabled the capture of images at much higher rates and the decay constants on phosphor screens became an issue that was important to consider in determining the speed at which patterns could be obtained. There have been other detectors used including micro-channel plate detectors [27] and even a rather costly single-crystal YAG detector [28] instead of a phosphor screen. These have mostly given way to the CMOS and phosphor screen system because of the higher speeds that can be obtained. The CMOS system may be used for both rapid scan rate imaging and for high-quality EBSD image collection. The practical indexing rate limit for modern detector systems is currently in the range of 5000–7000 images per second, but that number is likely to continue increasing with more powerful computers and optimized image-handling algorithms.

One of the latest developments in EBSD pattern collection is to use a direct electron detector system. At present, this system obtains patterns at lower accelerating voltages giving an increase in spatial resolution of the system because of a lower specimen/beam interaction volume. These systems are slower in performance than the CMOS systems, but efforts are underway to improve the speed and lower the required accelerating potentials. This will likely be an area of progress in the future for EBSD pattern acquisition [29].

2.4 Brief Overview of EBSD

There have been numerous papers and texts written on the physics of EBSD [cf. 30–32]. In short, the process for forming EBSD patterns is based upon Bragg's law (as are all diffraction processes):

$$\lambda = 2d\sin\theta$$

where λ is the wavelength of the electron beam, d is the interplanar spacing for a given set of lattice planes, and θ is the Bragg angle. The electron beam interacts with the specimen and electrons are scattered in all directions. Those that are elastically scattered or channeled and that move out of the specimen with similar energy to the incident beam energy form EBSD patterns. Considering that electrons are moving in all directions, the Bragg angle is satisfied for all planes in a given crystal that satisfy the diffraction selection criteria. Kikuchi bands form that are coincident with the geometric projection of the given diffracting plane. (Actually, the bands are formed by two diffracting cones with the width between the cones being proportional to the d-spacing of the diffracting plane.) Since all diffraction conditions are satisfied simultaneously, the intricate Kikuchi patterns form. Indexing of the patterns is determined by measuring the interplanar angles for several planes and referencing them against the angles in known Bravais lattices. The intersection of two or more Kikuchi bands corresponds to a zone axis, which is a direction common to multiple diffracting planes. For a more complete discussion of the physics of EBSD, the reader is referred to [30–32].

2.5 Pattern Indexing

The process of determining the crystallite lattice orientation from an EBSD pattern is referred to as indexing. The mathematics of indexing will not be detailed here. It will suffice to state that elaborate schemes for detecting Kikuchi bands have been developed. The most common is the Hough transform or mathematically similar Radon transform. The term "Hough Indexing" refers to the automatic determination of diffraction band positions and the use of the geometry of these band positions to identify the phase and orientation of the crystallite lattice from that point. The reader is referred to the discussion of Wright, Krieger-Lassen, and Schwarzer [15, 33, 34] for a complete description of the calculations required. Typically, with Hough indexing, there is a confidence index [35] or hit rate that is an indication of the data quality. Several open-source and publicly available algorithms have been produced over the years for Hough indexing. Some of the more prominent systems include EMSoft (https://github.com/EMsoft-org/EMsoft), the MatLab-based MTEX (https://mtex-toolbox.github.io/), ATEX (http://www.atex-software.eu/), and more recently Kikuchipy (https://github.com/pyxem/kikuchipy) to name a few. The main

EBSD system vendors (EDAX, Oxford Instruments, and Bruker) and other vendors and developers of specialty codes such as BLGVantage (CrossCourt) or BlueQuartz Software (Dream3D) are also reliable resources for information and tools for EBSD.

For weak EBSD images due to either poor specimen surface preparation, oxidation, or other defects near the specimen surface, the Hough approach generally works reasonably well. It has been demonstrated, however, that "dictionary" indexing or "spherical" indexing are more robust in many cases. In dictionary indexing, a "master pattern" is created—this is essentially a model of diffraction for a full sphere typically done using a dynamical diffraction model. Individual patterns at a given orientation and geometrical configuration of the EBSD detection system can be extrapolated from the master pattern. A dictionary of simulated patterns is created from the master pattern by generating a simulated EBSD pattern at each point in a regular sampling of in orientation space [36]. Pattern indexing is achieved by comparing a detected pattern against the dictionary of images for the given system geometry and assigned a phase and orientation based upon this comparison. This technique is slower than Hough indexing and generally requires that the user store all EBSD patterns in a file that is subsequently analyzed to determine the desired information. There is open-source code for EBSD dictionary indexing strategies, EMSoft (https://github.com/EMsoft-org/EMsoft), but the primary vendors of EBSD systems are also incorporating dictionary indexing into their commercial systems. GPU-based implementation of dictionary indexing has made the approach much more tractable than earlier CPU implementations. The maximum spacing between grid points is approximately 3 degrees. As the spacing between grid points gets smaller, the size of the dictionary increases rapidly. Dictionary size also increases with decreasing point group symmetry. This imposed very large memory requirements on the computational hardware.

Spherical indexing is similar in that it uses a pre-determined pattern formed on a spherical surface (the master pattern) but dramatically improves the computation efficiency over dictionary indexing. Spherical indexing uses a back-projection of the experimental pattern and a spherical harmonic transform technique to compare experimental patterns against the calculated master pattern. The orientation is determined with high precision using a spherical cross correlation between the experimentally obtained pattern and the calculated pattern [37, 38].

One particular challenge that arises in indexing EBSD patterns is that of pseudo-symmetry. A common example is martensite in steel samples. Martensite is body-centered tetragonal (BCT). The degree of tetragonality is typically small depending on carbon content. The tetragonality is often imperceptible using conventional EBSD indexing and thus martensite is often indexed as if it were body-centered cubic. Various techniques have been employed to overcome these pseudo-symmetric challenges, but the advent of dictionary and spherical indexing appear to hold much more promise than was achievable using the Hough transform-based approaches. This is partly due to the fact that the Hough transform only considers the geometrical arrangement of the bands in a pattern, whereas the dictionary and spherical methods also consider the variation of intensity within the bands which can be critical to resolving the pseudo-symmetry [39, 40].

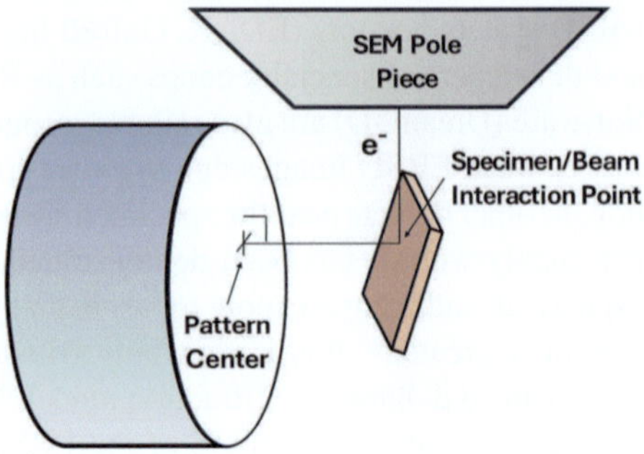

Fig. 3 Schematic showing the geometry of the "pattern center" that is an important calibration value for EBSD characterization. The pattern center is at a position in the plane of the EBSD detector where a line segment normal to the detector plane intersects the specimen/beam interaction point

2.6 EBSD Image Calibration

Regardless of the indexing strategy used, a proper calibration of the system is important in order to obtain reliable results. The pattern center position is the primary calibration value that must be determined. The pattern center can be defined as the point in the plane of the EBSD detector that is nearest to the specimen/beam interaction position. In other words, it is the point in the detector plane at which a line that is normal to that plane will intersect the specimen/beam interaction point (Fig. 3). The specimen to detector distance along this line must also be known and is easily calculated from the geometry of the diffraction patterns. We will not discuss EBSD calibration procedures herein. A discussion of calibration techniques for EBSD analysis was given by Day [41] and Krieger-Lassen and Bilde-Sorensen [42]. With modern EBSD systems the calibration values are optimized automatically using a captured pattern.

2.7 Multi-phase Analysis

Multi-phase analysis can present some difficulties when using Hough indexing for EBSD data collection, especially if the candidate phases are not known well. For mapping structures on multi-phase materials using any EBSD indexing strategy, it is best to first identify all possible phases. These phases can then be scrutinized for the best solution out of all possible candidate phases and orientations in order to obtain the results. Many complex-phase structures have been successfully characterized using EBSD. Some examples of results for titanium and steel alloys are included in the following references [43–48].

A couple of challenges that often exist with multi-phase analysis involve difficulties in obtaining high-quality diffraction patterns from all phases. It is common for the "background" image to have differing intensities for different phases. Typically, the background is collected as an average of all patterns that will be analyzed, and this is often adequate. At times, specialized background correction strategies would have to be employed. These generally include some form of low-pass filtering of the input image to obtain a background that is then used to manipulate the original image in order to pull out the Kikuchi patterns. Another challenge that is sometimes impossible to overcome is that of specimen preparation. Polishing techniques that are optimal for one phase may be inadequate for a neighboring phase. Again, this is beyond the scope of the current discussion. There are many resources available to the interested reader as referenced earlier in this section.

Another challenge in indexing multiphase materials is that some phases are not reliably differentiated. A phase may be slightly tetragonal, such as martensite in a ferrite matrix. This difference in the BCT and BCC phase has been distinguished using image quality, but not Hough indexing. A deformed BCC ferrite phase is virtually impossible to distinguish from a BCT martensite phase using Hough indexing strategies. Some success has been found in these situations by modifying the band identification algorithms and using higher resolution in the Hough accumulator array [28, 49]. Dictionary and spherical indexing procedures have been shown to address this issue much more rigorously than even higher resolution Hough indexing. This same type of situation occurs when phases are distinguished only by the difference of one or two low-intensity diffracting planes in the EBSD pattern. Again, dictionary or spherical indexing holds an advantage for these situations.

Figure 4 shows an orientation map and a phase map for a dual-phase titanium structure consisting of both alpha and beta phases (hexagonal close-packed and body-centered cubic, respectively). The image shows a cross section of the metal after shot peening. The right side of the maps is the shot-peened surface. Black regions show low confidence data and the other colors indicate poles aligned with the specimen surface normal orientation (at top) and the measured phase (bottom images). Also shown are representative low-quality EBSD images for both phases. The column indicating data obtained using spherical indexing has significantly more higher quality data than were obtained by the more conventional Hough indexing.

3 Extensions of EBSD

3.1 High-Resolution (Cross-Correlation) EBSD

Another EBSD-related technique that is now commonly used is generally referred to as High-Resolution EBSD (HR-EBSD) or cross-correlation EBSD. High resolution refers to the fact that standard EBSD has an angular resolution on the order of 0.25–0.5 degrees for standard Hough indexing procedures. HR-EBSD relies upon the comparison of two similar EBSD images using cross-correlation techniques to measure the distortion of one EBSD pattern compared with the other. If it is assumed that one of the patterns comes from a perfect crystal of a given orientation, the algorithms can determine with high accuracy how the distorted patterns is stretched and/or rotated in comparison to the first. There are multiple algorithms and strategies used to perform HR-EBSD measurements. There are also various approaches to

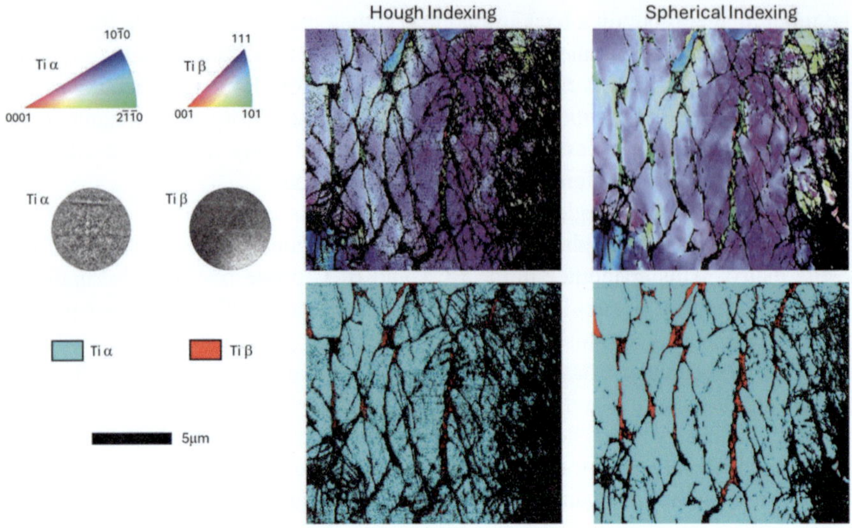

Fig. 4 Surface shot-peened dual-phase titanium sample showing lower quality EBSD data from standard Hough indexing as compared with spherical indexing (shot-peened surface is on the right side of the images). Sample courtesy of Mr. Prathompoom Newyawong, NSTDA Characterization and Testing Service Center (NCTC), with EBSD measurements performed by Dr. Rene de Kloe, EDAX

determining the required strain-free reference EBSD pattern both from measurement and from simulation. Some useful references for the reader are provided [50–53]. HR-EBSD is primarily focused on the investigation of the development of strain and stress that evolved during deformation.

Finally, using dual-beam focused-ion-beam and SEM (FIB/SEM) systems, three-dimensional EBSD data can be obtained. Data are obtained over a regular array on a surface before the FIB beam is used to remove a thin surface layer and EBSD measurements are again obtained. This leads to destructively obtained data from a full 3D volume of the specimen. Several examples of 3D EBSD have been demonstrated over the past couple of decades [54–56].

4 Summary

SEM-based orientation imaging microscopy, with electron backscatter diffraction (EBSD) at its core, represents a transformative approach to the study of crystallographic orientation and microstructure, offering unprecedented insights into the detailed characteristics of materials. The development of robust indexing algorithms, such as spherical and dictionary indexing, has effectively addressed critical limitations like pseudo-symmetry and data accuracy, enhancing the technique's reliability for complex analyses. Advanced adaptations, including High-Resolution EBSD (HR-EBSD), Transmission Kikuchi Diffraction (TKD), and three-dimensional EBSD (3D EBSD), have significantly extended their capabilities to high-resolution strain analysis, thin-film characterization, and volumetric microstructural analysis. These innovations not only facilitate the resolution of intricate microstructural details but also establish EBSD as a vital tool for cutting-edge research in materials science, engineering, and related disciplines. As technological advancements continue, EBSD is poised to address even more complex challenges, further solidifying its indispensable role in materials' characterization.

References

1. Nishikawa S, Kikuchi S (1928) The diffraction of cathode rays by calcite. Proc Imper Acad 4:475–477
2. Alam MN, Blackman M, Pashley DW (1954) High-angle kikuchi patterns. Proc R Soc Lond A221:224–242
3. Venables JA, Harland CJ (1973) Electron back-scattering patterns—a new technique for obtaining crystallographic information in the scanning electron microscope. Philos Magaz 27:1193–1200
4. Venables JA, Jaya RB (1977) Accurate microcrystallography using electron back-scattering patterns. Philos Magaz 35:1317–1328
5. Bunge HJ (1986) Experimental techniques of texture analysis. Informationsgesellschaft-Verlag, Germany, Oberursel
6. Alani R, Field DP, Ogura K (1999) SEM specimen preparation by broad beam etching for enhanced channeling and orientation imaging. Microsc Microanal 5(Suppl 2):910–911
7. Matteson TL, Schwarz SW, Houge EC, Kempshall BW, Giannuzzi LA (2002) EBSD investigation of focused ion beam surfaces. J Electr Matls 31
8. Troost KZ (1993) Submicron crystallography in the scanning electron microscope. Philips J Res 47:151–162
9. Trimby PW (2012) Orientation mapping of nanostructured materials using transmission kikuchi diffraction in the scanning electron microscope. Ultramicroscopy 120:16–24
10. Humphreys FJ (2001) Review–grain and subgrain characterisation by electron backscatter diffraction. J Mater Sci 36:3833–3854
11. Wright SI, Adams BL, Zhao J-Z (1991) Automated determination of lattice orientation from electron backscattered kikuchi diffraction patterns. Text Microstruct 13:123–131
12. Adams BL, Wright SI, Kunze K (1993) Orientation imaging, the emergence of a new microscopy. Metall Trans 24A:819–831
13. Wright SI (1993) A review of automated orientation imaging microscopy (OIM™). J Compu Assis Microsc 5:207–221
14. Jensen DJ, Schmidt NH (1990) Automatic recognition of electron backscattering patterns. In: Chandra TC (ed) Proceed Recrystallization '90. TMS, Warrendale, Pennsylvania, pp 219–224
15. Lassen NCK, Conradsen K, Jensen DJ (1992) (1992) Image processing procedures for analysis of electron back scattering patterns. Scan Microsc 6:115–121
16. Schwartz AJ, Kumar M, Adams BL, Field DP (2009) Electron backscatter diffraction in materials science 2. Springer, New York
17. Wright SI, De Graef M (2002) Electron backscatter diffraction. Inter Table Crystall C 1:6
18. Michael JR, Goehner RP (1993) Crystallographic phase identification in the scanning electron microscope: backscattered electron kikuchi patterns imaged with a CCD-based detector. MSA Bullet 23:168–175

19. Steinmetz DR, Zaefferer S (2010) Towards ultrahigh resolution EBSD by low accelerating voltage. 26(6)
20. Singh S, Guo Y, Winiarski B, Burnett TL, Withers PJ, De Graef M (2018) High resolution low kV EBSD of heavily deformed and nanocrystalline aluminium by dictionary-based indexing. Sci Rep 8:10991
21. Tripathi A, Zaefferer S (2019) On the resolution of EBSD across atomic density and accelerating voltage with a particular focus on the light metal magnesium. Ultramicroscopy 207:112828
22. Wright SI, Nowell MM, Kloe RD, Camus P, Rampton T (2015) Electron imaging with an EBSD detector. Ultramicroscopy 148:132–145
23. Payton EJ, Nolze G (2013) The backscatter electron signal as an additional tool for phase segmentation in electron backscatter diffraction. Microsc Microan 19:929–941
24. Brodusch N, Demers H, Gauvin R (2018) Imaging with a commercial electron backscatter diffraction (EBSD) camera in a scanning electron microscope: a review. J Imag 4:88
25. Dingley DJ, Kishi KZB, Randle V (1995) Atlas of backscattering kikuchi diffraction patterns. IOP Publishing, Bristol
26. Dingley DJ (1984) Diffraction from sub-micron areas using electron backscattering in a scanning electron microscope. Scan Electron Microsc 2:569–575
27. Eades A, Deal A, Bhattacharyya A, Hooghan T (2009) Energy filtering in EBSD. Elect Backscat Diffrac Mater Sci 53–63
28. Michael JR, Eades JA (2000) Use of reciprocal lattice layer spacing in electron backscatter diffraction pattern analysis. Ultramicroscopy 81:67–81
29. Wang F, Echlin MP, Taylor AA, Shin J, Bammes B, Levin BD, Graef MD, Pollock TM, Gianola DS (2021) Electron backscattered diffraction using a new monolithic direct detector: high resolution and fast acquisition. Ultramicroscopy 220:113160
30. Reimer L (1985) Scanning electron microscopy, physics of image formation and microanalysis. Springer-Verlag, Berlin, pp 338–356
31. Randle V, Engler O (2000) Introduction to texture analysis, macrotexture, microtexture and orientation mapping. Gordon and Breach, New York
32. Callahan PG, Graef MD (2013) Dynamical electron backscatter diffraction patterns. Part I: pattern simulations. Micro. Microanal. 19:1255–1265
33. Wright SI (1992) Individual lattice orientation measurements development and application of a fully automatic technique. PhD dissertation, Yale University, New Haven, CT
34. Schwarzer RA (1997) Automated crystal lattice orientation mapping using a computer-controlled SEM. Micron 28:249–265
35. Field DP (1997) Recent advances in the application of orientation imaging. Ultramicroscopy 67:1–9
36. Singh S, Graef MD (2016) Orientation sampling for dictionary-based diffraction pattern indexing methods. Modell Simul Mater Sci Eng 24:085013
37. Lenthe WC, Singh S, Graef MD (2019) A spherical harmonic transform approach to the indexing of electron backscattered diffraction patterns. Ultramicroscopy 207:112841
38. Hielscher R, Bartel F, Britton TB (2019) Gazing at crystal balls: Electron backscatter diffraction pattern analysis and cross-correlation on the sphere. Ultramicroscopy 207:112836
39. Pang EL, Schuh CA (2021) Crystal orientation and detector distance effects on resolving pseudosymmetry by electron backscatter diffraction. J Appl Crystall 54:513–522
40. Marquardt K, Graef MD, Singh S, Marquardt H, Rosenthal A, Koizuimi S (2017) Quantitative electron backscatter diffraction (EBSD) data analyses using the dictionary indexing (DI) approach: overcoming indexing difficulties on geological materials. Am Miner 102:1843–1855
41. Day A (1993) Developments in the EBSP technique & their application to grain imaging. PhD dissertation, University of Bristol, Bristol, England
42. Lassen NCK, Sorensen JBB (1993) Calibration of an electron back scattering pattern Set-up. J Microsc 170:125–129
43. Field DP, Wright SI, Dingley DJ (1996) Multi-phase texture analysis by orientation imaging microscopy. In: Liang Z, Zuo L, Chu Y (eds) Proceedings of the eleventh international conference on textures of materials (ICOTOM-11), Xi'an, China, Sept. 16–20. pp 94–99

44. Dupont ML, Baudin T, Penelle R, Naka S, Thomas M (1996) Evolution of texture and microstructure of gamma/alpha(2) biphase titanium as function of heat treatments. J Phys IV 6(C2):235–240
45. Levin L, Tokar A, Berner A (2000) Microstructural characterization of gamma-TiAl base alloy by electron probe x-ray microanalysis and electron backscatter diffraction. J Matls Sci 35:3923–3929
46. Schwarzer RA, Singh AK, Sukkau J (2000) Discrimination and mapping of phase distributions by automated crystal orientation measurement. Mater Sci Technol Ser 16:1389–1392
47. Paransky YM, Gotman IY, Gutmanas EY (2000) Reactive phase formation at AlN-Ti and AlN-TiAl interfaces. Matls Sci Eng 277A:83–94
48. Wilson AW, Spanos G (2001) Application of orientation imaging microscopy to study phase transformations in steels. Matls Charac 46:407–418
49. Kishi KZB (1998) Measurement of crystal parameters on backscatter kikuchi diffraction patterns. Scanning 20:17–127
50. Wilkinson AJ, Meaden G, Dingley DJ (2006) High-resolution elastic strain measurement from electron backscatter diffraction patterns: new levels of sensitivity, Ultramicroscopy 106(4–5)
51. Britton TB, Maurice C, Fortunier R, Driver JH, Day AP, Meaden G, Dingley DJ, Mingard K, Wilkinson AJ (2010) Factors affecting the accuracy of high resolution electron backscatter diffraction when using simulated patterns. Ultramicroscopy 110:1443–1453
52. Fullwood D, Vaudin M, Daniels C, Ruggles T, Wright SI (2015) Validation of kinematically simulated pattern HR-EBSD for measuring absolute strains and lattice tetragonality. Mater Charac 107:270–277
53. Ernould C, Beausir B, Fundenberger JJ, Taupin V, Bouzy E (2020) Global DIC approach guided by a cross-correlation based initial guess for HR-EBSD and on-axis HR-TKD. Acta Mater 191:131–148
54. Zaefferer S, Wright SI (2009) Three-dimensional orientation microscopy by serial sectioning and EBSD-based orientation mapping in a FIB-SEM. In: Electron backscatter diffraction in materials science. Springer US, Boston, MA, pp 109–122
55. Echlin MP, Burnett TL, Polonsky AT, Pollock TM, Withers PJ (2020) Serial sectioning in the SEM for three dimensional materials science. Curr Opin Solid State Mater Sci 24:100817
56. Rowenhorst DJ, Nguyen L, Leonard ADM, Fonda RW (2020) Characterization of microstructure in additively manufactured 316L using automated serial sectioning. Curr Opin Solid State Mater Sci 24:100819

Chapter 3
Electron Backscatter Diffraction-Based Strain Analysis in the Scanning Electron Microscope

Stuart I. Wright, Josh Kacher, and Tim Ruggles

1 Introduction

Electron Backscatter Diffraction (EBSD) systems have become commonplace in materials science and geology research microscopy laboratories. EBSD is a Scanning Electron Microscope (SEM)-based technique capable of measuring crystallographic orientation at sub-micron spatial resolution. Some of the very first applications of EBSD after its automation were to investigate the small changes in orientation that occur when a polycrystal is deformed [1, 2] and the study of deformation in polycrystals continues to be one of the most common applications of EBSD [3–10]. The most common way for crystalline materials to accommodate deformation is to slip along crystallographic planes. This slip results in a small rotation of the crystal lattice. While the lattice in a grain would like to rotate as a rigid body, the rotation is constrained by the crystal lattices of the neighboring grains. Without this constraint, the microstructure would develop voids and the material would consequently fail. To maintain compatibility at the boundaries between grains, orientation gradients develop within grains and eventually a substructure develops as grains fragment into sub-grains. Various metrics have been developed to characterize these local changes in lattice orientation using EBSD measurements in an effort to gain understanding into the active mechanisms involved in the deformation process and their role in microstructural evolution [3–10]. To understand the basis of these metrics it is helpful to understand the influence of strain and defects on EBSD patterns.

S. I. Wright (✉)
Gatan, Pleasanton, USA
e-mail: stuart.wright@ametek.com

J. Kacher
School of Materials Science and Engineering, Georgia Institute of Technology, Atlanta, USA

T. Ruggles
Sandia National Laboratories, Albuquerque, USA

S. Suwas and D. P. Field (eds.), *Advances in Texture, Microtexture, and Allied Techniques*, Materials Horizons: From Nature to Nanomaterials,
https://doi.org/10.1007/978-981-96-5346-1_3

2 Terminology

As this chapter discusses strain and deformation measurements, a brief discussion of terminology is appropriate. At the macroscopic scale, strain can be divided between elastic strain, or those strains that are recovered on unloading, and plastic strains, which are not recovered. However, at the mesoscale and below, it is more convenient to define elastic strain as any deformation associated with shear or dilation of the crystal lattice, which includes recoverable strain due to macroscopic loading as before, but also strains associated with residual stress, such as the strain fields around dislocations, grain boundaries, or other defects. Plastic strain is then any deformation not associated with a change in the lattice, such as the change in shape of a single crystal when a dislocation moves through it, leaving perfect crystal structure in its wake. As will be shown in this chapter, elastic strain can be qualitatively and quantitatively characterized via EBSD-based techniques. Plastic strain has no direct effect on EBSD patterns. However, plastic strain in crystalline materials is most commonly accompanied by the propagation and accumulation of dislocations. Although the passage of dislocations cannot be detected, it leaves no permanent change to the crystal lattice, the accumulation of dislocations can be detected by resultant curvature of the crystal lattice and by the degradation of EBSD pattern quality. These measurements will be discussed in further detail in the following sections.

3 What is the Effect of Strain and Defects on Individual EBSD Patterns?

The local strain and defect state play a vital role in governing a range of material responses, ranging from crack formation processes under mechanical loads to charge transport in functional materials. Elastic strains produce observable but typically subtle effects on EBSD patterns that are not always visible to the naked eye. Plastic strain cannot be directly detected via EBSD analysis, but defect populations, which are often correlated with local plasticity, can be detected and quantified.

3.1 Elastic Strain

Elastic strains cause a distortion of the crystal lattice which is reflected in the EBSD patterns. Consider, for example, a cubic lattice strained uniaxially in one of the principal directions of the unit cell as shown schematically in Fig. 1 along with simulated patterns for the unstrained and strained crystals [11, 12]. The distortion is reflected in the EBSD pattern by a shift in the positions of some of the zone axes as well as by a change in width of some of the diffraction bands. For the case of

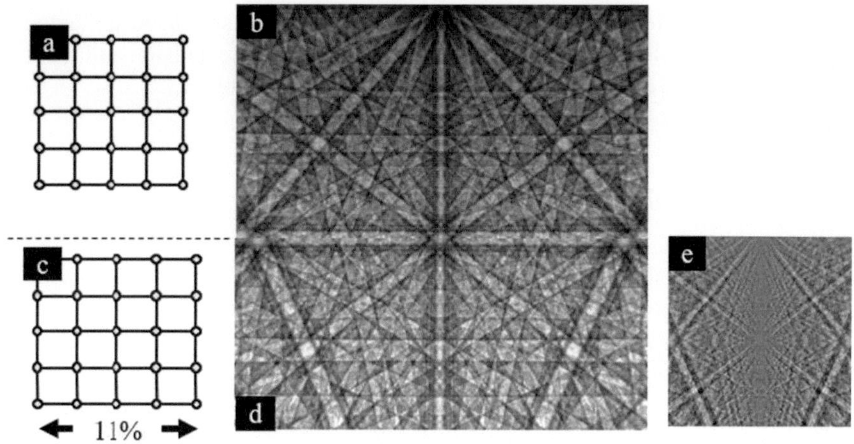

Fig. 1 Two-dimensional schematic of an **a** unstrained lattice and **c** 11% strained lattice. Dynamically simulated patterns for **b** the unstrained lattice and **d** the strained lattice. **e** Pattern **b** subtracted by pattern **d**

uniform dilation, the zone axes would remain in the same locations, only the band widths would change.

Another form of elastic distortion is a "bent" lattice as shown schematically in Fig. 2. For a "bent" crystal, not only do the zone axes shift as in the uniaxial extension case, but there would be a slight degradation in pattern quality as the planes within the diffraction volume are no longer exactly parallel. These slight deviations from parallel result in deviations of the Bragg angle along the length of the planes which produces the blurring that can be observed in Fig. 2 [13].

The elastic distortions used in Figs. 1 and 2 are for exaggerated degrees of strain in order that the effects in the patterns are observable at the scale of the figures. Elastic strains in real materials are much smaller and tend to produce only sub-pixel shifts even in patterns captured using high-resolution EBSD cameras. Very careful image analysis of high-resolution images is needed to make meaningful measurements of elastic strain which will be covered in a later section. The blurring shown in

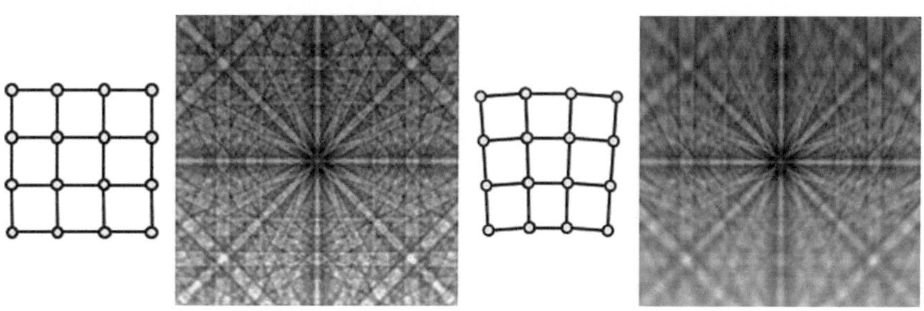

Fig. 2 Schematic and dynamically simulated EBSD patterns for unstrained and a "bent" crystal

Fig. 2 is not modeled but only approximated by summing simulated patterns for each individual "unit cell" in the bent lattice both in terms of extension and rotation but at a tenth of the extension and rotation shown in the diagram.

3.2 Plastic Deformation

Plastic strain in crystalline materials is accommodated by the nucleation and motion of defects such as dislocations and deformation twins. While twins can be directly characterized via orientation relationships, dislocations can only be indirectly characterized by EBSD. Figure 3 shows an example of the effect of high dislocation content on pattern quality in patterns obtained from a partially recrystallized sample of ferritic steel.

The diagram in Fig. 4 illustrates two types of dislocations. (1) There are regions in the material with significant dislocation density with a net Burgers vector of zero. Such dislocations are often referred to as "statistically stored" dislocations (SSDs). (2) There are also regions with net non-zero Burgers vectors resulting in a change in crystallographic orientation or lattice curvature. Dislocations that produce local variations in crystallography orientation are often termed "geometrically necessary" dislocations (GNDs). A sub-grain boundary is composed of an array of GNDs. Note that the designation of the two is length scale dependent as, at a sufficiently small length scale, all dislocations are geometrically necessary.

If the EBSD interaction volume contains dislocations with a net Burgers vector of zero (i.e., SSDs), the resulting pattern will be degraded. This is due to local perturbations of the diffracting lattice planes as shown schematically in Fig. 5 for a single pair of opposing edge dislocations. The local perturbation leads to incoherent scattering reducing the quality of the corresponding diffraction pattern—the higher the dislocation density, the greater the degradation in pattern quality.

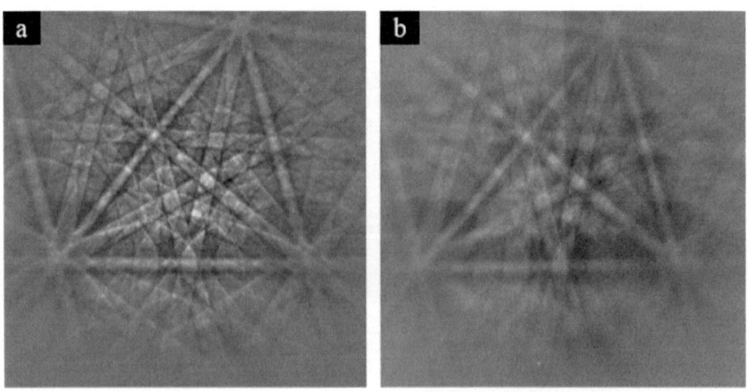

Fig. 3 EBSD patterns from a partially recrystallized grain in a ferritic steel obtained using a direct electron detector: **a** recrystallized and **b** deformed. Patterns courtesy of René de Kloe, Gatan

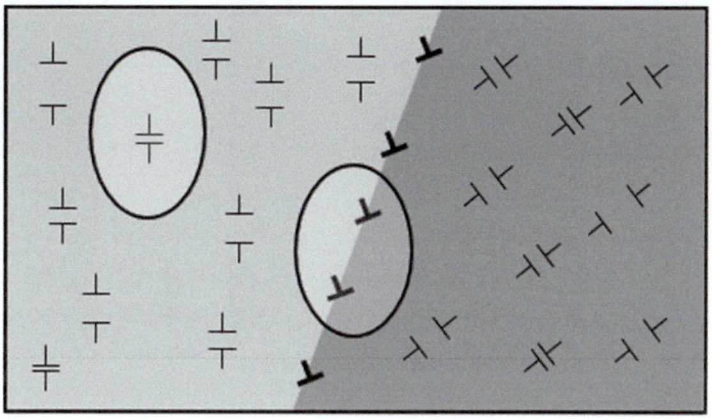

Fig. 4 Diagram of geometrically necessary and statistically stored dislocations (in bold)

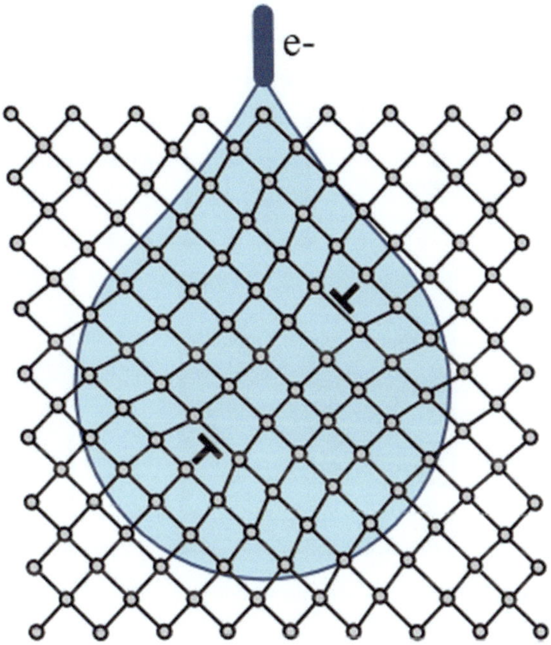

Fig. 5 Diagram showing the perturbation in the crystal lattice due to two opposing edge dislocations within an interaction volume

EBSD pattern quality is also degraded when the diffraction volume contains GNDs such as those indicated in Fig. 4. The resulting pattern is essentially a superposition of the patterns from each individual sub-grain within the diffraction volume as shown in Fig. 6. As with SSDs, the degree of degradation depends on the density of dislocations within the interaction volume.

The interaction volume may contain both SSDs and GNDs. The size of the diffraction volume is a function of beam size (typically a function of beam current) and accelerating voltage. For a given dislocation structure, the larger the volume the more likely it is to contain dislocations. Thus, pattern degradation is linked to spatial resolution and operating conditions of the SEM as well as the material itself. If

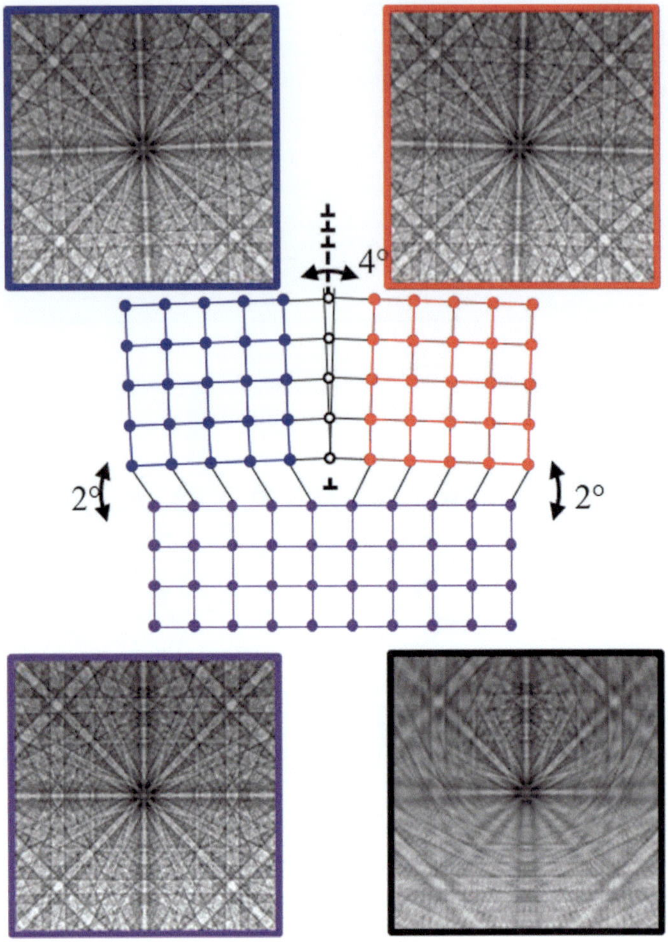

Fig. 6 Schematic showing the creation of three sub-grains due to a linear array of dislocations. Three corresponding simulated patterns and at bottom-right the average of the three simulated patterns

the spatial resolution is high and the material investigated forms a well-organized dislocation structure, then the degradation of the patterns may be minimal. Typical ductile metals and alloys have dislocation densities ranging from 10^{12} to 10^{16} m^{-2}, depending on the degree of deformation. As the diffraction volume generally spans tens of nanometers, each diffraction pattern can be expected to contain at least a few dislocations for moderately to heavily deformed materials.

When characterizing defect populations, we tend to focus on the misorientation between two orientation measurements. A note on terminology is useful in this context. The rotation required to bring one crystal lattice into coincidence with another of different orientation is referred to as a misorientation. Any two crystals will have a common axis between them. A rotation about that axis will bring the two crystals into coincidence. The misorientation is then composed of both the angle and axis of rotation. Because of crystal symmetry there are many symmetrically equivalent axis-angle pairs. For example, in two crystals with cubic symmetry there are $24^2 = 576$ symmetrically equivalent misorientations as there are 24 symmetry operators for cubic crystal symmetry. The disorientation is defined as the angle for the misorientation variant with the smallest angle of rotation regardless of axis. However, the term misorientation or misorientation angle is often synonymous with disorientation.

3.3 *Quantitative Analysis*

As noted, the effect of elastic strain and defect populations on EBSD patterns can be subtle. Standard approaches to indexing EBSD patterns to determine crystallographic orientation are capable of measurement rates of a few thousand points per second. The angular precision of such measurements is generally between 0.1 and 0.5 degrees depending on the quality of the patterns as well as both SEM and EBSD operating conditions. The indexing algorithms give some feedback on the fit of the indexing to the patterns which in principle could give some information on the magnitude of elastic strain in the material but in practice this has rarely been done as the perturbations due to elastic strain are too small to capture using standard indexing. Unfortunately, the angular precision of standard orientation measurements is generally not high enough to capture the local variations in orientation associated with small dislocation populations. However, as strain increases and dislocations multiply, standard EBSD measurements can reliably capture these small crystal rotations albeit with some degree of noise. As deformation increases so do local crystal rotations but pattern degradation also decreases which can increase the amount of noise in the orientation measurements.

To overcome the limitations in the accurate determination of misorientations critical to the study of deformation, methods based on the direct comparison of the pattern themselves (without the intermediate step of indexing for orientation determination) have been developed. The method that has gained the most acceptance in the EBSD community is a cross-correlation method pioneered by Wilkinson and co-workers [14, 15] and has come to be known as HR-EBSD where HR refers to high (angular) resolution. A coarse flow chart is shown in Fig. 7.

Two patterns are compared by first identifying matching sets of regions of interest (ROIs) in the two patterns. The cross correlation between each pair of ROIs is calculated to determine the shift between the two ROIs (typically many more ROIs than shown in the figure). These shifts are typically sub-pixel and reflect strains and rotations between the two points in the material from which the diffraction patterns are collected. They can be used to determine the displacement gradient tensor, $\partial u_i/\partial x_j$. For small strains and rotations, the elastic strain ε_{ij} is the symmetric part of the displacement gradient tensor and the antisymmetric part is the rotation ω_{ij}:

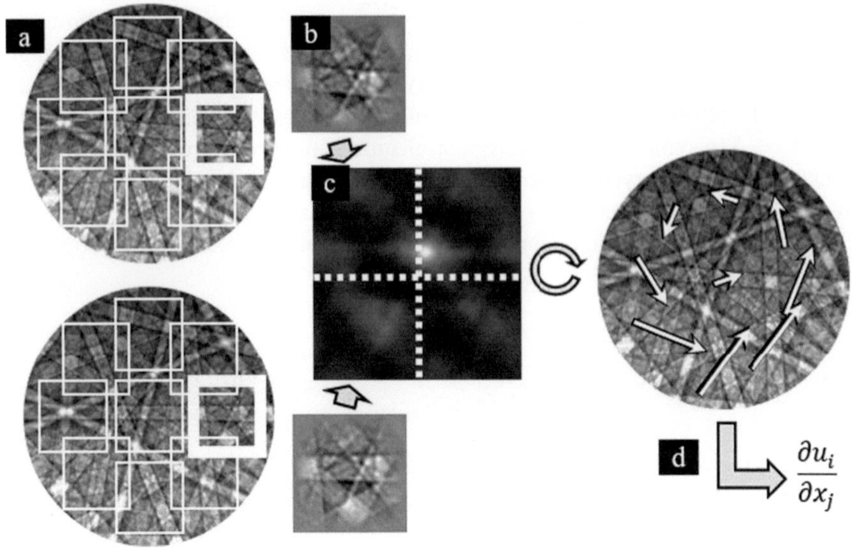

Fig. 7 Illustrated workflow for HR-EBSD: **a** define a paired sets of regions of interest (ROIs) in the reference pattern and the test pattern. **b** perform image processing on the ROI sub-images. **c** calculate the cross correlation between the ROI image pair to determine a vector describing the shift. **d** compile the vectors together to calculate the displacement gradient tensor. Vector shifts exaggerated. *Adapted from* original with permission of Cambridge [6]

$$\varepsilon_{ij} = \frac{1}{2}\left(\frac{\partial u_i}{\partial x_j} + \frac{\partial u_j}{\partial x_i}\right) \text{and}\ \omega_{ij} = \frac{1}{2}\left(\frac{\partial u_i}{\partial x_j} - \frac{\partial u_j}{\partial x_i}\right)$$

Because a hydrostatic dilation does not generate a shift in the strained pattern the individual diagonal terms of the displacement gradient tensor $\partial u_i/\partial x_j$ cannot be uniquely determined. If the stress σ_{33} normal to the free surface is assumed to be zero, then the diagonal terms can be determined using single-crystal stiffness constants for the material. The estimated resolution that can be achieved using these methods is $0.01°$ for angular precision (an order of magnitude improvement over misorientations calculated from orientation determined by conventional indexing) and 2×10^{-4} for elastic strain. If the elastic stiffness tensor (C_{ijkl}) for the material is known, the local stress state can also be calculated using the general form of Hooke's law:

$$\sigma_{ij} = C_{ijkl}\varepsilon_{kl}$$

HR-EBSD requires that the patterns used be of high quality, both in terms of the signal-to-noise ratio and also in their pixel resolution. Thus, pattern acquisition times are typically much longer than those needed for basic orientation mapping. Post-processing is also computationally intensive, both in terms of the cross-correlation calculations and image filtering required to achieve optimal results. The strain measured by HR-EBSD is relative between the two points from which the two patterns were obtained. If one of the patterns is from a known strain-free region (of the same, or nearly the same, crystal orientation as the other pattern) then the absolute strain can be obtained. Efforts to use simulated patterns to overcome this limitation have been proposed [16–18].

Recent complementary work appears promising for using the framework employed for Digital Image Correlation (DIC) for performing the cross-correlation measurements in a full-field mode [19–24]. Two other approaches proposed for measuring strain in individual EBSD patterns are (1) through sensitive measurements of the width of the Kikuchi bands [25, 26] and (2) analysis of the higher order Laue zone (HOLZ) rings that appear in EBSD patterns for some materials [27].

4 What Can Be Observed in OIM™ Maps?

Automated EBSD, where patterns are obtained at points on regular grids and the orientation determined from the acquired patterns has become the standard tool for characterizing the spatial distribution of crystallographic orientation in polycrystalline microstructures. As the orientations are measured on regular grids, it is possible to map the measurements to color or gray scale to form maps of the microstructure. The technique is often termed Orientation Imaging Microscopy or OIM™ [28]. Not only can orientation be mapped, but other metrics derived from the EBSD patterns such as pattern quality can also be mapped.

One source of confusion in mapping is that the terms pixel and scan point are used interchangeably. EBSD maps are constructed by assigning a color or grayscale value using a mapping scheme based on orientation, pattern quality, or other factors. As each point has an x, y coordinate, an image (or map) can be formed by assigning the color associated with each point to the corresponding pixel (or pixels) in the image. There is not necessarily a one-to-one correlation between the scan points on the sample and the pixels in the reconstructed image, particularly when a hexagonal scan grid is used. However, the term pixel in EBSD maps is well understood to refer to a specific scan point in the measurement grid.

4.1 Elastic Strain

As mentioned previously, several parameters are recorded during scanning, three of which are theoretically relevant to the measurement of elastic strain: pattern quality (PQ), fit, and confidence index (CI) [29]. The fit (or mean angular deviation) describes the average angular deviation between the location of the bands as determined by indexing and the locations of the bands as detected by the band detection algorithm (the Hough transform). CI describes the confidence the indexing algorithm has in its determination of the correct orientation from the detected bands. To maximize speed during mapping, the captured image size of the patterns can be small (i.e.,

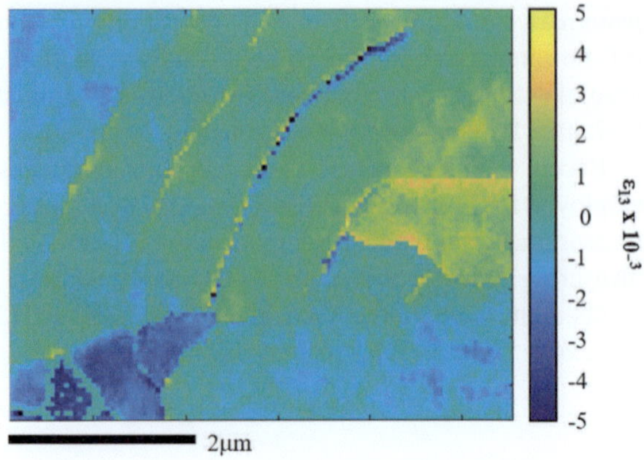

Fig. 8 Elastic strain map from an additive manufactured austenitic stainless-steel sample

100 × 100 pixels). In principle, the PQ would decrease with increasing elastic strain due to blurring of the patterns; and the fit would increase due to elastic strain as the distortions would modify the location of the bands in the pattern from the expected strain-free locations. The CI would be expected to decrease with elastic strain due to a combination of the pattern quality and shifting bands. In general, the effects of elastic strain are simply too small to be detected in the patterns used for conventional EBSD work. Nonetheless, positive mapping results using PQ have been obtained on elastically strained materials processed in a conventional manner [13]. Typically, HR-EBSD is required for mapping elastic strain [30, 31]. An example of elastic strain map is shown in Fig. 8 for one component of the elastic strain tensor from an additive manufactured austenitic steel sample. Here the variations in elastic strain arise both from dislocations structures and from deformed regions near grain boundaries.

4.2 Plastic Deformation

Considerable work has been published on OIM™-based analysis of the local deformation state. The two approaches for analyzing the effects of plastic deformation using OIM™ measurements are (1) pattern quality based [32] and (2) the analysis of local variations in orientation [33].

4.2.1 Pattern Quality Approach

Maps can be generated by mapping the pattern quality (PQ) value at each point in the scan to a grayscale as shown in Fig. 9. In this case, the PQ is defined as the average height of detected peaks in the Hough transform of the pattern which is a measure of the prominence of the bands in the EBSD pattern [34]. Since defects in materials lower PQ, regions with high defect content appear darker than pristine regions in

PQ maps. Thus, a PQ map provides, at the very least, a qualitative description of the distribution of defects in the material, which often correlates with the plastic strain distribution [35].

However, PQ is not solely dependent on defect content; other factors that impact PQ include grain boundaries, surface topology, second phases, beam conditions, sample preparation, camera settings, and others. Some of these factors can be observed in Fig. 9. When the interaction volume encompasses a grain boundary the diffraction volume includes the superposed contribution of both crystal lattices leading to a poorer quality pattern in the same vein as that described for an interaction volume containing a sub-grain boundary. Surface topology will also affect PQ as the surface will deviate from the optimal tilt condition for EBSD and, if severe enough, can lead to shadowing in the patterns. This can be helpful in some cases, as slip traces can be illuminated as sharp linear features in PQ maps. Some planes in the crystal lattice produce higher contrast diffraction bands relative to others. A pattern from an orientation containing many of these high-intensity reflections will have a higher PQ than a pattern with less high-intensity reflectors. Consequently, there will be some grain-to-grain variation in PQ simply due to lattice orientation [34]. The difficulty in separating these contributing factors makes it difficult to correlate absolute values of PQ with defect content.

Some of these effects can be mitigated through post-processing of the OIM™ data. For example, grain boundary effects can be mitigated by averaging PQs within the individual grains as shown in Fig. 10. The variation in PQ within a grain is lost, but the ability to make grain-to-grain comparisons of defect content would be improved and boundaries would no longer give the false impression as sites of defect accumulation. However, as can be seen in Fig. 10, this approach is not always effective. For the sample in Fig. 10, the misorientations associated with defect contents are better captured by the deviation of PQ within grains than the grain-averaged PQ.

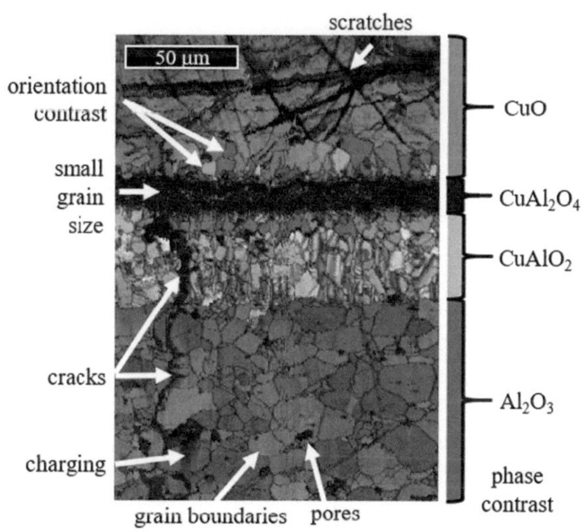

Fig. 9 Pattern quality map of a reaction zone between copper and aluminum with various microstructural features influencing pattern quality highlighted. *Modified from* original with permission of Cambridge [6]

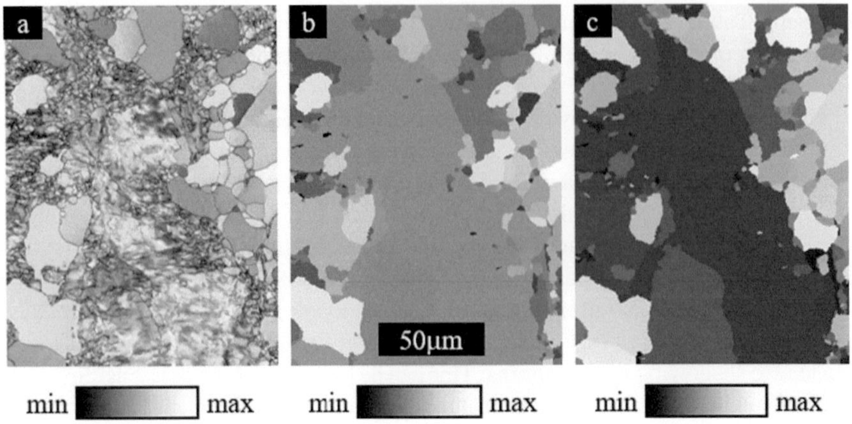

Fig. 10 Maps of partially recrystallized steel derived from EBSD pattern quality. **a** Pattern quality, **b** grain-averaged pattern quality, and **c** standard deviation of pattern quality within each grain

Other measures of PQ [36] have been proposed to reduce such effects but have found to provide only minimal improvement [34]. It should be emphasized that, in general, strain and defect content are the dominant contrast mechanism for PQ maps on well-prepared engineering materials using conventional OIM™ measurements. Other PQ-type maps [37] can be used for microstructural imaging including the distribution of strain within the microstructure, but, to date, not in a quantified manner.

4.2.2 Local Misorientation Approach

One manifestation of residual strain in the microstructure are local variations in lattice orientation due to dislocation accumulation. As OIM™ is well suited to the quantified measurement of orientation, these local variations provide a more quantitative avenue for characterizing defect populations via EBSD than PQ. There are many different methods based on the mapping of misorientation data. They can be broadly categorized into three types: (1) grain based, (2) point based, and (3) a hybrid of the first two. For grain-based metrics, there will not be any point-to-point variation in the metric, but the value will vary from grain to grain. For point-based metrics, each point in the map will have its own individual value. The hybrid approach combines elements of grain-based maps, each point will have its own value, but that value will have a relationship to the grain to which the individual scan point belongs. To provide some guidance on these different map types, data from an in situ tensile experiment on medium-carbon steel are used [38]. The original sample was fully recrystallized and then deformed under uniaxial tension to 20% strain. The tensile motion was stopped at the yield point and an EBSD scan performed. This point is denoted as the 0% strain position. The tensile motion was stopped after every 100 m of elongation and an EBSD scan performed up to 600μm elongation (6% strain) after which the motion was stopped every 200μm to 1200μm elongation (12% strain). Additional scans were performed at 15% and 20% strain but only the results out to 10% strain will be shown.

Grain-Based Maps

Two grain-based maps are the *Grain Average Misorientation* (GAM), which is the average disorientation between each neighboring pair of measurement points within the grain and the Grain Orientation Spread (GOS) which is the average disorientation between each point in a grain and the average orientation of the grain [39]. The GOS maps are the more common of the two, partly, because it is less sensitive to the step size of the measurement grid. The GOS is given by

$$\mathrm{GOS}(j) = \frac{1}{N(j)} \sum_{n=1}^{N(j)} \omega(j)_n$$

where $N(j)$ is the number of pixels in grain j and $\omega(j)_n$ is the disorientation between the average orientation [40] of grain j and the nth constituent pixel of the grain. Each grain in the GOS map is assigned a color based on the GOS value. All pixels in the grain are mapped to the assigned grain color. GOS can also be calculated by determining the average misorientation between each point in the grain with every other point in the grain [41]. Several parameters can affect the GOS calculations including those that control the algorithm for grouping pixels into grains [6].

Point-Based Maps

Several point-based maps that can be constructed from the EBSD data [6]. The most prevalent is the Kernel Average Misorientation (KAM) map. KAM is the average disorientation between a given point in the scan and its neighbors, i.e., a kernel, and provides a qualitative measure of the local GND density. For a given point, i, in the scan, the KAM is given by

$$\mathrm{KAM}(i) = \frac{1}{N} \sum_{n=1}^{N} \omega(i)_n$$

where N is the number of neighboring pixels (six for orientations measurements collected on a hexagonal grid and four for data collected on a square grid) and $\omega(i)_n$ is the disorientation between the point i and the nth point in the kernel or window of neighboring points. For a KAM map, a color is assigned to each point in the scan grid based on its KAM value. An additional criterion is also used in the KAM calculation, namely, a cutoff value for disorientation. If the disorientation

$\omega(i)_n$ exceeds this cutoff value the disorientation is excluded from the averaging calculation [41]. This methodology can be extended to larger kernels; for example, a second-order kernel would include the next nearest neighbors. While each point in the kernel can be used in the calculation, it is more common to use the points at the perimeter of the kernel in the calculation. This provides a means for accounting for differences in step size between scans.

Hybrid Maps

The Grain Reference Orientation Deviation (GROD) map merges pixel-to-pixel mapping into a grain-based framework [1, 42–44]. Each pixel is colored according to the disorientation of the pixel relative to a grain-based reference orientation. The reference orientation is typically the average orientation for a grain, but several other possibilities have been proposed as well [38]. This is a point-based map where the GROD value at point i in the scan grid is given by

$$\mathrm{GROD}(i) = \omega(i, \overline{g}(i))$$

where the function $\omega(i, \overline{g}(i))$ denotes the misorientation between the orientation at point i and the average orientation of the grain to which point i belongs, $\overline{g}(i)$. The GROD map is then the deviation within the grain with respect to the average orientation of the grain.

Mapping Example

An example of GROD, KAM, and GOS maps for an in situ tensile sample of medium-carbon steel is shown in Fig. 11 at several steps in the strain sequence along with a plot of the average GOS and KAM values as a function of strain shown in Fig. 12. As expected, Figs. 11 and 12 show that KAM and GOS increase with increasing strain. KAM and GOS have been compared in deformation studies [7, 45].

In Fig. 11, the color maps are overlaid on grayscale PQ maps. The PQ map was included partly to show the location of the grain boundaries in the map but also to show the general decrease in pattern quality with increasing strain. As expected, PQ decreases steadily with increasing strain. While some of the decrease in PQ with strain is due to increased dislocation density leading to more diffuse or noisy patterns, two other factors also will have decreased the PQ in these in situ measurements: (1) the development of surface topography with increasing strain and (2) an increase in the buildup of hydrocarbon contamination with repeated scanning of the electron beam over the surface of the sample both of which are challenges of in situ EBSD work.

While GROD maps provide a visualization of the spatial variation in crystallographic orientation within individual grains, a GROD map does not provide any

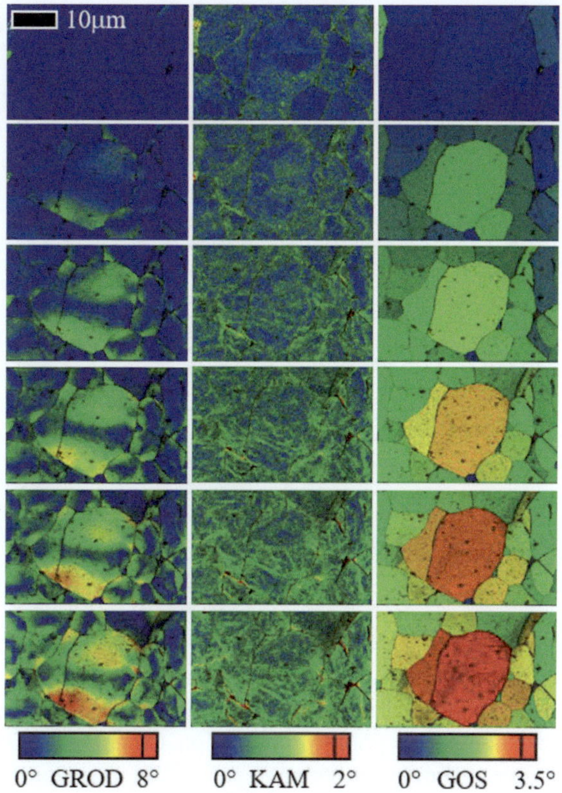

Fig. 11 GOS, KAM, and GROD maps for an in situ steel tensile specimen at 0, 2, 4, 6, 8, and 10% strain. Tensile axis is vertical. GROD and KAM maps adapted from originals by permission of TMS [38]

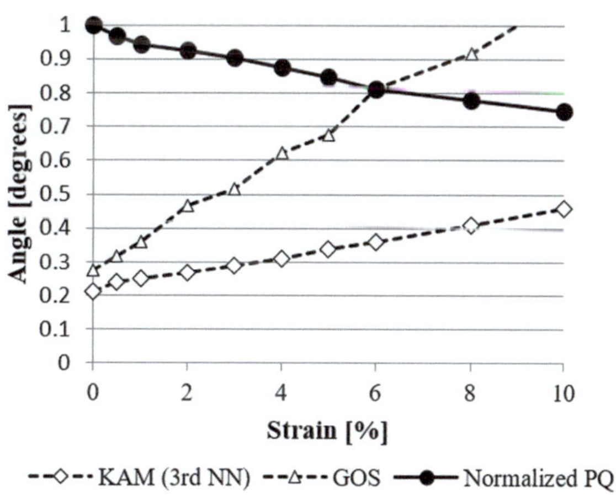

Fig. 12 Plots of the average values for GOS, KAM, and PQ (normalized to the maximum) for an in situ steel tensile specimen as a function of strain for an in situ steel tensile specimen. *Adapted from* original by permission of TMS [38]

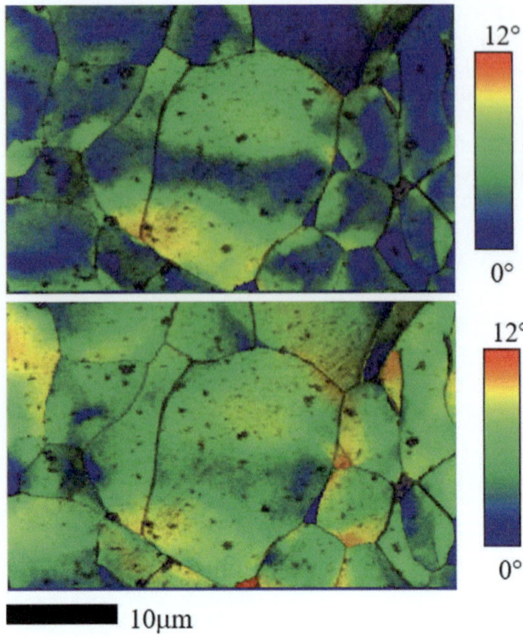

Fig. 13 **a** $GROD_A$ and **b** $GROD_0$ maps for in situ tensile steel sample at 10% strain. *Adapted from* originals by permission of TMS [38]

indication of how the orientations within the grain have evolved relative to their original orientations prior to deformation. Coupling the GROD idea with analysis of the in situ tensile data provides a picture of the evolution of orientation with deformation. First, the average orientation of each grain within the scan area for the initial 0% strain scan is calculated. These average orientations are then used as the reference orientations for each of the grains in the 10% strain scan to create the GROD map. This variant of the GROD map is denoted $GROD_0$ to differentiate it from the original GROD map which, hereafter, will be labeled $GROD_A$. An example for the in situ tensile data is shown in Fig. 13.

A comparison of these two maps with a focus on the large grain at the center of the scan elucidates the twofold nature of strain accommodation in a polycrystal: (1) $GROD_0$ is dominated by the rigid body rotation of the grain as a whole and (2) $GROD_A$ captures the orientation gradients that develop within the grain. $GROD_0$ correlates with the uniform strain "Taylor" model of deformation where the grain rotates independently of its neighbors and $GROD_A$ correlates with the iso-stress "Sachs" model where domains within the grain rotate to preserve structural compatibility with neighboring grains. Another feature of note is that a GROD map derived from a single post-deformation scan ($GROD_A$) provides a description of the current state of the material but does not provide the evolutionary context for the current state that $GROD_0$ can provide. The in situ data shows that it is invalid to assume that orientations near the average grain orientation in $GROD_A$ (blue in the color mapping scheme used here) represent regions that remain stationary during deformation and the surrounding points are rotating away from or around these stationary points.

Fig. 14 $GROD_A$ maps for in situ tensile steel sample at 10% strain showing the axis of misorientation relative to the average orientation for each grain: **a** in crystal space, **b** the in-plane direction of the misorientation axis relative to the sample coordinates, and **c** the out-of-plane direction of the misorientation axis relative to the sample surface. **a** modified from original by permission of TMS [38]

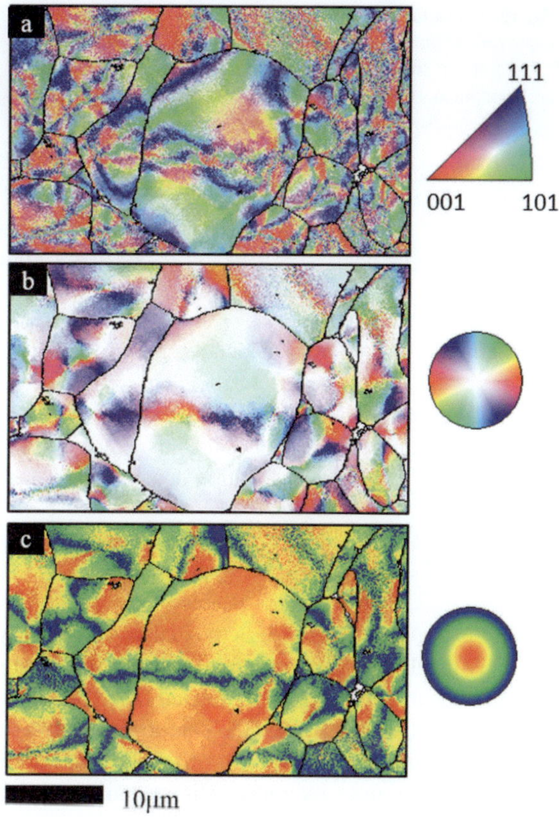

EBSD-based strain studies have been performed considering both the axis and angle of rotation [46, 47]. The foundation of the GROD map can be extended to illustrate the role the misorientation axis plays in the orientation deviation. Figure 14 shows a GROD map that expresses the misorientation not in terms of disorientation but in terms of the misorientation axis in both the crystal and in sample reference frames.

Note that regions in the GROD disorientation map with small disorientation tend to be speckled in the GROD misorientation axis map. Speckling is due to ambiguity as the disorientation approaches zero. This is in part due to lack of angular precision in the orientation measurements but is also due to the fact that as the disorientation approaches zero, the axis of rotation is not well defined. The fragmentation of the grain into domains is emphasized in the GROD axis maps, particularly for the map in crystal space. Both the axis and angle of misorientation can be coupled together into a single description using Rodrigues vectors [48].

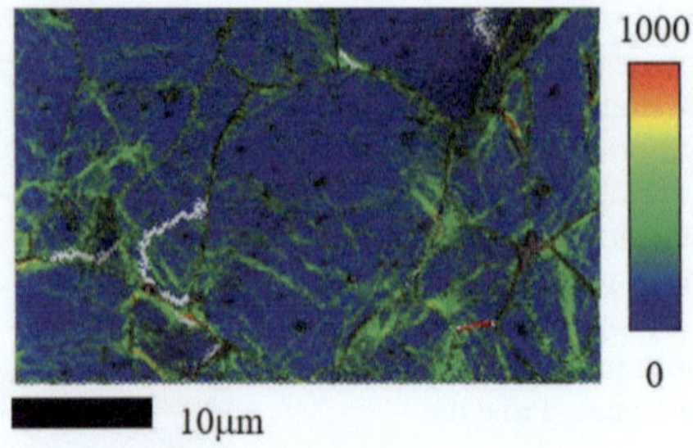

Fig. 15 Color GND map overlaid on a gray scale PQ map for in situ tensile steel sample at 10% strain

Geometrically Necessary Dislocation Maps

The GND density can be quantified from the orientation data using a Nye's-tensor-based approach. Nye's tensor is a statistical representation of the local dislocation density and is related to dislocation density by

$$\alpha_{ij} = \sum_t \rho^t b_i^t l_j^t$$

where t is the slip system, ρ^t is the dislocation density associated with that system, and l_j^t is the line direction associated with the system. If elastic strain gradients are ignored, this tensor can be related directly to the curvature tensor, or the curl of the orientation field [49, 50]. Due to the number of available slip systems and the unavailability of gradient information in the z-direction, Nye's tensor cannot be mapped directly onto individual slip systems without additional assumptions [51], so it is often used to estimate bulk dislocation density irrespective of slip system. However, work by multiple groups has shown that this approach can provide accurate dislocation densities for individual slip systems, particularly when coupled with the increased orientation accuracy of HR-EBSD [51–54]. It should be reiterated that the GND density provides only a partial description of the dislocation density in a material as it does not capture the dislocation density associated with the SSDs. EBSD-based GND calculations are very sensitive to the distance between the pairs of points on the EBSD scan grid and the precision of the orientation measurements [49]. A GND map for the in situ steel data is presented in Fig. 15.

4.2.3 Orientation Precision

The impact of the precision of the EBSD orientation measurements has been alluded to several times. Typically, with decreasing step size the disorientation between points on the scan grid also decreases. This has two effects on gradient-based GND density measurements. First, GND density measurements rely on measuring spatial derivatives. Decreasing the step size decreases the signal-to-noise ratio in the measurements and increases the minimum measurable GND density. This effect is captured in Fig. 16 [55].

Secondly, very small step sizes can lead to misorientations nearing the angular resolution limit of EBSD. As the metrics presented give an indication of strain via point-to-point misorientations, the strain will be overestimated because some misorientations due to noise in the orientation measurements will contribute. Thus, for example, on a scan with a very fine measurement grid third nearest-neighbor KAMs may give more reliable results than first nearest-neighbor KAMs as the distance between the point pairs used in the KAM calculations will be larger and hence the misorientations are more likely to fall within the range that EBSD can accurately measure [56, 57]. As deformation increases the point-to-point misorientations increase and are easier for EBSD to reliably measure. But simultaneously, pattern quality decreases (due to both GNDs and SSDs) which can hamper orientation precision. Nonetheless, it is possible, even in severely deformed materials or in materials deformed at high rates (i.e., shock), to obtain EBSD patterns of high enough quality to get useful results [58, 59]. Noise effects can be mitigated by adapting image filtering techniques to orientation data [60]. Orientation precision can also be improved through Neighbor Pattern Averaging (NPAR) [61] and Dictionary Indexing (DI) with refinement even on data obtained at very rapid collection rates [62]. Figure 17 compares results obtained from a nickel sample using standard Hough-based indexing, Hough-based indexing after NPAR, and dictionary indexing with orientation refinement. Clearly some improvement in the orientation precision is achieved even with the low-resolution patterns used (60 × 60 pixels).

Greater improvement can be achieved using HR-EBSD for the misorientation calculations at the core of local misorientation mapping. As noted previously, HR-EBSD is best suited to high-resolution patterns (i.e., large pixel dimensions) and thus acquisition times are considerably longer compared to those typical of conventional Hough-based measurements of orientation. Figure 18 compares GND results derived from conventional EBSD measurements and HR-EBSD measurements. It is clear that

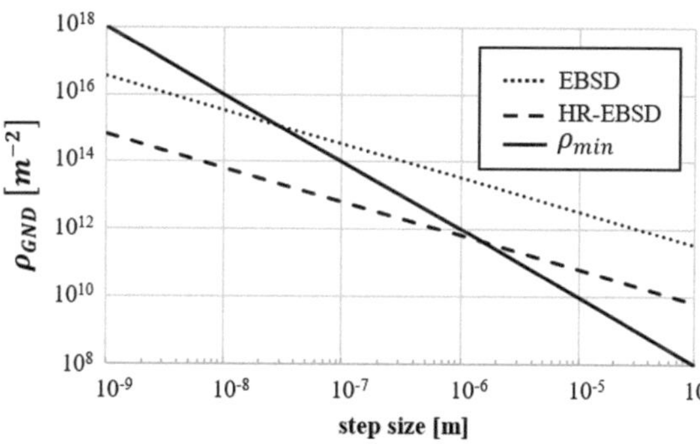

Fig. 16 Dislocation density resolution estimates as a function of EBSD scan grid step size

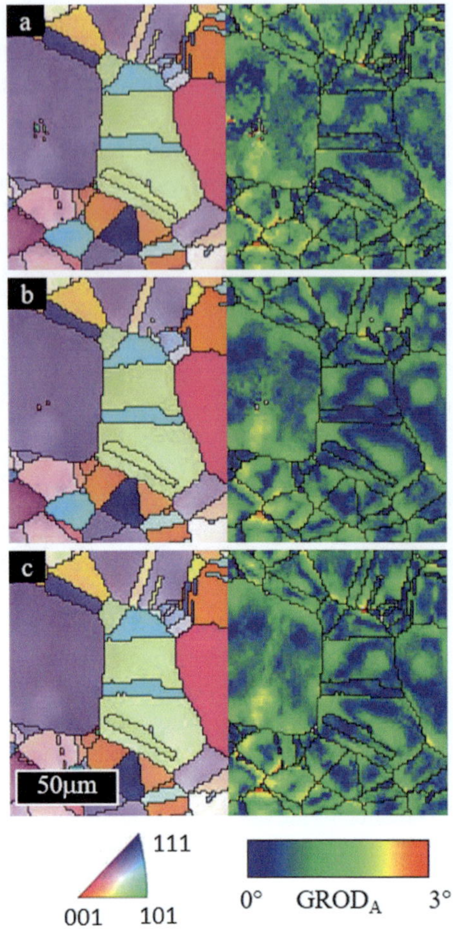

Fig. 17 Orientation maps colored according to the crystal direction parallel to the sample normal and $GROD_A$ maps for nickel using EBSD measurements obtained **a** using standard Hough-based indexing, **b** Hough-based indexing on patterns after NPAR, and **c** dictionary indexing followed by orientation refinement

some of the structure in the conventional measurements is an artifact as it is absent from the HR-EBSD measurements.

4.2.4 Scaling to Bulk Strain

If the average values from the local misorientation maps described are assumed to correspond to the measured overall strain, then the local strain can be approximated by assuming a linear relationship between the local misorientation metric and the strain. However, it should be remembered that the misorientation metrics are based solely on GNDs and provide no information on SSDs. Since EBSD can only detect GNDs, one would have to make the gross assumption that the SSDs are uniformly distributed within the microstructure to scale the EBSD results to a bulk strain. Another obstacle to scaling is that the strain distribution in the material is inherently three dimensional

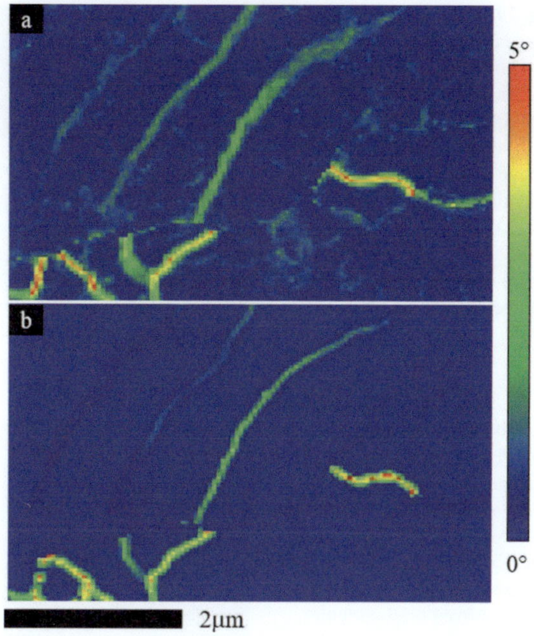

Fig. 18 KAM maps based on **a** conventional EBSD and **b** HR-EBSD for a sample of additive manufactured 316 stainless steel

while OIM™ maps are obtained on two-dimensional section planes. Furthermore, materials that lack internal obstacles to dislocation propagation such as single crystals can be strained without any dislocation accumulation. In this case, local crystal rotations would not develop, and the strain would be undetectable via orientation gradient-based measurements.

5 Complementary Techniques

Correlating EBSD results with other characterization techniques can provide a more comprehensive understanding of the accommodation of strain within the microstructure. Individual dislocations can be imaged using the transmission electron microscope (TEM) and/or at the same SEM spatial resolution as EBSD using electron channeling contrast imaging (ECCI) as shown in Figs. 19 and 20 [51].

Micro-digital image correlation (DIC) in the SEM can be correlated directly with EBSD measurements [63, 64]. The example shown in Fig. 21 shows how the correlation between the two techniques enables the varying modes of slip within individual grains to be identified and also highlights the difference between dislocation motion (the narrow slip bands shown in DIC) and dislocation/stress accumulation (shown in $GROD_A$). DIC provides a description of the total strain and EBSD provides a description of the material state.

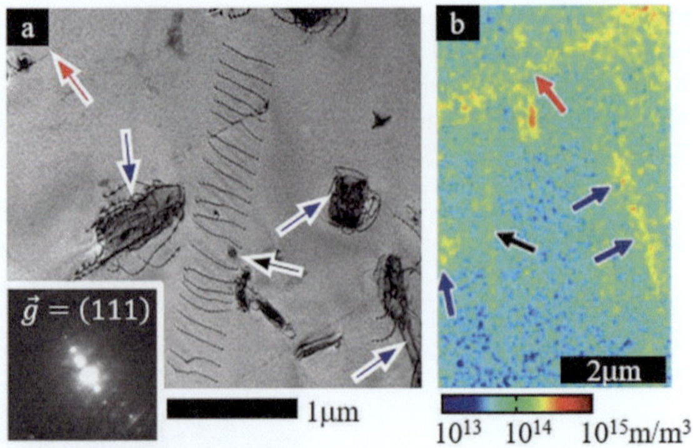

Fig. 19 **a** Bright field TEM image and **b** HR-EBSD GND map from an austenitic 302 stainless-steel sample highlighted to show dislocations around a second-phase particle (blue), individual dislocations (black), and twin boundary (red) in both techniques. Figure adapted from original with permission of Elsevier [51]

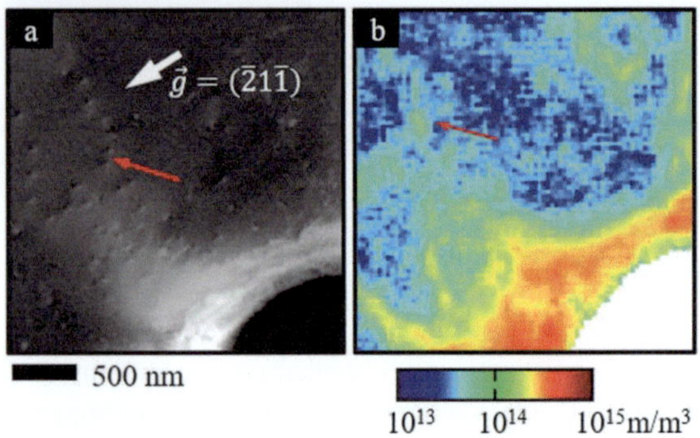

Fig. 20 **a** ECCI image and **b** HR-EBSD GND map from an austenitic 302 stainless-steel sample. The red arrow highlights the same pair of dislocations in both images for reference. Figure modified from original with permission of Elsevier [51]

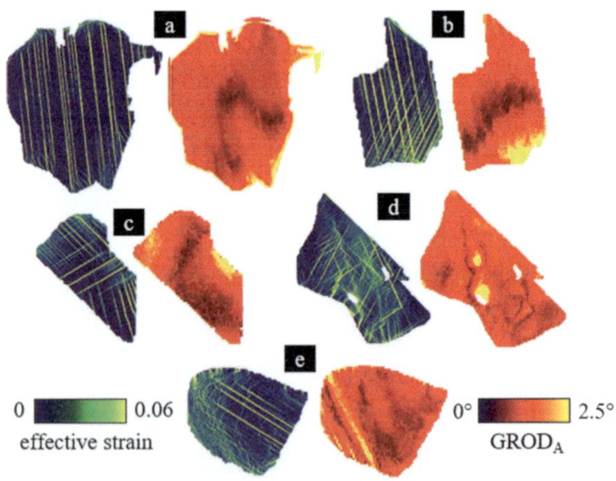

Fig. 21 Micro-DIC and GROD maps from a polycrystalline sample of Ni-based superalloy RR1000 deformed in uniaxial tension (the tensile axis is horizontal). Maps for individual grains exhibiting **a** single slip, **b** double slip, **c** triple slip, **d** cross slip, and **e** diffuse slip. Adapted from original courtesy of Elsevier and A. Harte [63]

EBSD is a two-dimensional surface-focused characterization technique but it can be extended to three dimensions through serial sectioning [65–67]. It can also be illuminating to compare EBSD results in both two and three dimensions with three-dimensional results obtained by high-energy X-ray diffraction (HEDM), i.e., obtained at synchrotron beam lines [68].

6 Conclusions

EBSD-based strain and deformation analysis remains a very active research arena both in terms of technical improvements and in application particularly in connecting EBSD-based strain characterization with deformation models linked with complementary techniques. Local misorientation analysis based on EBSD orientation measurement is very capable of identifying areas of concentrated strain within a microstructure due to dislocation accumulation but is not as well suited to identifying the actual magnitude of strain in those areas. As EBSD is very much a surface-sensitive technique careful preparation of the sample surface for EBSD characterization is imperative [69]. Any deformation introduced during preparation of the sample can mask the true deformation structure [70]. The metrics introduced here provide only a portion of the comprehensive characterization of strain in polycrystalline materials [71]; integrating EBSD with other complementary techniques can help complete the picture. The strain analyses described here focus on strain being accommodated through dislocations. However, dislocation generation can be accompanied by grain boundary sliding [72] and/or twinning [73, 74].

Acknowledgements This work represents the compilation of the efforts of many outstanding colleagues. In particular, the authors acknowledge David Field of Washington State University and Matt Nowell and René de Kloe of Gatan for their many helpful insights in the arena of EBSD strain analysis. Seiichi Suzuki of TSL Solutions in Japan is acknowledged for his help with the in situ data presented. Thanks to Allan Harte of UK Atom Energy Authority and João Quinta da Fonseca of the University of Manchester for sharing their DIC results. JK acknowledges support from the U.S. National Science Foundataion (DMR-2043264). Sandia National Laboratories is a multi-mission laboratory managed and operated by National Technology and Engineering Solutions of Sandia, LLC, a wholly owned subsidiary of Honeywell International, Inc., for the US Department of Energy's National Nuclear Security Administration under contract DE-NA0003525. This paper describes objective technical results and analysis. Any subjective views or opinions that might be expressed in the paper do not necessarily represent the views of the US Department of Energy or the United States Government.

References

1. Wright SI (1993) A review of automated orientation imaging microscopy (OIM^{TM}). J Comput Assit Microsc 5(3):207–221
2. Wright SI, Gray GT, Rollett AD (1994) Textural and microstructural gradient effects on the mechanical behavior of a tantalum plate. Metall Mater Trans A 25:1025–1031
3. Schwartz AJ, Kumar M, Adams BL, Field DP (2009) Electron backscatter diffraction in materials science, 2nd edn. Springer, New York

4. Kimura H, Wang Y, Akiniwa Y, Tanaka K (2005) Misorientation analysis of plastic deformation of austenitic stainless steel by EBSD and X-ray diffraction methods. Trans Jpn Soc Mech Eng A 71(712):1722–1728
5. Dillien S, Seefeldt M, Allain S, Bouaziz O, Houtte PV (2010) EBSD study of the substructure development with cold deformation of dual phase steel. Mater Sci Eng A 527(4–5):947–953
6. Wright SI, Nowell MM, Field DP (2011) A review of strain analysis using electron backscatter diffraction. Microsc Microanal 17(3):316–329
7. Allain-Bonasso N, Wagner F, Berbenni S, Field DP (2012) A study of the heterogeneity of plastic deformation in IF steel by EBSD. Mater Sci Eng A 548:56–63
8. Kamaya M (2012) Assessment of local deformation using EBSD, quantification of local damage at grain boundaries. Mater Charac 66:56–67
9. Jedrychowski M, Tarasiuk J, Bacroix B, Wronski S (2013) Electron backscatter diffraction investigation of local misorientations and orientation gradients in connection with evolution of grain boundary structures in deformed and annealed zirconium. A new approach in grain boundary analysis. J Appl Cryst 46(2):483–492
10. Schayes C, Bouquerel J, Vog JB, Palleschi F, Zaefferer S (2016) A comparison of EBSD based strain indicators for the study of Fe-3Si steel subjected to cyclic loading. Mater Charac 115:61–70
11. Callahan PG, Graef MD (2013) Dynamical electron backscatter diffraction patterns, Part I: Pattern simulations. Microsc Microanal 19(5):1255
12. Winkelmann A, Cowan CT, Sweeney F, Day AP, Parbrook P (2007) Many-beam dynamical simulation of electron backscatter diffraction patterns. Ultramicroscopy 107:414–421
13. Keller RR, Roshko A, Geiss RH, Bertness KA, Quinn TP (2004) EBSD measurement of strains in GaAs due to oxidation of buried AlGaAs layers. Microelectron Eng 75:96–102
14. Wilkinson AJ, Meaden G, Dingley DJ (2006) High-resolution elastic strain measurement from electron backscatter diffraction patterns: new levels of sensitivity. Ultramicroscopy 106(4–5):307–313
15. Wilkinson AJ, Britton TB (2012) Strains, planes, and EBSD in materials science. Mater Today 15(9):366–376
16. Kacher J, Landon C, Adams BL, Fullwood D (2009) Bragg's Law diffraction simulations for electron backscatter diffraction analysis. Ultramicroscopy 109(9):1148–1156
17. Jackson BE, Christensen JJ, Singh S, Graef MD, Fullwood DT, Homer ER, Wagoner RH (2016) Performance of dynamically simulated reference patterns for cross-correlation electron backscatter diffraction. Microsc Microanaal 22(4):789–802
18. Tanaka T, Wilkinson AJ (2019) Pattern matching analysis of electron backscatter diffraction patterns for pattern centre, crystal orientation and absolute elastic strain determination–accuracy and precision assessment. Ultramicroscopy 202:87–99
19. Alkorta J, Marteleur M, Jacques PJ (2017) Improved simulation based HR-EBSD procedure using image gradient based DIC techniques. Ultramicroscopy 182:17–27
20. Vermeij T, Hoefnagels J (2018) A consistent full-field integrated DIC framework for HR-EBSD. Ultramicroscopy 191:44–50
21. Ruggles T, Bomarito G, Qiu R, Hochhalter J (2018) New levels of high angular resolution EBSD performance via inverse compositional Gauss-Newton based digital image correlation. Ultramicroscopy 195:85–92
22. Shi Q, Roux S, Latourte F, Hild F (2019) Estimation of elastic strain by integrated image correlation on electron diffraction patterns. Ultramicroscopy 199:16–33
23. Ernould C, Beausir B, Fundenberger JJ, Taupin V, Bouzy E (2020) Global DIC approach guided by a cross-correlation based initial guess for HR-EBSD and on-axis HR-TKD. Acta Mater 191:131–148
24. Zhu C, Kaufmann K, Vecchio KS (2020) Novel remapping approach for HR-EBSD based on demons registration. Ultramicroscopy 208:112851
25. Fodchuk I, Balovsyak S, Borcha M, Garabazhiv Y, Tkach V (2011) Determination of structural inhomogeneity of synthesized diamonds by backscattering electron diffraction. Phys Status Solidi 208:2591–2596

26. Ram F, Zaefferer S, Raabe D (2014) Kikuchi bandlet method for the accurate deconvolution and localization of Kikuchi bands in Kikuchi diffraction patterns. J Appl Cryst 47:264–275
27. Michael J, Eades J (2000) Use of reciprocal lattice layer spacing in electron backscatter diffraction pattern analysis. Ultramicroscopy 81:67–81
28. Adams BL, Wright SI, Kunze K (1993) Orientation imaging: the emergence of a new microscopy. Metal Trans A 24:819–831
29. Nowell MM, Wright SI (2005) Orientation effects on indexing of electron backscatter diffraction patterns. Ultramicroscopy 103:41–58
30. McLean MJ, Osborn WA (2018) In-situ elastic strain mapping during micromechanical testing using EBSD. Ultramicroscopy 185:21–26
31. Miyamoto G, Shibata A, Maki T, Furuhara T (2009) Precise measurement of strain accommodation in austenite matrix surrounding martensite in ferrous alloys by electron backscatter diffraction analysis. Acta Mater 57:1120–1131
32. Lassen NCK, Jensen DJ, Conradsen K (1994) Automatic recognition of deformed and recrystallized regions in partly recrystallized samples using electron backscattering patterns. Mat Sci For 157–162:149–158
33. Field DP (1995) Quantification of partially recrystallized polycrystals using electron backscatter diffraction. Mat Sci Eng A 190:241–246
34. Wright SI, Nowell MM (2006) EBSD image quality mapping. Microsc Microanal 12:72–84
35. Wardle ST, Lin LS, Cetel AD, Adams BL (1994) Orientation imaging microscopy: monitoring residual stress profiles in single crystals using and imaging quality parameter. IQ. In: Bailey GW, Reed AJG (eds) Proceedings of the 52nd Annual Meeting of the Microscopy Society of America. San Francisco Press, San Francisco, pp 680–681
36. Tao X, Eades A (2005) Errors, artifacts and improvements in EBSD processing and mapping. Microsc Microanal 11:79–87
37. Wright SI, Nowell MM, Kloe RD, Camus P, Rampton T (2015) Electron imaging with an EBSD detector. Ultramicroscopy 148:132–145
38. Wright SI, Suzuki S, Nowell MM (2016) In situ EBSD observations of the evolution in crystallographic orientation with deformation. JOM 68(11):2730–2736
39. Wright SI (1999) Quantification of recrystallized fraction from orientation imaging scans. In: Szpunar JA (ed) Proceedings of the twelfth international conference on textures of materials (ICOTOM 12), vol 1. NRC Research Press, Ottawa, pp 104–109
40. Kunze K, Wright SI, Adams BL, Dingley DJ (1993) Advances in automatic EBSP single orientation measurements. Text Microsc 20:41–54
41. Lehockey EM, Lin YP, Lepik OE (2000) Mapping residual plastic strain in materials using electron backscatter diffraction. In: Schwartz AJ, Kumar M, Adams BL (eds) Electron backscatter diffraction in materials science. Kluwer Academic/Plenum Pub, New York, pp 247–264
42. Brewer LN, Othon MA, Young LM, Angeliu TM (2002) Misorientation mapping for visualization of plastic strain via electron back-scattered diffraction In: Voelkel E, Piston D, Gauvin R, Lockley AJ, Bailey GW, McKernan S (eds) Microscopy and Microanalysis. Cambridge University Press, Québec City, Québec, Canada, pp 684CD–685CD
43. Kysar JW, Briant CL (2002) Crack tip deformation fields in ductile single crystals. Acta Mater 50:2367–2380
44. Li BL, Godfrey A, Liu Q (2002) Investigation of macroscopic grain sub-division of an IF-steel during cold-rolling. In: Lee DN (ed) ICOTOM 13, vol Vols. 408–412. Textures of Materials. Trans Tech Publications Inc., Seoul, Korea, pp 1185–1190
45. Gandarilla FC, Garrido AMS, Bolmaro RE, Baudin T, Natalia S, Avalos M, Moreno JGC, Leon HM (2016) Microstructural evolution and mechanical properties on an ARB processed IF steel studied by X-ray diffraction and EBSD. Mater Char 118:332–339
46. Chun Y, Davies C (2011) Investigation of prism< a> slip in warm-rolled AZ31 alloy. Metal Mater Trans A 42:4113–4125
47. Michels ZD, Kruckenberg SC, Davis JR, Tikoff B (2015) Determining grain-scale vorticity axes from crystallographic orientation data. In: AGU fall meeting abstracts

48. Field DP, Kumar M, Trivedi PB, Wright SI (2005) Analysis of local orientation gradients in deformed single crystals. Ultramicroscopy 103:33–39
49. Field D, Merriman C, Bonasso NA, Wagner F (2012) Quantification of dislocation structure heterogeneity in deformed polycrystals by EBSD. Model Simul Mater Sci Eng 20:024007
50. Wilkinson AJ, Randman D (2010) Determination of elastic strain fields and geometrically necessary dislocation distributions near nanoindents using electron back scatter diffraction. Philos Magaz 90:1159–1177
51. Ruggles T, Yoo Y, Dunlap B, Crimp M, Kacher J (2020) Correlating results from high resolution EBSD with TEM-and ECCI-based dislocation microscopy: approaching single dislocation sensitivity via noise reduction. Ultramicroscopy 210:112927
52. Ruggles T, Fullwood D, Kysar J (2016) Resolving geometrically necessary dislocation density onto individual dislocation types using EBSD-based continuum dislocation microscopy. Inter J Plast 76:231–243
53. Yoo YSJ, Book TA, Sangid MD, Kacher J (2018) Identifying strain localization and dislocation processes in fatigued Inconel 718 manufactured from selective laser melting. Mater Sci Eng A 724:444–451
54. Wallis D, Hansen LN, Britton TB, Wilkinson AJ (2017) Dislocation interactions in olivine revealed by HR-EBSD. J Geophys Res Solid Ea 122:7659–7678
55. Adams BL, Kacher J (2009) EBSD-based microscopy: Resolution of dislocation density. Comput Mater Cont 14:185–196
56. Ram F, Zaefferer S, Jäpel T, Raabe D (2015) Error analysis of the crystal orientations and disorientations obtained by the classical electron backscatter diffraction technique. J Appl Cryst 48:797–813
57. Wright SI, Nowell MM, Kloe RD, Chan L (2014) Orientation precision of electron backscatter diffraction measurements near grain boundaries. Microsc Microanal 20:852–863
58. Bingert JF, Livescu V, Cerreta EK (2009) Characterization of shear localization and shock damage with EBSD, In: Electron backscatter diffraction in materials science. Springer, pp 301–315
59. McNelley TR, Zhilyaev AP, Swaminathan S, Su J, Menon ES (2009) Application of EBSD methods to severe plastic deformation (SPD) and related processing methods. In: Electron Backscatter Diffraction in Materials Science. Springer, pp 277–289
60. Wright S, Field D, Nowell M (2015) Post processing effects on GND calculations from EBSD-based orientation measurements. IOP Conf Ser Mat Sci Eng 1:012049
61. Wright SI, Nowell MM, Lindeman SP, Camus PP, Graef MD, Jackson MA (2015) Introduction and comparison of new EBSD post-processing methodologies. Ultramicroscopy 159:81–94
62. Singh S, Ram F, Graef MD (2017) Application of forward models to crystal orientation refinement. J Appl Cryst 50:1664–1676
63. Harte A, Atkinson M, Preuss M, Fonseca JQD (2020) A statistical study of the relationship between plastic strain and lattice misorientation on the surface of a deformed Ni-based superalloy. Acta Mater 195:555–570
64. Kammers AD, Daly S (2013) Self-assembled nanoparticle surface patterning for improved digital image correlation in a scanning electron microscope. Exp Mech 53:1333–1341
65. Echlin MP, Mottura A, Torbet CJ, Pollock TM (2012) A new TriBeam system for three-dimensional multimodal materials analysis. Rev Sci Inst 83:023701
66. Groeber MA, Rowenhorst DJ, Uchic MD (2009) Collection, processing, and analysis of three-dimensional EBSD data sets. In: Electron Backscatter Diffraction in Materials Science. Springer, pp 123–137
67. Zaefferer S, Wright SI, Konjinenberg PJ (2015) 3D EBSD. In: Brissett F (ed) EBSD: Analyse par diffraction des électrons rétrodiffusés. Groupment National de Microscopie Electronique a Balayage et de Microanalyses, Paris, pp 159–184
68. Nygren K, Pagan D, Miller M (2019) Sub-grain orientation resolution during continuous loading using only far-field HEDM. IOP Conf Ser Mat Sci Eng, IOP Publishing. 1:012018
69. Katrakova D, Mucklich F (2001) Specimen preparation and electron backscatter diffraction–part I: metals. Pract Metall 38:547–565

70. Field D, Magid K, Mastorakos I, Florando J, Lassila D, Morris JJ (2010) Mesoscale strain measurement in deformed crystals. A comparison of X-ray microdiffraction with electron backscatter diffraction. Philos Magaz 90:1451–1464
71. Jiang J, Zhang T, Dunne FP, Britton TB (2016) Deformation compatibility in a single crystalline Ni superalloy. Proc R Soc A Math Phys Eng Sci 472:20150690
72. Han J, Kang SH, Lee SJ, Kawasaki M, Lee HJ, Ponge D, Raabe D, Lee YK (2017) Superplasticity in a lean Fe-Mn-Al steel. Nat Commun 8:1–6
73. Urrutia IG, Zaefferer S, Raabe D (2010) The effect of grain size and grain orientation on deformation twinning in a Fe–22 wt.% Mn–0.6 wt.% C TWIP steel. Mater Sci Eng A 527:3552–3560
74. Yu Q, Wang J, Jiang Y, McCabe RJ, Tomé CN (2014) Co-zone 1 012 twin interaction in magnesium single crystal. Mater Res Lett 2:82–88

Chapter 4
In Situ EBSD: A High Throughput Technique Across the Real and Reciprocal Space to Decipher Micro-Mechanisms in Materials

Subhasis Sinha, Reshma Sonkusare, and N. P. Gurao

1 Introduction

Microstructure dependence of mechanical response and thermal fields is a fundamental area of interest that constitutes microstructure-processing-property relationships used for materials design. Therefore, the study of microstructural evolution during deformation and heat treatment is essential to decipher the underlying micro-mechanisms of plasticity, damage evolution, fracture, recovery, recrystallization, grain growth and phase transformation. Traditionally, post-mortem analyses of deformed, fractured and heat-treated specimens are used to correlate the microstructural evolution in the sample subjected to thermomechanical treatment or mechanical testing with operative micro-mechanisms. While the latter remains the workhorse for materials science research, the quest for detailed investigation into a problem has led to the development of newer techniques such as in situ experimentation, to supplement the information and insight gained from traditional post-mortem microscopy of deformed and heat-treated samples. The advantage with in situ methods is that microstructural evolution can be captured as it happens with the application of load

S. Sinha
Department of Metallurgical Engineering, Indian Institute of Technology (BHU) Varanasi, Varanasi, India
e-mail: ssinha.met@iitbhu.ac.in

R. Sonkusare
Institute of Materials and Process Design, Helmholtz Zentrum Hereon, Geesthacht, Germany
e-mail: reshma.sonkusare@hereon.de

N. P. Gurao (✉)
Department of Materials Science and Engineering, Indian Institute of Technology Kanpur, Kanpur, India
e-mail: npgurao@iitk.ac.in

S. Suwas and D. P. Field (eds.), *Advances in Texture, Microtexture, and Allied Techniques*, Materials Horizons: From Nature to Nanomaterials,
https://doi.org/10.1007/978-981-96-5346-1_4

or thermal fields on the specimen. In situ EBSD can provide evolution of microstructure in terms of size, shape and distribution of different phases, their composition and crystallographic orientation as well defect structure like dislocation density and damage accumulation from in operando observation of the real microstructural and compositional space as well as the reciprocal diffraction space. The evolution of microstructure and micro-texture as a function of external field captured during this high throughput experimental technique can be used to tune crystal plasticity simulations for predicting deformation behaviour or phase field models to capture recrystallization and phase transformation. Thus, in situ EBSD forms a very important part of the ICME paradigm that relies on multi-scale computational simulations complimented by high throughput experiments for rapid materials and process development.

In situ electron back scatter diffraction (in situ EBSD) experiments consist of the use of a tensile stage that is inserted into the scanning electron microscope (SEM) chamber for carrying out deformation. The first in situ EBSD experiments were performed in 1996 [1]. Subsequently, simultaneous advancement in both automated EBSD technology and SEM technology (with the advent of field emission gun SEMs) contributed to the development of the in situ EBSD method. It must be mentioned here that before in situ stages became widespread, a few researchers also performed *pseudo* in situ experiments to achieve similar goals, by tracking the microstructural evolution within a region on the specimen marked by microhardness indentations [2, 3]. The primary aim of these experiments (both pseudo in situ and in situ deformation) is to capture the lattice rotations of individual grains, development of intragranular orientation gradients, grain fragmentation or subgrain formation caused by deformation of the polycrystalline material.

Another type of in situ EBSD experiments is in situ heating. The in situ stage that is inserted into the SEM chamber could be a dedicated tensile or only heating stage or a combined tensile/heating stage. For the in situ heating experiments, the stage is connected to an external temperature controller. In situ heating experiments are usually carried out to study recrystallization and grain growth or phase transformation.

With the development of in situ EBSD into a standard technique, many researchers used this method to study microstructural and micro-textural evolution in various metals and alloys [4–11]. Buchheit et al. carried out an in situ tensile EBSD study of annealed polycrystalline tantalum to reveal the deformation-induced development of orientation gradients and substructure within individual grains [6]. Interestingly, they could distinguish the macro-textural evolution from the micro-textural evolution, wherein the former evolved slowly but the intragranular orientation gradients comprising the latter developed rapidly with deformation. Through this study, they also identified that significant orientation changes occurred at the grain boundaries making them the preferred sites for subgrain formation.

Donoghue et al. carried out in situ EBSD with heating of additively manufactured Ti-6Al-4 V, where they showed that when slight in-process deformation is applied, the α colonies undergo deformation twinning [7]. These twinned α colonies gives rise to new β orientations upon reheating near the transus temperature, which is a possible mechanism for obtaining structurally refined β grains in the additively manufactured Ti-6Al-4 V alloy. On the other hand, many researchers also carried out in situ tensile EBSD studies of Ti-6Al-4 V [10, 12].

A recent investigation by Hémery and Villechaise discussed deformation mechanisms and strain partitioning in Ti-6Al-4 V in great detail based on analysis of lattice rotations through in situ EBSD method [12]. The lattice rotations were tracked through in situ EBSD and used as an indicator of slip activity. The rotation axis associated with plastic deformation was used to analyze active slip systems. Combined with slip trace analysis, the slip plane and direction of active slip systems could both be determined. Crystal plasticity was used to interpret plastic shear magnitude from the lattice rotation data and quantitatively obtain the micro-scale plastic strain. From the latter estimate, strain partitioning based on microstructure and active slip modes was explained. Once again, this is an example where in situ EBSD enabled to distinguish between the local (intragranular) and global deformation behaviours. The study revealed that intragranular plastic slip contributed significantly to global lattice rotation behaviour, while rigid body rotations were less significant.

Thus, the aforementioned studies highlight the advantage of in situ EBSD deformation studies with respect to the micro-mechanisms of deformation and their contribution to the macroscopic deformation behaviour. This technique enables detailed discussion of finer aspects of deformation mechanisms that are camouflaged by the bulk mechanical response, yet crucial to develop principles of microstructure tailoring for enhanced mechanical properties.

2 Scope

This chapter discusses the study of deformation induced microstructural evolution and underlying micro-mechanisms through in situ EBSD method. The plan of this chapter is as follows. A detailed description of the in situ EBSD method is presented to acquaint the general reader with the scope and limitations of this technique. This includes an overview of the various types of in situ ESBD experiments that are performed to obtain a direct correlation between applied deformation and observed microstructural evolution. After covering this background about the technique, the chapter deals with a thorough description of in situ deformation induced microstructural evolution in specific metals or alloys. A detailed discussion is presented using hexagonal close packed (HCP) structured commercially pure titanium (cp-Ti) as a model material. The results of in situ experiments performed in the present work are used to discuss the micro-mechanisms of deformation, including some experiments carried out to gain insight on microstructural evolution during in situ annealing of the deformed sample. The chapter also includes some discussion about in situ EBSD deformation studies in other face/body centred cubic (FCC/BCC) metals and alloys. Finally, the technological importance is discussed and the insight gained is summarized in the concluding remarks.

3 Process Details

Figure 1a shows a photograph of the in situ tensile-heating stage that was used in the present study. The in situ stage has a set of 70° pre-tilted jaws where the specimen is mounted to enable EBSD acquisition. The setup is provided with suitable cables to connect the stage to the controller for movement of the jaws (to facilitate tension/compression deformation) and to the temperature controller unit for heating experiments. The typical design and dimensions of the in situ specimen is shown in Fig. 1b. There are four through holes in the grip regions that are used to clamp the specimen onto the holder. The thickness of the specimen is usually 1–2 mm, while the gage length and width are 4.25 and 3 mm, respectively. Some variation in specimen dimensions is possible, depending on the capacity of the load cell of the in situ stage and the material being tested. For example, the present in situ stage has a load capacity of 2 kN. Therefore, for testing titanium that has UTS of the order of 400–500 MPa, the specimen dimensions must be such that the cross-sectional area at the gage is within 4 mm^2. On the other hand, these dimensions can be slightly increased if a 5 kN in situ tensile stage is used or if a softer material (with lower UTS) is being tested.

The surface of the in situ specimen is metallographically polished for EBSD acquisition prior to inserting the sample stage inside the SEM chamber. The amount of tensile or compressive strain that can be applied is determined by the limits of the jaw movement, relative to the starting position of the jaws. At any specified amount of strain, the orientation imaging may be performed by setting up an EBSD scan in the same way as ex-situ EBSD. The only difference is that, for EBSD mapping at each stage of an in situ test (e.g. at different levels of strain), the controller (for movement of jaws) is stopped. Therefore, it is very important to optimize the EBSD scan parameters to obtain best pattern quality in minimum acquisition time.

In the present work, three types of in situ experiments were performed (1) in situ tensile tests till failure of the specimen (2) in situ tension–compression load reversal tests and (3) in situ annealing (of in situ tensile deformed specimen) up to 773 K.

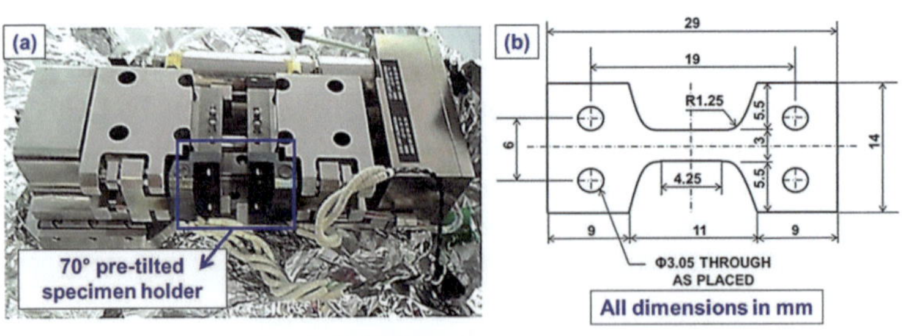

Fig. 1 **a** Photograph of Gatan 2 kN in situ tensile-heating EBSD stage **b** In situ tensile specimen design and dimensions

4 In Situ Microstructural Evolution During Deformation of HCP Titanium

The majority of the results presented in this chapter deal with the microstructural evolution during deformation of HCP structured cp-Ti. Therefore, before discussing the in situ EBSD results for cp-Ti, it is necessary to briefly discuss the microstructure-texture-mechanical response relationship of cp-Ti to provide the background for the in situ investigation.

HCP structured cp-Ti (consisting of α-Ti phase) has a c/a ratio of 1.587 (lower than ideal 1.633) and deforms by a combination of slip and deformation twinning [13–19]. The importance of deformation twinning as a strain accommodation mechanism in the low-symmetry HCP metals, the crystallography and various modes of deformation twinning are discussed in detail in the textbook of Hirth and Lothe [20] and the classical review by Christian and Mahajan [21]. While there are 7 theoretically possible twin systems in HCP metals, $\{10\overline{1}2\}$ and $\{11\overline{2}2\}$ twins are most commonly observed [22, 23]. For titanium, the $\{10\overline{1}2\}$ twin is the extension twinning system and $\{11\overline{2}2\}$ is the contraction twinning system. Prismatic slip is favoured due to lower than ideal c/a ratio (unlike Mg where basal slip is favoured). Basal and pyramidal $< c + a >$ slip comprise the other major deformation modes.

In the present study, the cold rolled and annealed (as-received) plate of cp-Ti had characteristic split transverse direction (split-TD) initial texture as seen from {0002} pole figure in (Fig. 2). Tensile specimens of three different initial orientations were obtained from the plate (Fig. 2) [24]. The initial textures of these three different specimens A, B and C, as determined by X-ray texture measurements, are also depicted on tensile axis inverse pole figures (IPFs) in Fig. 2. The tensile properties of these orientations (A, B and C) were obtained and correlated with microstructure from post-mortem EBSD characterization. From this study, the following salient observations were made.

The samples A and C showed prismatic-pyramidal (away from basal) texture along the tensile axis, while sample B was basal textured along the tensile axis [24]. The basal textured sample B showed distinctly higher strength but lower ductility compared to orientations A and C (while the latter two exhibited similar properties). This difference in mechanical response was attributed to the difference in initial texture and the hypothesis was confirmed from microstructural evaluation that revealed significant difference in the amounts and nature of deformation twinning in the sample B, as compared to samples A and C. Specifically, thick single variant extension twins ($\{10\overline{1}2\}$) were observed in sample B, while samples A and C showed thin, intersecting, multi-variant extension and contraction ($\{11\overline{2}2\}$ type) twinning that increased strain hardening and delayed failure, resulting in higher tensile elongation [24, 25].

Since these aforementioned results clearly indicated that the propensity of deformation twinning influenced the tensile response of HCP Ti, we sought to investigate further into the activity of twinning during load reversal. In situ EBSD was especially useful to gain insight about twin formation in cp-Ti during load reversal. The

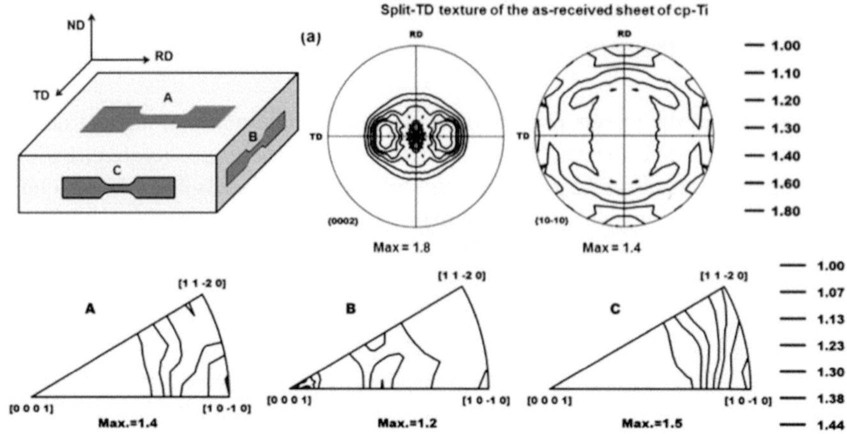

Fig. 2 Tensile specimen orientations and initial textures of cp-Ti [24]

following sections discuss the observations from in situ EBSD deformation experiments carried out on cp-Ti. First, in situ tensile-EBSD test till failure was performed to obtain the twinning propensity as a function of strain. Subsequently, various in situ load reversal tests were performed to observe the propensity of twinning and detwinning micro-mechanisms and use the insight to explain cyclic deformation mechanisms.

4.1 In Situ Tensile Deformation of HCP Ti

Figure 3 shows the engineering stress–strain curve for an in situ tensile test till failure. The microstructural evolution with strain is represented as crystal orientation or inverse pole figure (IPF) maps captured at various stages of the test. The amount of twinning increases with strain but it reaches a maximum after which twin saturation occurs, due to which the overall fraction of twin boundaries in the microstructure decreases. At high strain near fracture, the quality of EBSD pattern also diminishes severely, therefore it is not possible to acquire a sufficiently indexed EBSD map. The key insight obtained from the in situ tensile test is that from the onset of twinning, the increase in twin fraction increases the strain hardening rate (dσ/dε) until twin saturation occurs (Fig. 3).

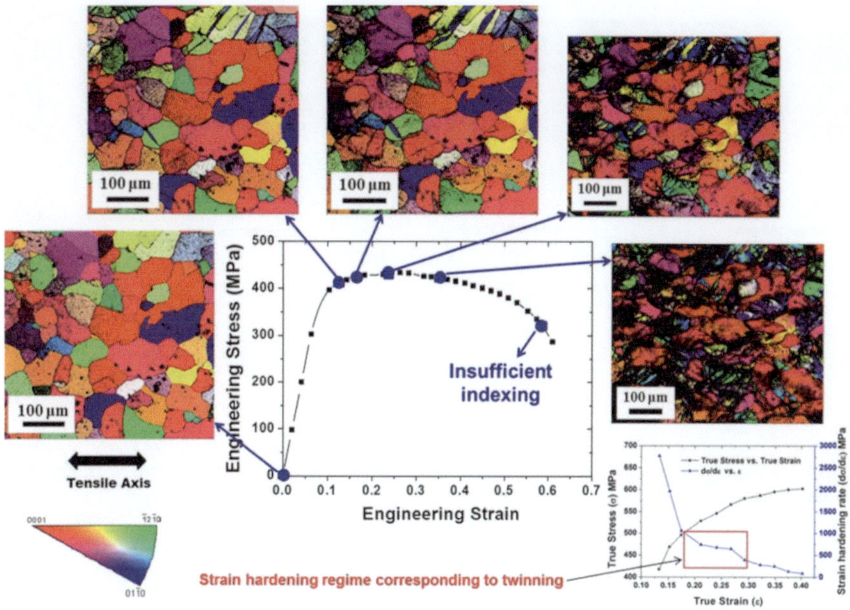

Fig. 3 In situ tensile EBSD results of cp-Ti [24]

4.2 *In Situ Load Reversal of HCP Ti*

Figure 4 shows the deformation stages I to VI marked on a schematic stress–strain curve for in situ test involving load reversal carried out in load control mode. IPF maps obtained from EBSD at each stage of the in situ load reversal experiment are also shown in Fig. 4 [26]. This experiment clearly depicts the initiation of deformation twinning with loading, from the starting microstructure (stage I) to the stage II when the first twin forms in grain 1, appearing merely as unindexed points (since it is not yet thick enough for its orientation to be resolved in EBSD). The twin orientation is clearly visible (yellow in IPF map) in stage III (the twin is marked by the number 2).

At stage III, twins of other variants are also observed within the same grain, i.e. multiple variant twinning occurs with increase in stress just above yielding. The present experiment focuses on twin formation within a single grain. In line with this, loading to above the yield stress but sufficiently below UTS ensures that surface roughness does not hamper EBSD indexing significantly. On the other hand, loading to higher stress would have probably captured twinning in multiple grains but also twinning induced surface roughness would increase. Unloading to 0 N in stage IV causes decrease in un-indexed points within twin variant 2 compared to stage III which may be due to annihilation of some dislocations. Subsequent compression (stage V) to similar level of stress as the previous forward tension (stage III) leads to thinning of the already formed twins due to reversal in loading direction suggesting

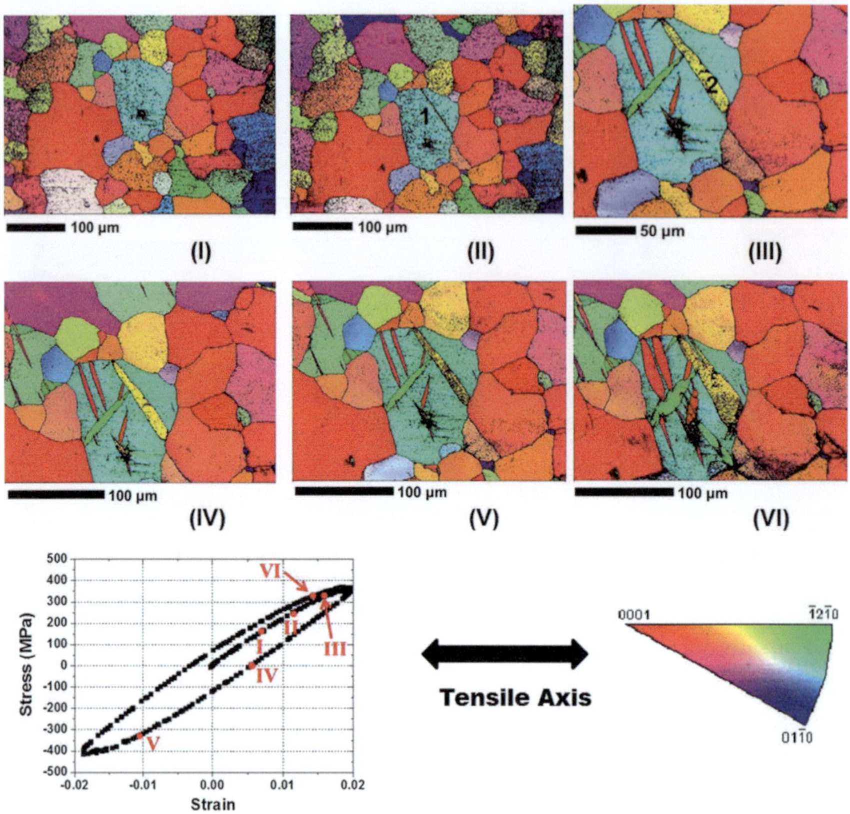

Fig. 4 In situ load reversal test in load control mode [26]

some degree of twin reversibility. The variations in un-indexed points within the twins at these different stages are probably related to changes in dislocation density.

Reloading in tension (stage VI) changes the twin morphology. The previously formed twins are pinned by grain boundaries or twin boundaries at both ends, where they nucleated and impinged upon. Reloading in tension under these circumstances causes barrelling out of the twins. Also, the twin variant 2 now shows a tapered shape with fragmentation at the narrow end. The impinging twin boundaries at the narrow end of the twin impose a constraint that hinders lateral thickening of the twin. Therefore, the differential degree of lateral thickening of the twin at its two ends causes fragmentation of the twin. Another interesting observation in stage VI is that more twins belonging to the previously formed variants are nucleated, mainly at the boundaries of the already formed twins.

This experiment provides valuable insight about the deformation dependence of twinning propensity in HCP Ti. First, crystal orientation and grain size are of utmost importance for twin nucleation. These two factors determine the favourable parent grain for twin nucleation within a specific microstructural region as captured in the present experiment (Fig. 4). Second, deformation twinning being an important strain accommodation mechanism in HCP Ti, the amount of twinning is strongly dependent on the applied stress as well as the loading mode (monotonic or cyclic). Third, the impact of applied stress or load reversal on deformation twinning is primarily related to the various stages of twin formation—nucleation, propagation and growth [27]. Thus, the overall twinning propensity will be dependent on how the loading conditions influence the nucleation and growth tendency of individual variants; and the latter is determined by the sum effect of all the influencing factors (e.g. crystal orientation, amount and direction of applied stress, local stress state and interactions between twin and grain boundaries). Finally, the importance of this experiment (Fig. 4) is that twin formation propensity is explored through in situ load reversal experiment.

Several in situ load reversal tests were performed in displacement control mode to study the microstructural evolution and nature of twinning in cp-Ti during cyclic deformation. Figure 5 shows the results from one such test. Evidently, twinning is less pronounced in the displacement controlled test than the load controlled test for comparable amount of strain. For instance, multiple variant twinning and sufficient thickening of the twins occurred progressively within a single grain during consecutive stages of the previous load controlled test (Fig. 5); while, in the present displacement controlled experiment (Fig. 5), a few thin twins are formed within some grains but without pronounced twin growth (lateral thickening). At the same time, this is also the reason why such displacement controlled in situ load reversal test is appropriate for studying and predicting the mechanisms of low cycle fatigue (LCF), where displacement controlled low strain amplitude cyclic deformation is applied.

The microstructural evolution with tension compression load reversal in (Fig. 5) provides some useful insight based on the changes in parent grain orientations at each stage. The grains undergoing changes in orientation during successive tension–compression stages are numbered 1–4 in Fig. 5b. No twins are observed in grains 1 and 3, while grains 2 and 4 show twin nucleation. From the starting stage Fig. 5a to the tension stage Fig. 5b, orientation change is visible in grains 1 and 4 (along with some intragranular strain gradient); while grains 2 and 3 do not undergo change in orientation (although some intragranular strain gradient is present). However, during subsequent compression (Fig. 5c), all the four grains undergo orientation change. Subsequently, these grains also revert back to the orientation in (Fig. 5b) after reloading in tension (Fig. 5d). This clearly illustrates that load reversal induced microstructural reversibility occurs in HCP Ti.

The interesting point of this discussion is that the aforementioned microstructural reversibility in Fig. 5 is not specific to the deformation twinning phenomenon. Now, since slip dislocations can revert their motion easily under the influence of load reversal, this observed microstructural reversibility is somewhat similar to microstructural Bauschinger effect observed in other common FCC metals such as copper [28]. With respect to partial reversibility of plastic deformation (coupled with microstructural reversibility) in HCP Ti, twinning plays a significant part. Twinning

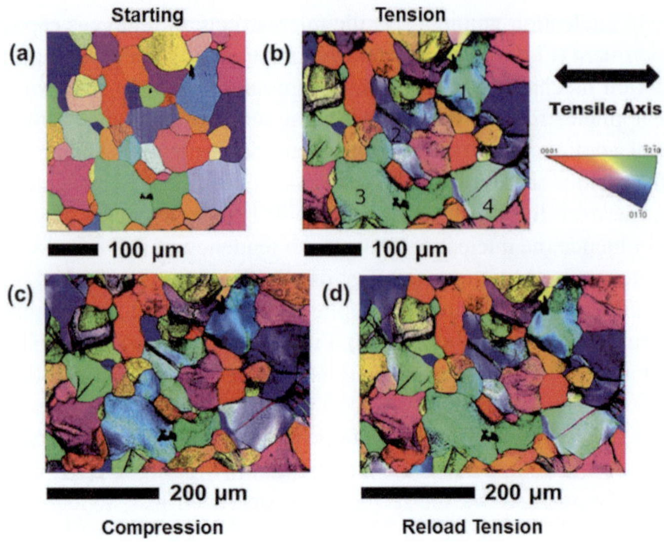

Fig. 5 In situ load reversal test in displacement control mode [30]

dislocations are polar in nature and cannot revert their motion as easily as slip dislocations [29]. This is a fundamental factor that influences the partial microstructural and mechanical plastic deformation reversibility in HCP Ti. Therefore, the subsequent in situ load reversal experiments aim to explore the reversibility of twinning in the microstructure of cp-Ti during cyclic deformation.

In order to focus specifically on reversibility of twinning, particular grains where twins nucleated were selected and tracked for the subsequent stages of in situ load reversal tests. Figure 6 shows the microstructural evolution during tension followed by compression. Twinning is first observed (after tensile loading) in a grain with $< 01\overline{1}2 >$ IPF normal direction (purple coloured grain 1 in Fig. 6). Twins (numbered 2 and 3) are observed along with a few other thin, small twins. Twins 2 and 3 are of the same variant although they are nucleated at different grain boundaries and their propagation directions are almost perpendicular to each other. Twin 3 is visibly impinging on the grain boundary opposite to the one where it nucleated and thicker, while twin 2 shows incomplete propagation (advancement of the twin front) and growth (lateral thickening). Subsequently, twin 2 (whose propagation direction is aligned close to the tensile axis) gets shortened due to compression deformation. This firmly establishes that the extent of twin propagation and growth is vital for its response to load reversal and the reversibility of twin formation is sensitive to its orientation with respect to the loading direction (loading axis).

Finally, upon reloading in tension to higher strain, both twins 2 and 3 grow (laterally thicken), while another twin 4 of the same variant is nucleated parallel to twin 3 (and the twin fronts of 2 and 4 impinge upon each other restricting each other's propagation). The nucleation of twin 4 supports the previous observation that increasing

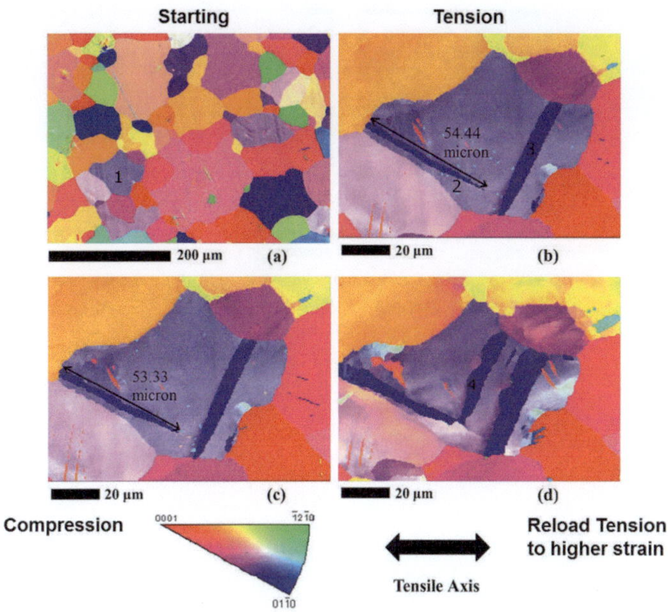

Fig. 6 Displacement controlled in situ load reversal test showing twin formation [30]

strain and reversal in loading direction enhance the twinning probability. While the effect of strain (that higher strain promotes twinning) was already evident from the in situ tensile test, the effect of load cycling is more complex. On the one hand, load cycling coupled with increasing accumulated strain favours twin formation; while on the other hand, the extent of twin propagation and growth decides their microstructural reversibility, giving rise to the possibility of detwinning during cyclic deformation.

Figure 7 shows the microstructural evolution during another similar in situ load reversal test; the only difference from the previous test is that first compression is applied, followed by tension. Deforming the starting microstructure (Fig. 7a) in compression results in the nucleation of the twin 6 (pink colour) within the purple coloured grain 5 (Fig. 7b). The IPF normal direction orientation of grain 5 is also $< 01\bar{1}2 >$, similar to grain 1 in (Fig. 6a). Thus, twinning always initiates first in grains with $< 01\bar{1}2 >$ IPF normal direction orientation irrespective of whether the first deformation step is in tension or compression. This proves that the twinning propensity of individual grains is strongly dependent upon their crystallographic orientation even in the case of cyclic deformation.

Subsequent unloading results in partial detwinning due to which, the twin is visible only as unindexed points (Fig. 7c). This indicates partial microstructural and plastic deformation reversibility associated with twinning. During tensile loading in the next step, once again the twin orientation is clearly visible (pink colour) in (Fig. 7d), although the twin is now thinner than it was in (Fig. 7b). Once again, this is an

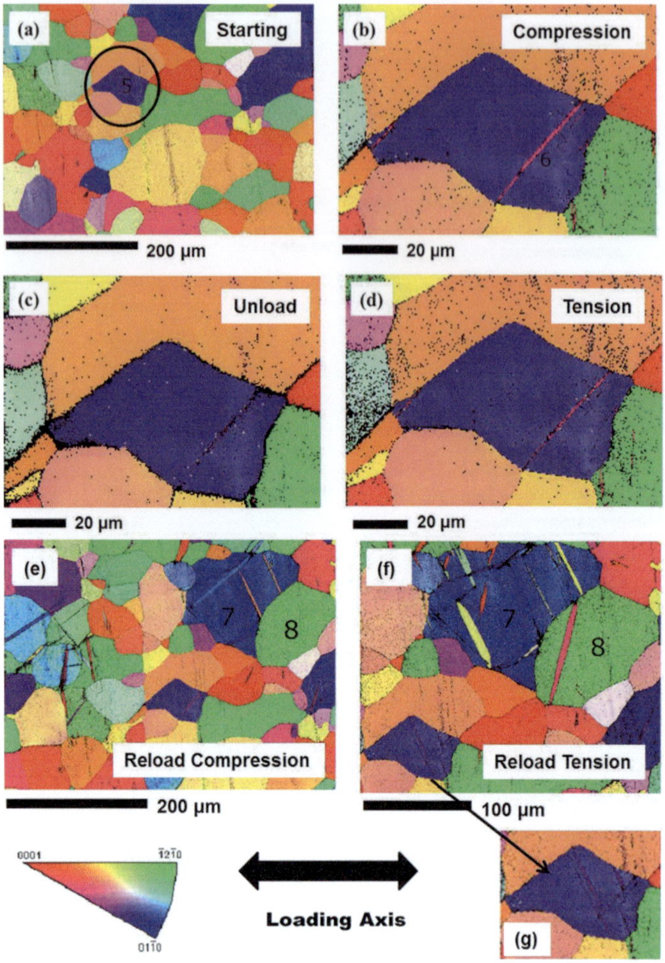

Fig. 7 Displacement controlled in situ load reversal test showing twinning-detwinning [31]

indication of load reversal induced partial detwinning, rather than complete detwinning. Partial detwinning is particularly relevant to cyclic deformation (LCF) where small strain amplitude is applied in each half-cycle between tension–compression load reversals. Such detwinning tendency influences the cyclic hardening/softening behaviour and therefore, the cyclic life.

Figure 7e and f show the microstructure after reloading in compression and subsequently reloading in tension to a higher strain, respectively. Also, Fig. 7e and f include a larger microstructural area surrounding grain 5 (while Fig. 7b–d only focused on grain 5). Complete detwinning of a few twins and partial detwinning of some other twins are observed in the nearby grain 7 during load reversal from Fig. 7e, f. The application of higher strain in the tensile reloading stage (Fig. 7f) also results in the formation of new twins of different variants in grains 7 and 8. The grain 5 also shows interesting twinning activity in (Fig. 7f), shown enlarged in (Fig. 7g). Once again, there is partial detwinning of the pink coloured twin during load reversal from Fig. 7e, f, while new twins of the same variant (pink colour denotes IPF normal direction close to $< 01\bar{1}1 >$) are nucleated in grain 5, whose in-plane orientation with respect to the loading axis is different (at an obtuse angle although not completely perpendicular to) from the previous pink twin (that has undergone partial detwinning).

Therefore, this experiment unambiguously confirms that reversal in loading direction and the accumulated strain during cyclic deformation strongly influence the extent of detwinning. Some twins completely disappear while others are partially detwinned. Strain accumulation during cycling can also promote nucleation of new twins of the same or different variants, especially because twin nucleation is dependent on the local stress state. In short, small strain amplitude coupled with reversal of loading direction in each cycle definitely impacts the lateral growth of twins. Consequently, thinner twins are observed in cyclic loading. The interplay between twinning and detwinning is what makes the cyclic hardening/softening behaviour more complicated and interesting than the twinning induced increase in strain hardening rate in quasi-static tension. In other words, while strain hardening in quasi-static tension depends only on the twinning propensity of a particular crystallographic texture, the cyclic hardening/softening mechanisms are determined by the net effect of twinning and detwinning propensities of the particular crystallographic orientation (texture).

4.3 *Micro-Mechanisms of Deformation*

The in situ EBSD results discussed in the previous section throw light on the deformation micro-mechanisms of cp-Ti that are essential to explain the initial texture dependence of mechanical response in the present study. Specifically, the utility of the in situ EBSD results may be explained as follows. The initial texture dependence of tensile properties was first discussed to establish the background and purpose of in situ deformation experiments in the present study; while, the in situ EBSD results were analysed to obtain insight about deformation micro-mechanisms so that the initial texture dependence of cyclic (LCF) behaviour of cp-Ti can be explained. Therefore, this section is dedicated to the insight obtained from the in situ experiments in the preceding section, about the impact of twinning/detwinning micro-mechanisms on cyclic response of the different initial orientations of cp-Ti.

Here it must be mentioned that earlier, orientations A and C showed similar tensile behaviour, while orientation B was distinctly different (Fig. 2). Therefore, displacement controlled cyclic tests were performed only on orientations A and B to compare their LCF response [31]. In addition to these out-of-plane orientations A and B, a study of in-plane orientations (i.e. along the rolling and transverse directions in the same plane as orientation A) was also carried out for a thorough systematic investigation [32]. The salient points from the investigation of cyclic response are

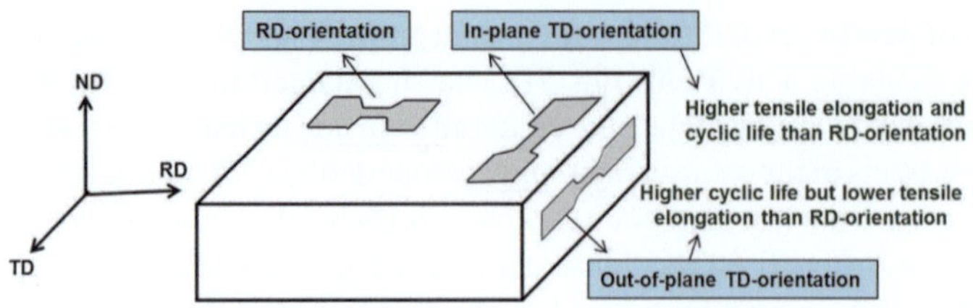

Fig. 8 Summary of tensile and cyclic behaviour study of in-plane and out-of-plane orientations of cp-Ti [30]

briefly summarized here to connect the dots between the inferences from the in situ EBSD study and the crystallographic texture dependence of cyclic properties.

The combined investigation of tensile and cyclic properties for both the in-plane and out-of-plane orientations revealed that the transverse direction (TD) oriented specimens show higher cyclic life than rolling direction (RD) oriented specimens, irrespective of whether the TD-oriented specimen is in-plane or out-of-plane with the RD-oriented specimen [30–32]. On the other hand, the in-plane TD-oriented specimen showed higher tensile elongation than the RD-oriented specimen; while, the out-of-plane TD-oriented specimen showed lower tensile elongation than the RD-oriented specimen (Fig. 8). These results are explained in this section from the underlying concepts of twinning-detwinning micro-mechanisms that were established from the in situ EBSD studies, as follows.

4.3.1 Twinning-Detwinning Propensity

The in situ EBSD results in the previous section clearly established that twinning propensity is important during quasi-static tension, while both twinning and detwinning tendencies are of paramount importance during load reversal. The microstructural analysis was supplemented with post-mortem ex-situ EBSD characterization of fractured tensile and cyclic specimens of the RD and TD-orientations for in-depth analysis of deformation mechanisms and contribution of slip/twin modes [30–32], which made it possible to justify the initial orientation dependence of mechanical response (Fig. 8).

Crystallographic texture involves a plane and a direction (e.g. a {hkl} < uvw > texture component denotes that the plane normal to {hkl} lies along the ND and the RD lies along< <uvw>). Therefore, the initial texture of the in-plane TD-orientation is obviously closer to the RD-orientation than the out-of-plane TD-orientation. This is because for the latter, both the texture plane and direction are different from the RD-orientation. The result of initial texture dependence on deformation mechanisms is that the in-plane TD-orientation shows intersecting, multi-variant extension and contraction twinning in quasi-static tension, similar to the RD-orientation. This nature of twinning increases strain hardening and delays failure. Hence, the in-plane TD-orientation shows higher tensile elongation than the RD-orientation. The out-of-plane

TD-orientation is in contrast with the RD-orientation because the initial texture is distinctly different (basal along the tensile axis in the former and away from basal along the tensile axis in the latter). Consequently, the nature of twinning is also markedly different. The out-of-plane orientation in Fig. 8 (same as orientation B in Fig. 2) shows thick, single variant extension twins accounting for lower tensile elongation.

For the cyclic deformation case, the aforementioned texture dependence of twinning propensity is a factor influencing twin nucleation, but other factors also come into play to decide the overall twin activity. First, as highlighted by the in situ EBSD load reversal experiments, application of small strain in each half-cycle coupled with cyclic reversal of loading direction restricts lateral thickening of twins, irrespective of the initial texture/orientation. In other words, RD, in-plane TD or out-of-plane TD orientations would all exhibit the formation of thin, small twins with limited propagation and growth during low cycle fatigue, unlike in quasi-static tension. Additionally, the initial orientation strongly influences the detwinning tendency of twins over a large number of tension–compression cycles.

Now, lateral thickening of twins may be influenced by orientation of individual variants in case of quasi-static tension too. However, in cyclic loading, the orientation effect is coupled with the effect of loading direction. In other words, twins that are oriented with their lateral growth direction (thickness) parallel to the loading direction are susceptible to alternate micro-scale thickening and thinning during tension–compression load reversal (for example, as observed in Fig. 7). Similarly, twins with their twin fronts oriented along the loading axis would undergo restricted propagation (for example, as observed in Fig. 6). Through this mechanism, a lot of thin and small twins are formed during cyclic deformation that exerts competitive effects on the cyclic hardening/softening behaviour. Also, the probability of detwinning during successive cycling is higher for such thin, small twins. This is precisely what was observed in the present study. The TD-orientations (both in-plane and out-of-plane) showed almost complete detwinning of contraction twins [32]. Thus, the detwinning tendency is more sensitive to one particular direction, while the twinning tendency was found to be more sensitive to the texture plane. Higher detwinning tendency (of contraction twins) during cyclic loading in both the in-plane and out-of-plane TD-oriented specimens evidently influenced the cyclic life. Hence, higher cyclic life was obtained for both the TD-orientations than the RD-orientation. This detwinning induced increase in cyclic life is explained in the following.

4.3.2 Impact of Partial Reversibility of Plastic Deformation

From a microstructural standpoint, partial detwinning is a signature of partial reversibility of plastic deformation in HCP titanium. The significance of this partial microstructural reversibility is that the cyclic life exhibits texture dependence based on the relative detwinning tendency of the different initial orientations of cp-Ti. The aforementioned dependence is due to the fact that detwinning relieves backstress and thereby prolongs cyclic life. The mechanism is explained as follows.

Twin boundaries serve as obstacles to dislocation motion. Therefore, just like at grain boundaries, dislocations also pile up at twin boundaries, giving rise to backstress. During cyclic loading, the dislocation pile-ups formed during forward loading aid reverse loading. Bauschinger effect is more as more dislocations are annihilated. However, the backstress causes hardening during subsequent forward loading since the pile-ups act as barriers to dislocation motion. If only slip was operative, this backstress induced hardening would result in early failure; but in case of combined slip and twinning (such as in HCP Ti), detwinning helps to relieve backstress by releasing dislocation pile-ups formed at twin boundaries. Complete detwinning of a twin would cause all the dislocations in that pile-up to be released, while partial detwinning could also release some of the dislocations in the pile-up and reduce the length of the pile-up. Thus the backstress induced hardening effect is reduced and the cyclic life is increased. In our study, the detwinning mediated release of backstress was qualitatively estimated from X-ray peak asymmetry and it was found that the TD-oriented specimens exhibited decreased peak asymmetry during cyclic loading, which is consistent with the higher detwinning tendency obtained in the TD-oriented specimens [30]. Hence, the TD-oriented specimens exhibited higher cyclic life.

5 In Situ Microstructural Evolution During Annealing of HCP Titanium

A few in situ annealing experiments were also performed on cp-Ti, using the heating setup. Figure 9 shows the microstructural evolution upon in situ annealing of the in situ deformed specimen from Fig. 4. After reloading in tension (Stage VI), the specimen was subjected to annealing at 773 K for 1 h (Stage VI → VII shown in Fig. 9). Note that, since α-Ti phase is stable up to 1155 K, a temperature above 573 K should be sufficient for recrystallization to occur. Now, Nourbakhsh and Brien carried out a classical investigation on the textural reversibility of cp-Ti during cold rolling and annealing and discussed that annealing at 848 K causes significant changes in the microstructure of α-Ti [33]. Therefore, annealing was carried out at the upper limit of the present in-situ heating stage (773 K) to get as close to 848 K as possible.

Significant microstructural changes due to annealing are observed in Fig. 9. The parent grain and the twin 2 had not shown any change in IPF normal direction orientation throughout the deformation stages (Fig. 4) but upon annealing they undergo reorientation towards the same orientation (near $< 01\bar{1}2 >$). Many of the other previous deformation twins undergo complete detwinning and disappear, while a few new basal $< 0001 >$ IPF normal direction oriented annealing twins are nucleated. Thus, in situ annealing in the present study leads to microstructural evolution that encompasses multiple events—(1) reorientation of parent grain and twin, (2) detwinning of previously formed deformation twins (both complete and partial detwinning are observed) and (3) nucleation of new annealing twins. Clearly, the recrystallization annealing of titanium is complicated and requires detailed further study.

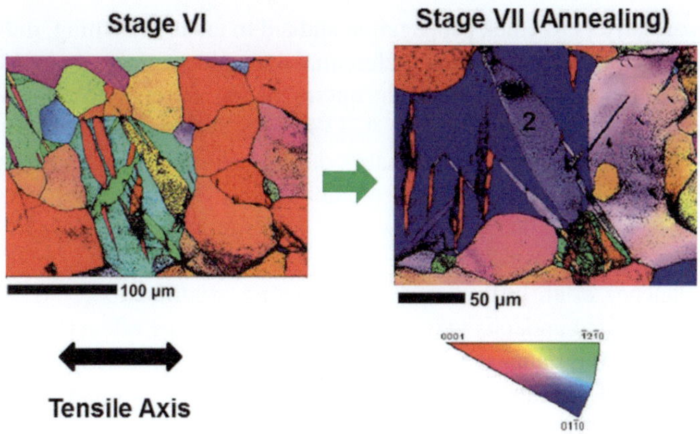

Fig. 9 Effect of in situ annealing of the deformed microstructure (stage VI) in Fig. 4 [26]

6 In Situ EBSD Deformation and Annealing Studies of Other Metals and Alloys

The in situ EBSD study of deformation (and annealing) in HCP Ti showed some really interesting microstructural changes, especially because deformation twinning involves sudden reorientation of grains and such orientation changes can be tracked and captured as they happen. Hence, in situ EBSD has also been used effectively for studying deformation twinning in other HCP metals (and their alloys) such as Mg/Zr [34, 35]. For the same reason, in situ EBSD experiments also provide very useful insight in FCC materials that exhibit deformation twinning, such as austenitic stainless steels or TWIP steels. In situ EBSD observation of twin evolution with strain in 316L austenitic stainless steel was discussed in one of our earlier studies [11].

Apart from materials that undergo deformation twinning, in situ EBSD is also useful to throw light on deformation and strain hardening mechanisms in other FCC metals and alloys. Mishra et al. carried out in situ EBSD experiments to characterize the microstructural evolution in a study of the strain hardening behaviour of age hardening Al alloys [36]. In situ EBSD was used in the study to experimentally validate the simulation results of slip activity based strain hardening (SASH) model, wherein peak aged samples showed single slip at low strain and secondary slip was operative only at higher strain; whereas in solutionized samples, multiple slip was activated at low strain itself. This was achieved by tracking the grain orientation evolution of several individual grains (relative to the loading axis) during in situ quasi-static tension or tension–compression load reversal.

Ghassemali et al. used in situ EBSD to study fracture mechanisms (crack propagation) and elevated temperature behaviour of dual-phase (FCC + BCC) AlCoCrFeNi high entropy alloys [37, 38]. Room temperature in situ tensile-EBSD results showed that the precipitation of tougher FCC phase along the grain boundaries of BCC phase acts as barriers to crack propagation and aid in crack branching, deflection or restricting crack growth [37]. Also, preferential dynamic precipitation of FCC phase in some BCC grains was observed in the microstructure during in situ tensile testing at an intermediate temperature of 550 °C [38]. This dynamically precipitated FCC phase was deemed to relax the BCC matrix during deformation and it was proposed from this study that enhanced dynamic precipitation of FCC phase at higher temperatures (700–800 °C) could be a pathway to increase the ductility of AlCoCrFeNi alloys.

Lopez-Sanchez et al. [39] carried out annealing experiments ($\sim 0.6T_m$) using quasi in situ EBSD to understand deformation microstructure effect on static recrystallization of AZ31B magnesium alloy. Samples were deformed under different temperature and strain rate to obtain microstructures with different defect density and dislocation arrangements. Recrystallization occurred via nucleation and growth of sub-structure free grains and developed rapidly. The recrystallization kinetics was found to be controlled by dislocation stored energy rather than grain boundary energy, which also led to higher grain growth exponent (n = 7–18). Figure 10 shows the steps utilized to analyse the annealed microstructure. A positive correlation between the recrystallized fraction and the initial average KAM was observed. The recrystallization rate evolution during annealing revealed that recrystallization occurred faster in samples which had highest initial average KAM value. Lee et al. [40] studied deformation behaviour and static recrystallization behaviour of magnesium alloy (AZ31) having an off-basal texture, using quasi in situ EBSD. The cold rolled sample had multiple shear bands throughout the material, along with twins and their interactions in the non-shear-band regions, which acted as nucleation sites for recrystallized grains. The non-shear-band region included $\{10\bar{1}2\}$ tension twins, $\{10\bar{1}2\}$ –$\{10\bar{1}2\}$ double twins, twin-twin and twin-grain boundary intersections. During annealing, recrystallization occurred in both the regions of the cold rolled sample but with different recrystallization rates and grain growth behaviour. At the start of annealing, the recrystallization took place quickly in the shear band region due to the large internal strain energy, but the new formed grains got constrained in that region. In the non-shear-band region, recrystallization rate was slow. However, the new grains pursued to form and grow with annealing, leading to the final recrystallized microstructure and texture.

Zhao et al. [41] investigated mechanisms of nucleation and grain growth in ultrafine surface rolled CoCrNi medium entropy alloy, with gradient nanograined structure, using quasi in situ EBSD. The alloy had gradient depth of 450 μm, with nano grains at the surface and multiple deformation twins at the subsurface, due to low stacking fault energy. The initial microstructure before annealing had a mixture of mostly recrystallized grains, few grains with a dominant dislocation sub-structure and almost no deformed grains. Figure 11 shows microstructural evolution during annealing of CoCrNi alloy. It was observed that the proportion of Σ3 coincidence site lattice boundary increases with grain coarsening, along with slight lattice rotation. The thermal stability of nano-grained top surface layer was found to be relatively higher than nano-twined subsurface layer, revealed from activation energy calculation and layer by layer grain growth kinetics. The nano-grains on the top surface

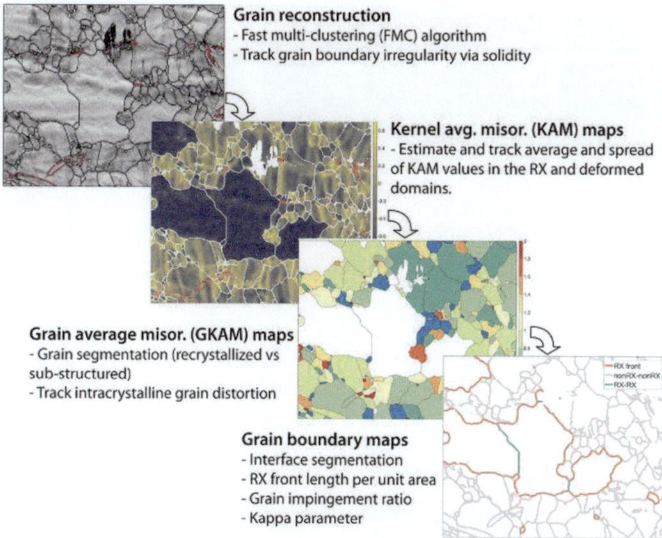

Fig. 10 Steps used for analysing the annealed microstructure of AZ31B alloy, using MTEX software [39]

encountered grain boundary relaxation and dynamic recrystallization, leading to improvement of thermal stability.

Nakamichi et al. [42] studied in situ EBSD analysis of recrystallizing interstitial free steels at different temperatures and deformed to different strain values, (a) sample I—80% rolled at 400 °C ($\varepsilon = 1.6$) and (b) Sample II—50% rolled at room temperature ($\varepsilon = 0.75$). It was observed that recrystallization behaviour depends on the deformation history. The amount of recrystallization increased gradually and saturated at 65% and 50% for sample I and II, respectively. Sample II had lower stored energy of deformation and hence low recrystallization fraction. It was observed in sample I that 70% recrystallized grains had 'γ texture fibre', which includes $\{111\} < 121 >$, $\{111\} < 123 >$ and $\{111\} < 110 >$. In the sample II, 25% grains recrystallized from $\{111\} < 123 >$, but no strong texture evolution was detected. Tokita et al. [43] examined microstructural evolution in 304 austenitic stainless steel, subjected to solution treatment followed by 3% cold rolling and then annealing at 1220 K, employed for grain boundary engineering. The in situ EBSD observations during annealing with different time, revealed heterogeneous growth of clusters of grains with high coincident site lattice (CSL) boundaries, evident from the grain boundary character distribution (GBCD) maps. A rise in the frequency of CSL boundaries and reduction in the percolation probability of random boundaries was observed along with the evolution of clusters of grains. It was found that the strain-induced grain growth led to the growth of clusters, from the grain orientation spread (GOS) analysis. The twin variant selection using in situ EBSD also confirmed disconnection of random boundaries between large clusters due to formation of annealing twins. Annealing behaviour of copper-added bake-hardenable (BH) steel was studied by Kim et al.

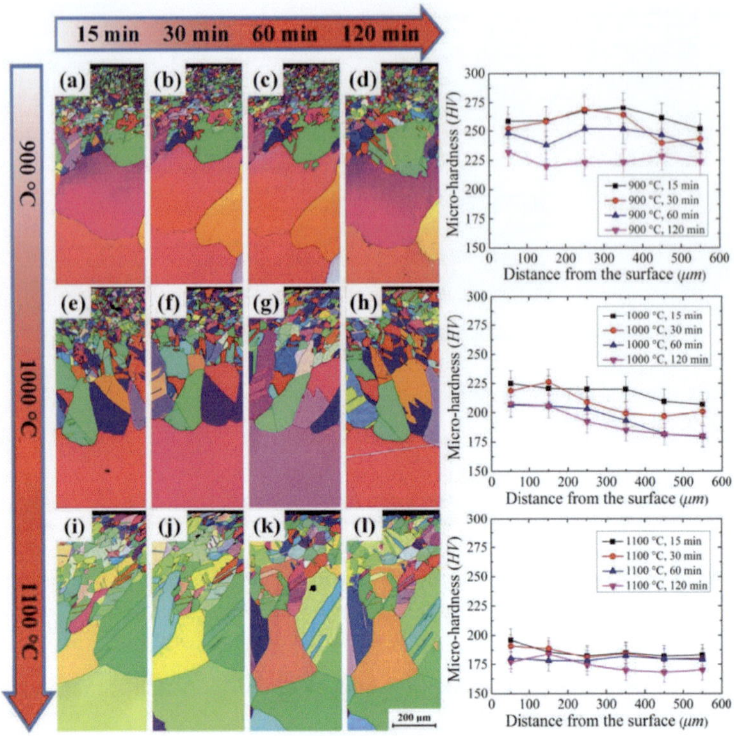

Fig. 11 Inverse pole figure (IPF) maps showing microstructure evolution during annealing at different time and temperatures and corresponding micro-hardness of the CoCrNi medium entropy alloy [41]

[44], using in situ EBSD. The deformed grains in the alloy were divided into three categories based on grain average misorientation (GAM) method: (I) average orientation of unique grain had single γ-fiber component, (II) multiple γ-fiber component and (III) texture components could not be assigned to γ-fiber or α-fiber component. In type I deformed grains, the recrystallization started in the later stages and less number of recrystallized grains were observed. But the average grain size increased substantially once the recrystallization began. In II, the number of recrystallized grains increased continuously with annealing time. The new grains nucleated at the shear bands and high angle grain boundaries (HAGBs). In III, recrystallization kinetics was found to be similar to that in II. In this case, the new grains nucleated at the highly misoriented regions inside the deformed grains. However, microtexture analysis indicated that type III was minor deformed grains and type I and II were the principal feature governing the overall kinetics of recrystallization in this steel.

7 In Situ EBSD Phase Transformation Studies

Seward et al. [45] studied HCP (α) to BCC (β) phase transformation in commercially pure titanium at 900 °C using in situ SEM-EBSD. The β phase nucleates and grow as plates within the α grains and as allotriomorphs along the α-α grain boundaries. The intragranular β possessed a habit plane near to specific {334} for a particular OR variant. The growth mechanism of the β phase present within the grain was found to be a military transformation. Whereas, the allotriomorphic β showed civilian growth mechanism and does not have a specific habit plane. In the final transformation stage, there is a competitive growth between the two β forms, and the allotriomorphic β dominated due to its faster moving α/β interfaces. The grain growth of β was found to be an order of magnitude faster than α, at the same temperatures. The dominant {001} < 100 > texture component of the β phase was found to be consistent with the transformation.

Ubhi et al. [46] reported in situ EBSD results of austenite to ferrite phase transformation in plain carbon steel and recrystallization behaviour in bronze strip casted on a ferritic steel substrate. The recrystallization starts first in bronze and then in steel, but the recrystallization rate is faster in steel. It was observed that the recrystallization starts at the grain boundaries in bronze and not at the bronze–steel interface. In the plain carbon steel, 20 μm grains of austenite phase transforms to 200 μm grains of ferrite phase. Lischewski et al. [47] carried out high temperature in situ EBSD of micro-alloyed low carbon steels to capture microstructure and orientation changes during phase transformation, recovery, recrystallization and grain growth. The α-γ-α transformation (886–900 °C) takes place via nucleation and growth and, texture development and variant selection plays a vital role during the transformation. Figure 12 shows phase transformation obtained using in situ EBSD measurement. The grey colour grains represent low temperature BCC α phase and coloured grains are high temperature FCC γ phase. During α to γ transformation, the new phase nucleates at grain boundaries and triple points. The new grains are homogeneous and isotropic and causes grain refinement. The FCC texture changes slightly in 5 to 50% of transformation, except for the copper component on the β-fiber during the first transformation step at 897 ∘C (5% FCC). The γ to α microstructure development was found to be inhomogeneous and its growth kinetics was faster than α to γ transformation. In this transformation, grain coarsening was observed in the microstructure and no texture could be calculated due to poor grain statistics. It was also found that the mobility of the transformation front is dependent on proximity to Kurdjumov–Sachs relationship.

Fukino et al. [48] investigated dynamics of α/γ phase transformation and orientation relationship in Fe-9.5at%Ni alloy. The γ phase nucleated preferentially at random grain boundaries and triple junction points, and not inside the grains. It was observed that around 90% of the allotriomorphic γ phase had either a Kurdjumov-Sachs (K-S) or a Nishiyama-Wasserman (N-W) orientation relationship (OR) with the α parent grains (Fig. 13). The γ phase grew into adjoining grain if it had special OR with that grain and grew into both adjoining grains symmetrically if it had special OR with

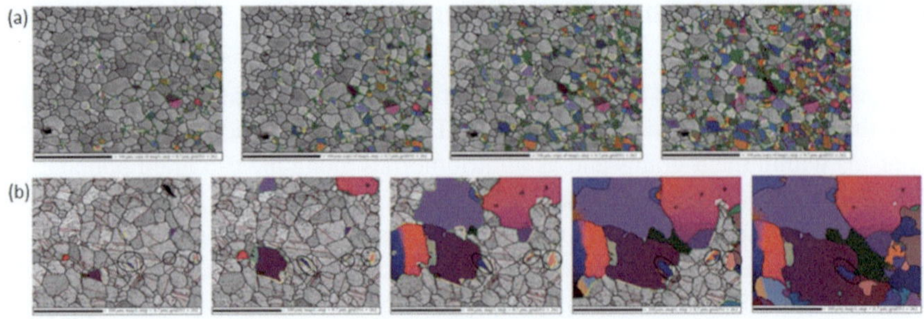

Fig. 12 Phase transformation **a** α to γ from 897 to 900 °C, **b** γ to α from 893 to 886 °C, in low carbon steel [47]

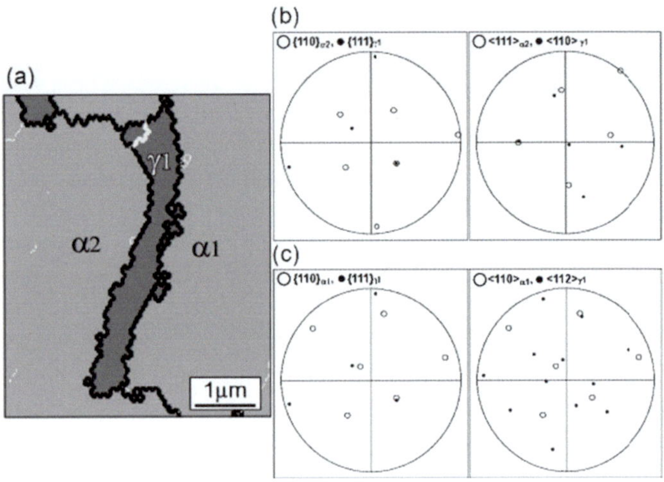

Fig. 13 **a** Microstructure of Fe-9.5at%Ni alloy at 883 K showing formation of γ_1 at the grain boundary of α grains. Stereographic orientation showing **b** K-S relationship between α_2 and γ_1, **c** N-W relationship between α_1 and γ_1 [48]

both the grains. The grain growth of γ phase was sometimes accompanied by Σ3 boundary formation, by a two-dimensional nucleation process on the {111} plane.

Wang et al. [49] studied tensile behaviour of Fe-30Mn-10Co-10Cr-0.5C (at. %) interstitial high entropy alloy using in situ EBSD to understand the phase transformation and twinning mechanisms in the alloy. It was observed that the alloy deform by formation and multiplication of stacking faults along {111} microbands. When there is sufficient overlap of stacking faults within the microbands, it leads to intrinsic nucleation of HCP ε phase. The incoherent annealing twin boundaries act as sites for ε phase nucleation. Upon further straining, the ε phase grows into adjacent FCC γ matrix. The stress required to activate twinning in the FCC γ grains with a size of around 10 μm was found to be approximately 736 MPa, at room temperature.

With further deformation, new twin lamellae started to nucleate continuously and grow in the direction of motion of mobile partial dislocations. The desired strain hardening profile in the alloy can be tuned and excellent combination of strength and ductility can be achieved owing to the deformation-driven athermal phase transformation from FCC into HCP phase, stacking fault formation, precipitation hardening, and mechanical twinning.

8 In Situ EBSD Cryogenic Studies

Cryogenic temperature deformation is a helpful tool in comprehending the fundamental deformation behaviour of various alloys, since thermally activated modes will be repressed. It is also important from industrial point of view if unique textures can be obtained using cryogenic deformation. For example, the strong basal textures produced in magnesium alloys (e.g. AZ) during standard thermo-mechanical processing leads to poor formability at room temperature. Zhang et al. [50] studied compression of commercial cast magnesium alloy AZ31 at room temperature (RT) and cryogenic temperature (CT) and traced evolution of microstructure and plasticity behaviour using a quasi in situ EBSD method. The CT deformed sample showed higher flow stress and higher strain hardening rate compared to the RT deformed sample. The quasi in situ EBSD revealed increased fraction of finer twins, twin-twin interactions and lower average KAM (kernel average misorientation) in CT deformed sample, leading to hardening effect, more effective than the stored tangled dislocation hardening. It was observed that CT leads to multiple twin variants in a single grain (Fig. 14).

Cheng et al. [51] analysed tensile behaviour of solution treated Al-Cu-Mn alloy at RT and CT, via quasi in situ EBSD. At CT, the sample showed resistance to localized thinning and crack propagation, leading to enhanced ductility and strain hardening,

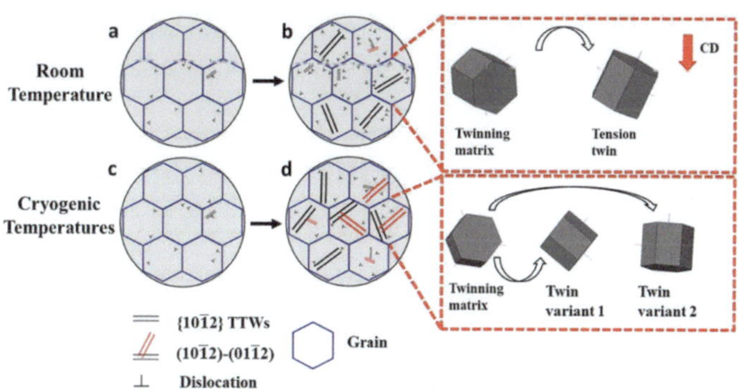

Fig. 14 Schematic of microstructure of AZ31 alloy before and after deformation at room temperature and cryogenic temperature [50]

compared to RT deformation. The ductility is enhanced due to increased dislocation storage capacity inside the grains and reduced number of movable dislocations along the grain boundaries. The strain hardening is enhanced due to reduced relative slip distance of activated dislocations and the constraint collapse of dislocation tangles and cells. Chen et al. [52] employed quasi in situ EBSD to study microstructural heredity in 6.5%Ni-containing cryogenic steel. The alloy was cast and then subjected to homogenization treatment at 1473 K for 2 h, followed by hot rolling and then quenching. During hot rolling, different as-rolled lath martensite structures were developed, depending on the type of rolling and prior austenite grain morphologies. During the inter-critical quenching, the as-rolled lath martensite go through austenite reversion and re-martensitic transformation, which results in the formation of lath-like fresh martensite (LFM) and blocky fresh martensite (BFM) along the lath boundaries and prior austenite grain boundaries, respectively. It was observed that the LFM maintains K-S orientation relationship with the parent austenite grain, leading to memory effect. Additionally, the as-rolled lath martensite gets refined and hence promotes BFM formation to further refine the inter-critically quenched microstructure. The heredity effect of as-rolled lath martensitic refinement resulted in excellent cryogenic toughness. Wang et al. [53] studied the effect of secondary twinning in high manganese austenitic steel at 77 K, by employing quasi in situ EBSD. Sample with fine grains (26.8 μm) showed excellent tensile properties (tensile strength = 1344 MPa, total elongation = 57%) at cryogenic temperature, compared to sample with coarse grains (78.8 μm) (tensile strength = 1238 MPa, total elongation = 48%). Both the samples reveal primary and secondary twins in the microstructure, and volume fraction of secondary twins decrease with increase in grain size. The grain refinement due to primary twins is 'one dimensional', which indicates activation of secondary twins become difficult with increase in primary twins. In the CG steel, large number of primary twin formation at an early stage prevents the formation of secondary twins, leading to decrease in strain hardening capacity. However, in the FG steel, sufficient secondary twin formation leads to increase in yield strength and strain hardening capacity due to strong twin-twin interaction. This indicates that secondary twin activation depends strongly on primary twins, which are influenced by the grain size.

9 Technological Importance as High Throughput Technique

The key significance of in situ EBSD deformation is that microstructural evolution can be tracked and the information extracted from the microstructural evolution helps to unravel the underlying deformation micro-mechanisms.

For example, in the present study on cp-Ti, in situ load reversal experiments enabled us to obtain valuable insight about the relative importance of the various factors influencing twinning and detwinning propensity. These various factors are

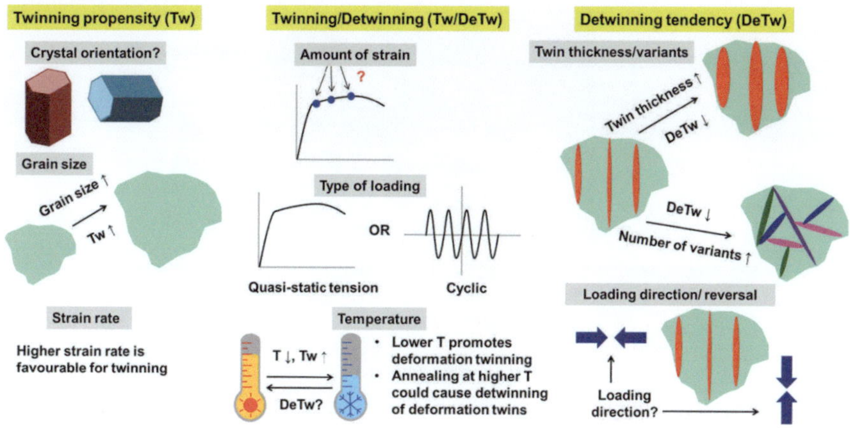

Fig. 15 Factors influencing twinning and detwinning in HCP Ti

summarized in the schematic in Fig. 15. The factors influencing twinning propensity such as initial orientation, strain/strain rate, temperature and grain size are quite well established [54–56]. The mechanisms of twin variant selection in titanium were also discussed in detail by Qin and Jonas [57]. In comparison, detwinning is a more complicated phenomenon. Although a few earlier studies showed in situ evidence of detwinning [58–60], there is a lot of scope for further discussion on detwinning mechanisms.

The use of a real-time microstructural characterization technique such as in situ EBSD in the present study enabled an in-depth study of detwinning characteristics and its impact on the microstructural evolution and deformation behaviour. The present study provided some unambiguous evidence of the factors influencing detwinning in HCP Ti, especially the dependence on loading mode and direction, twin thickness and variants (Fig. 15). Thus, it was clearly established that detwinning tendency is strongly dependent on the type of twins present in the microstructure and the extent of their growth. Also, the number of twin variants formed influences the detwinning tendency. Similarly, the few other in situ EBSD studies on FCC/BCC alloys also provided valuable and new insight about deformation (work hardening) and fracture (crack growth) behaviour.

In one of our recent investigation [61], we have combined in situ EBSD study on tensile deformation of a solutionized Al–Mg alloy with crystal plasticity simulations using the elasto viscoplastic fast Fourier transform simulations using the Dusseldorf Advanced Materials Simulation Kit (DAMASK) software [62, 63] to understand the micro-mechanisms of evolution of texture and damage in the presence of second phase particles. Figure 16 shows the initial and final microstructure of the Al–Mg alloy in terms of tensile axis inverse pole figure map. The secondary electron image shows the occurrence of slip lines, surface roughness evolution, kink band formation and accumulation of damage on the deformed sample at failure. The initial microstructure of the sample was digitised and used to perform crystal

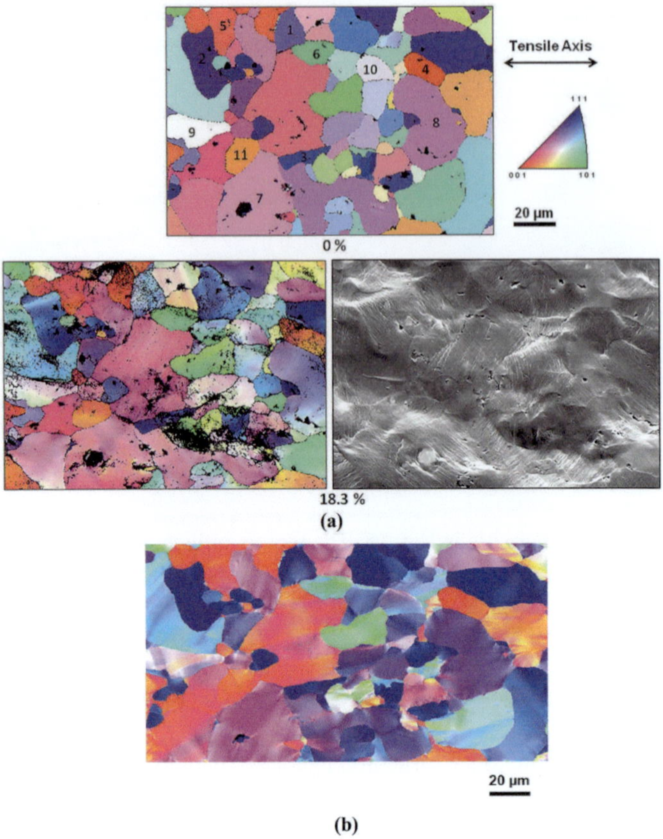

Fig. 16 Experimental tensile axis inverse pole figure map for initial sample **a** and experimental inverse pole figure map and secondary electron image after strain of 18.3% and **b** inverse pole figure map from elasto viscoplastic fast Fourier transform simulations using DAMASK software for Al–Mg alloy [61]

plasticity simulations using DAMASK assuming octahedral slip in the solid solution aluminium matrix while the dispersoid phase was considered fully elastic. Two dimensional crystal plasticity simulations performed with periodic boundary conditions yield a good match with the experimental stress–strain response (Fig. 17a) as well as the experimental texture (Fig. 17b), thus providing correct information about lattice rotations for different grains in the microstructure. The full field simulations also provided information about evolution of von Mises stress and strain (Fig. 18a and b, respectively) in the microstructure providing information on damage accumulation at the microstructural length scale. A mechanistic insight into the deformation process was obtained by plotting the micro-mechanical Taylor factor map (Fig. 18c) that links the local strain state with operative slip systems [64]. Thus, a thorough slip activity and deformation/damage correlation can be obtained which can help in choosing different crystal plasticity simulation parameters like critical resolved shear stress for the slip systems, hardening law as well as constitutive relationship. The high fidelity parameters used to fit in situ experimental data are expected to capture not just the stress–strain and bulk texture data but also the gain level intragranular misorientation data and local slip activity. Thus, full field crystal plasticity simulations performed with in situ experiments can reduce the number of experiments to provide better prediction of deformation and damage for different stress and strain boundary conditions. The above example clearly manifests the important contribution of in situ EBSD in the ICME paradigm which relies heavily on crystal plasticity simulations to link the microstructure with micromechanical component scale deformation behaviour. Similarly, in situ EBSD of recrystallization and phase transformation provides important information for industrial important processes and when coupled with phase field [65] or cellular automata modelling, help in process development under the ICME paradigm.

To summarize the main advantages, in situ EBSD deformation studies are of paramount importance for crystal orientation mapping based analysis of underlying deformation micro-mechanisms. A single cycle of a usual cyclic deformation test can be broken down further into steps (forward loading, unloading, reverse loading and unloading, reloading etc.) and after each stage, the crystal orientation mapping

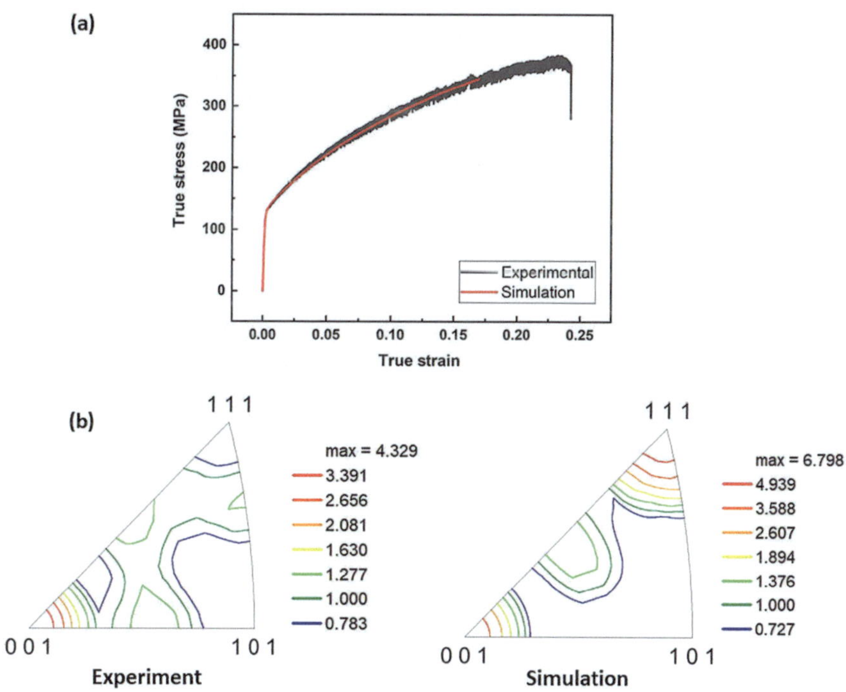

Fig. 17 Comparison of experimental and simulated **a** stress–strain response and **b** tensile axis inverse pole figure from in situ tensile test of Al–Mg alloy [61]

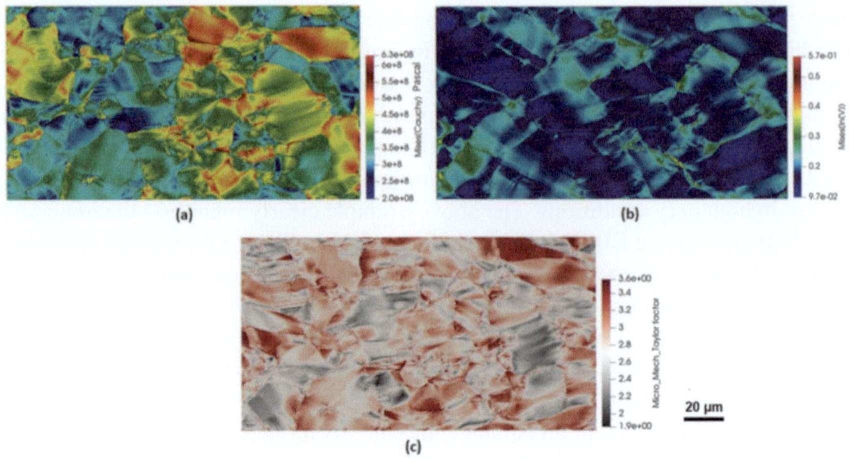

Fig. 18 Distribution of **a** von Mises stress **b** von Mises strain and **c** micro-mechanical Taylor factor map for in situ EBSD tested Al–Mg sample deformed till strain of 18.3% from elasto viscoplastic fast Fourier transform simulation using DAMASK software [61]

can be carried out, i.e. the microstructural changes caused by each step of the load reversal process can be captured.

Nevertheless, the following limitations and nuances of the in situ EBSD method must also be mentioned here to highlight the utility of the technique. First, the range of strain rates applicable in the in situ tensile stage is limited. For example, the stage used in the present study has five discrete crosshead velocity values available in the range of 0.05–0.4 mm/min. Second, the in situ EBSD deformation experiments carried out using a 2 kN stage are appropriate for observing the microstructure and crystal orientation evolution during deformation. This is sufficient for investigating deformation micro-mechanisms, but is not a reliable method if the objective is to obtain the mechanical properties of the metal/alloy. In other words, there could be a systematic error between the levels of stress/strain obtained from a bulk mechanical test and that obtained from an in situ EBSD test. Therefore, trial experiments are necessary to calibrate the applied load/displacement before conducting experiments to study deformation behaviour pertaining to any particular regime of the tensile flow curve.

9.1 Concluding Remarks

In situ EBSD studies have expanded the scope of detailed investigation of deformation, recrystallization and phase transformation micro-mechanisms in polycrystalline metals and alloys in the last 25 years. The two primary advantages of this method are that microstructural evolution with deformation and heating/cooling can be captured as it happens and the analysis makes it possible to distinguish the micro-scale crystallite orientation evolution and deformation, recrystallization and phase transformation behaviour from the global texture evolution or mechanical response. As a result of the latter, it is also possible to obtain the contribution of the local microstructural effects to the overall deformation, recrystallization and phase transformation behaviour. This is of utmost importance for microstructure tailoring to achieve enhanced mechanical properties. Presently, in situ EBSD deformation studies are widely used by researchers across the world to perform in-depth investigation of the micro-mechanisms of tensile and cyclic deformation behaviour of various metals and alloys in tension, compression, bending in quasistatic regime as well as phase transformation over a wide temperature range. Nevertheless, there is further scope to utilize this technique to unravel the micro-mechanisms cyclic hardening/softening during high cycle fatigue and high pressures. Specifically, many researchers have carried out in situ load reversal studies involving a few cycles; however, it would be interesting if the in situ method could be suitably developed to study various cyclic hardening/softening regimes at higher number of cycles as well as studying deformation and phase transformation at high pressures and temperature. There is also a scope to provide a wide range of strain rate covering the creep regime to dynamic deformation at different temperatures from cryogenic to 1273 K along with different atmosphere like hydrogen for in situ EBSD experiments so that material behaviour in actual engineering applications can be captured. Even with the present high speed EBSD cameras available these days, in situ EBSD experiments are not truly real time experiments as EBSD scan takes finite time to capture the data and hence efforts should be directed to capture the Kikuchi patterns at higher speed with better angular resolution and faster data analysis to be in sync with the actual processes happening in the microstructure. Nevertheless, the important contribution of in situ EBSD experiments to provide high fidelity microstructural evolution as a function of mechanical and thermal field provides a high throughput method that forms an integral part of ICME paradigm to develop robust crystal plasticity and phase field models for capturing material behaviour to enable accelerated materials and process development.

Acknowledgements The authors gratefully acknowledge funding support from Science and Engineering Research Board (SERB), DST, Govt. of India for carrying out this research. Thanks are due to Tata Consultancy Services Research for funding high throughput experimentation research at IIT Kanpur and the Indian Space Research Organization for their continued support.

References

1. Wright SI, Nowell MM (2009) A review of in situ EBSD studies. In: Schwartz A, Kumar M, Adams B, Field D (eds) Electron backscatter diffraction in materials science. Springer, Boston, MA, pp 329–337
2. Mishra SK, Pant P, Narasimhan K, Rollett AD, Samajdar I (2009) On the widths of orientation gradient zones adjacent to grain boundaries. Scr Mater 61:273–276

3. Gurao NP, Suwas S (2013) Deformation behaviour at macro- and nano-length scales: the development of orientation gradients. Mater Lett 99:81–85
4. Weidner A, Glage HB (2010) In-situ characterization of the microstructure evolution during cyclic deformation of novel cast TRIP steel. Procedia Eng 2:1961–1971
5. Borkar H, Seifeddine S, Jarfors AEW (2015) In-situ EBSD study of deformation behavior of Al–Si–Cu alloys during tensile testing. Mater Des 84:36–47
6. Buchheit TE, Carroll JD, Clark BG, Boyce BL (2015) Evaluating deformation-induced grain orientation change in a polycrystal during in situ tensile deformation using EBSD. Microsc Microanal 21:969–984
7. Donoghue J, Gholinia A, Fonseca JQD, Prangnell P (2016) In-situ high temperature EBSD analysis of the effect of a deformation step on the alpha to beta transition in additive manufactured Ti-6Al-4V. In: Venkatesh V, Pilchak AL, Allison JE, Ankem S, Boyer R, Christodoulou J, Fraser HL, Imam MA, Kosaka Y, Rack HJ, Chatterjee A, Woodfield A (eds) Proceedings of the 13th World Conference on Titanium, TMS, pp 1283–1288
8. Gussev MN, Leonard KJ (2019) In situ SEM-EBSD analysis of plastic deformation mechanisms in neutron-irradiated austenitic steel. J Nucl Mater 517:45–56
9. Li S, Guo C, Hao L, Kang Y, An Y (2019) In-situ EBSD study of deformation behaviour of 600 MPa grade dual phase steel during uniaxial tensile tests. Mater Sci Eng A 759:624–632
10. Li W, Yamasaki S, Mitsuhara M, Nakashima H (2020) In situ EBSD study of deformation behavior of primary α phase in a bimodal Ti-6Al-4V alloy during uniaxial tensile tests. Mater Char 163:110282
11. Sinha S, Szpunar JA, Kiran Kumar NAP, Gurao NP (2015) Tensile deformation of 316L austenitic stainless steel using in-situ electron backscatter diffraction and crystal plasticity simulations. Mater Sci Eng A 637:48–55
12. Hémery S, Villechaise P (2019) In situ EBSD investigation of deformation processes and strain partitioning in bi-modal Ti-6Al-4V using lattice rotations. Acta Mater 171:261–274
13. Lutjering G, Williams JC (2003) Titanium. Springer-Verlag, Berlin Heidelberg
14. Salem AA, Kalidindi SR, Doherty RD (2002) Strain hardening regimes and microstructure evolution during large strain compression of high purity titanium. Scr Mater 46:419–423
15. Salem AA, Kalidindi SR, Doherty RD (2003) Strain hardening of titanium: role of deformation twinning. Acta Mater 51:4225–4237
16. Salem AA, Kalidindi SR, Doherty RD, Semiatin SL (2006) Strain hardening due to deformation twinning in α-titanium: Mechanisms. Met Mater Trans A 37A:259–268
17. Salem AA, Kalidindi SR, Semiatin SL (2005) Strain hardening due to deformation twinning in α-titanium: constitutive relations and crystal-plasticity modeling. Acta Mater 53:3495–3502
18. Roth MA, Lebyodkin TA, Lebedkina J-S, Lecomte T, Richeton KEKA (2014) Mechanisms of anisotropy of mechanical properties of α-titanium in tension conditions. Mater Sci Eng A 596:236–243
19. Bozzolo N, Chan L, Rollett AD (2010) Misorientations induced by deformation twinning in titanium. J Appl Crystallogr 43:596–602
20. Hirth JP, Lothe J (1982) Theory of dislocations, 2nd edn. Wiley, New York, pp 811–834
21. Christian JW, Mahajan S (1995) Deformation twinning. Prog Mater Sci 39:1–157
22. Capolungo L, Beyerlein IJ (2008) Nucleation and stability of twins in hcp metals. Phys Rev B 78:024117–1–024117–19
23. Galindo-Nava EI, Rivera-Diaz-del-Castillo PEJ (2014) Thermostatistical modelling of deformation twinning in HCP metals. Int J Plast 55:25–42
24. Sinha S, Ghosh A, Gurao NP (2016) Effect of initial orientation on the tensile properties of commercially pure titanium. Philos Magaz 96:1485–1508
25. Sinha S, Pukenas A, Ghosh A, Singh A, Skrotzki W, Gurao NP (2017) Effect of initial orientation on twinning in commercially pure titanium. Philos Magaz 97:775–797
26. Sinha S, Gurao NP (2017) In situ electron backscatter diffraction study of twinning in commercially pure titanium during tension-compression deformation and annealing. Mater Des 116:686–693

27. Arul Kumar M, Kanjarla AK, Niezgoda SR, Lebensohn RA, Tome CN (2015) Numerical study of the stress state of a deformation twin in magnesium. Acta Mater 84:349–358
28. Demir E, Raabe D (2010) Mechanical and microstructural single-crystal Bauschinger effects: observation of reversible plasticity in copper during bending. Acta Mater 58:6055–6063
29. Mendelson S (1969) Zonal dislocations and twin lamellae in h.c.p. metals. Mater Sci Eng 4:231–242
30. Sinha S (2017) Effect of twinning on tensile and cyclic deformation behaviour of hexagonal close packed titanium. PhD thesis, Indian Institute of Technology Kanpur, India
31. Sinha S, Gurao NP (2017) The role of crystallographic texture on load reversal and low cycle fatigue performance of commercially pure titanium. Mater Sci Eng A 691:100–109
32. Sinha S, Gurao NP (2017) In-plane anisotropy in mechanical behaviour and microstructural evolution of commercially pure titanium in tensile and cyclic loading. Met Mater Trans A 48A:5813–5832
33. Nourbakhsh S, O'Brien TD (1988) Texture formation and transition in cold-rolled titanium. Mater Sci Eng 100:109–114
34. Yin SM, Yang HJ, Li SX, Wu SD, Yang F (2008) Cyclic deformation behavior of as-extruded Mg–3%Al–1%Zn. Scr Mater 58:751–754
35. Yang HL, Kano S, Chai LJ, Shen JJ, Zhao ZS, McGrady J, Duan ZG, Abe H (2019) Interaction between slip and 1012 tensile twinning in Zr alloy: quasi in situ electron backscatter diffraction study under uniaxial tensile test. J Alloys Compds 782:659–666
36. Mishra S, Yadava M, Kulkarni KN, Gurao NP (2019) A new phenomenological approach for modeling strain hardening behavior of face centered cubic materials. Acta Mater 178:99–113
37. Ghassemali E, Sonkusare R, Biswas K, Gurao NP (2017) In-situ study of crack initiation and propagation in a dual phase AlCoCrFeNi high entropy alloy. J Alloy Compd 710:539–546
38. Ghassemali E, Sonkusare R, Biswas K, Gurao NP (2018) Dynamic precipitation at elevated temperatures in a dual-phase AlCoCrFeNi high-entropy alloy: an in situ study. Philos Magaz Lett 98:400–409
39. Lopez-Sanchez MA, Tommasi A, Barou F, Quey R (2020) Dislocation-driven recrystallization in AZ31B magnesium alloy imaged by quasi-in situ EBSD in annealing experiments. Mater Charac 165:110382
40. Lee SW, Park SH (2020) Static recrystallization mechanism in cold-rolled magnesium alloy with off-basal texture based on quasi in situ EBSD observations. J Alloy Comp 844:156185
41. Zhao PC, Guan B, Tong YG, Wang RZ, Li X, Zhang XC, Tu ST (2022) A quasi-in-situ EBSD study of the thermal stability and grain growth mechanisms of CoCrNi medium 9entropy alloy with gradient-nanograined structure. J Mater Sci Technol 109:54–63
42. Nakamichi H, Humphreys FJ, Brough I (2008) Recrystallization phenomena in an IF steel observed by in situ EBSD experiments. J Microsc 230:464–471
43. Tokita S, Kokawa H, Sato YS, Fujii HT (2017) In situ EBSD observation of grain boundary character distribution evolution during thermomechanical process used for grain boundary engineering of 304 austenitic stainless steel. Mater Charac 131:31–38
44. Kim DI, Kim JS, Kim JH, Choi SH (2014) A study on the annealing behavior of Cu-added bake-hardenable steel using an in situ EBSD technique. Acta Mater 68:9–18
45. Seward GGE, Celotto S, Prior DJ, Wheeler J, Pond RC (2004) In situ SEM-EBSD observations of the hcp to bcc phase transformation in commercially pure titanium. Acta Mater 52:821–832
46. Ubhi HS, Parsons J, Othen N, Campbell S, Poole R, Gholinia A (2014) In-situ EBSD phase transformation and recrystallisation. J Phys Confer Ser 522:012011
47. Lischewski I, Kirch DM, Ziemons A, Gottstein G (2008) Investigation of the α-γ-α phase transformation in steel: high-temperature in situ EBSD measurements. Text Stress Microst 294508
48. Fukino T, Tsurekawa S (2008) In-situ SEM/EBSD Observation of α/γ phase transformation in Fe-Ni Alloy. Mater Trans 49:2770–2775
49. Wang M, Li Z, Raabe D (2018) In-situ SEM observation of phase transformation and twinning mechanisms in an interstitial high-entropy alloy. Acta Mater 147:236–246

50. Zhang K, Shao Z, Daniel CS, Turski M, Pruncu C, Lang L, Roboson J, Jiang J (2021) A comparative study of plastic deformation mechanisms in room-temperature and cryogenically deformed magnesium alloy AZ31. Mater Sci Eng A 807:140821
51. Cheng W, Liu W, Fan X, Yuan S (2020) Cooperative enhancements in ductility and strain hardening of a solution-treated Al–Cu–Mn alloy at cryogenic temperatures. Mater Sci Eng A 790:139707
52. Chen QY, Zhang WN, Tang S, Wang PJ, Chen J, Liu ZY (2022) Microstructural heredity of Ni-containing cryogenic steel and its effect on the toughness at 77 K. Mater Sci Eng A 832:142441
53. Wang P, Ren J, Chen Q, Chen J, Liu Z (2021) Effect of secondary twins on strain hardening behavior of a high manganese austenitic steel at 77 K by quasi in situ EBSD. Mater Charac 180:111428
54. Meyers MA, Vöhringer O, Lubarda VA (2001) The onset of twinning in metals: a constitutive description. Acta Mater 49:4025–4039
55. Fu HH, Benson DJ, Meyers MA (2001) Analytical and computational description of effect of grain size on yield stress of metals. Acta Mater 49:2567–2582
56. Meyers MA, Benson DJ, Vöhringer O, Kad BK, Xue Q, Fu H-H (2002) Constitutive description of dynamic deformation: physically-based mechanisms. Mater Sci Eng A 322:194–216
57. Qin H, Jonas JJ (2014) Variant selection during secondary and tertiary twinning in pure titanium. Acta Mater 75:198–211
58. Morrow BM, Cerreta EK, McCabe RJ, Tome CN (2014) Toward understanding twin–twin interactions in hcp metals: utilizing multiscale techniques to characterize deformation mechanisms in magnesium. Mater Sci Eng A 613:365–371
59. Wang L, Lind J, Phukan H, Kenesei P, Park J-S, Suter RM, Beaudoin AJ, Bieler TR (2014) Mechanical twinning and detwinning in pure Ti during loading and unloading–an in situ high-energy X-ray diffraction microscopy study. Scr Mater 92:35–38
60. Wang S, Zhang Y, Schuman C, Lecomte J-S, Zhao X, Zuo L, Philippe M-J, Esling C (2015) Study of twinning/detwinning behaviors of Ti by interrupted in situ tensile tests. Acta Mater 82:424–436
61. Nagarajan S, Jain R, Gurao NP (2021) Microstructural characteristics governing the lattice rotation in Al-Mg alloy using in situ EBSD. Mater Charac 180:111405
62. Roters F, Diehl M, Shanthraj P, Eisenlohr P, Reuber C, Wong SL, Maiti T, Ebrahimi A, Hochrainer T, Fabritius H-O, Nikolov S, Friák M, Fujita N, Grilli N, Janssens KGF, Jia N, Kok PJJ, Ma D, Meier F, Werner E, Stricker M, Weygand D, Raabe D (2019) DAMASK—the Düsseldorf advanced material simulation kit for modeling multi-physics crystal plasticity, thermal, and damage phenomena from the single crystal up to the component scale. Comput Mater Sci 158:420–478
63. Lebensohn RA, Brenner R, Castelnau O, Rollett AD (2008) Orientation image-based micromechanical modelling of subgrain texture evolution in polycrystalline copper. Acta Mater 56:3914–3926
64. Raabe D, Sachtleber M, Zhao Z, Roters F, Zaefferer S (2001) Micromechanical and macromechanical effects in grain scale polycrystal plasticity experimentation and simulation. Acta Mater 49:3433–3441
65. Luan Q, Lee J, Zheng JH, Hopper C, Jiang J (2020) Combining microstructural characterization with crystal plasticity and phase-field modelling for the study of static recrystallization in pure aluminium. Comput Mater Sci 173:109–119

Chapter 5
Electron Channeling and Electron Channeling Contrast Imaging (ECCI)

Martin A. Crimp

1 Introduction

The basic premise of the field of materials science is that the properties of a given material are controlled by its underlying structure. In many instances, the nature of crystal defects is critical to the overall performance of materials. Consequently, a wide range of techniques has been developed to characterize these defects, including scanning and transmission electron microscopies (SEM and TEM). Of particular note, since the middle of the twentieth century the application of diffraction contrast TEM to the study of crystalline materials has allowed the crystallographic details of defects (grain boundary orientation and structure, habit planes, dislocation line directions and Burgers vectors, stacking fault displacements and planes, twin orientations, etc.) and their interactions to be understood in relation to the broader prosperities of the materials.

Nevertheless, while being an extremely powerful tool, TEM is limited by the requisite thin foils. Preparation of these thin foils can be time consuming and difficult. Furthermore, thin foils are subject to artifacts associated with image forces from the two free surfaces in close proximity. Historically, such foils were prepared by electropolishing or ion milling, both of which have significant limitations, including difficulties in establishing thinning conditions and the potential for the introduction of artifacts. More recently, focused ion beam (FIB) and plasma FIB (PFIB) preparation of TEM thin samples has become the norm, but these approaches can still result in significant artifacts, including the introduction of point defects, re-deposition of material, and curtaining. These FIB produced TEM thin foils are further limited by an overall lack of thin area. Additionally, while FIB approaches allow more site-specific thin foil preparation than electropolishing or traditional ion milling, these

M. A. Crimp (✉)
Chemical Engineering and Materials Science, Michigan State University, East Lansing, MI, USA
e-mail: crimp@egr.msu.edu

S. Suwas and D. P. Field (eds.), *Advances in Texture, Microtexture, and Allied Techniques*, Materials Horizons: From Nature to Nanomaterials,
https://doi.org/10.1007/978-981-96-5346-1_5

approaches, like the others, are destructive. As a consequence, it is difficult to use TEM to study specific areas before and after various treatments. While this limitation has the potential to be overcome with in situ studies, TEM can be limited for this approach. For example, while significant numbers of in situ deformation studies of dislocation behavior have been carried out over the years, these are typically limited by ill-defined/complex stress states because of the irregular nature of the thin foil sample.

Many of the limitations of TEM, particularly for the study of dislocations, twins, and stacking faults, can be overcome using electron channeling contrast imaging (ECCI). This SEM based characterization technique allows the imaging of these crystal defects to be carried out in the near-surface region of bulk samples. Consequently, while it is still necessary to have a well-prepared surface on the bulk sample, and the single free surface may still impart image forces, it is not necessary to prepare a thin foil sample. This leads to a number of advantages of ECCI over TEM. First, because a bulk sample is being examined, observations can be made over a very large area. This can result in more statistically relevant data collection, such as for defect density determinations. Furthermore, because a large amount of material is available for analysis, specific regions of interest can be explored, for example specific grain boundary orientations or crack tip regions, all the while non-destructively. Additionally, the surface preparation can be carried out and the material examined prior to many material treatments, such as mechanical or radiation. This allows the surface to be examined for any potential sample preparation artifacts prior to subsequent treatment, allowing the final structure to be fully attributed to the treatment. For example, in deformation studies, it is possible to characterize the pre-existing dislocation structure using ECCI, carry out deformation (either in situ or ex situ), and then perform ECCI to put the deformation structure in context to the pre-existing structure. Multi-step characterization is thus possible. In many cases, in situ studies can be carried out (although, for image capture time reasons that will be discussed below, real time dynamic studies can be limited). While ECCI does have some limitations, which will be outlined later, overall it offers an important alternative to diffraction contrast TEM for the study of crystalline defects in a number of situations.

2 Fundamentals of Electron Channeling

In the same way that diffraction contrast TEM images depend on the underlying electron diffraction as manifested by the associated electron diffraction pattern, ECC images are critically dependent on the underlying electron channeling phenomenon, often illustrated by an accompanying electron channeling pattern. Thus, it is critical to understand the nature of electron channeling in order to understand the effective use of ECCI.

In 1967 Coates was the first to report the electron channeling phenomenon, with the observation of "orientation-dependent patterns, very similar in appearance to

Kikuch patterns, obtained using thick single-crystal specimens in a scanning electron microscope [1]. The ability to potentially use this phenomenon to image crystal defects was immediately recognized by Booker et al. [2] (indeed, in the very next paper in Philosophical Magazine), who stated "It should in principle be possible to use the scanning electron microscope to detect dislocations. It is only necessary to orientate the crystal at the Bragg position and the local bending of the crystallographic planes should produce the necessary contrast. Such contrast should be directional and lead to Burgers vector determination". Thus, the relationship between electron channeling and the ability to characterize dislocations was recognized early on, although practical implementation took a significant amount of time.

The observation by Coates was effectively that the backscattered electron yield was a function of the angle between the incident electron beam and the orientation of the crystal lattice [1]. The behavior was schematically illustrated by Newbury and Yakowitz [3], as shown in Fig. 1, where it is seen that the backscattered electron yield, η, varies as the angle between the incoming electron beam and crystal lattice changes. While this schematic shows the general phenomenon, it suggests that η will be low when the electron trajectory is nearly parallel to the crystal lattice planes. In fact, this is opposite to the actual case due to the bloch wave nature of the electrons, which was illustrated by Newbury and Yakowitz [3] as shown in Fig. 2. This figure shows the backscattered electron (BSE) intensity (in arbitrary units) as a function of angle from parallel to (110) planes in silicon for 30 keV incident electrons. Strong backscattering is revealed with the incident electron nearly parallel to the planes, but the BSE intensity rapidly decreases as the angle is increased to near that of the $(\bar{2}\bar{2}0)$ and (220) allowed Bragg reflections for Si. At angles greater than these Bragg angles, the BSE intensity increases again, but then dramatically decreases in the vicinity of the higher order {440} reflections, repeating the sequence in a weaker manner for the {660} Bragg reflections.

This electron channeling phenomenon is very commonly seen in backscattered electron images of polycrystals (and even secondary electron images of low topography samples, as the backscattered electrons contribute to the captured secondary

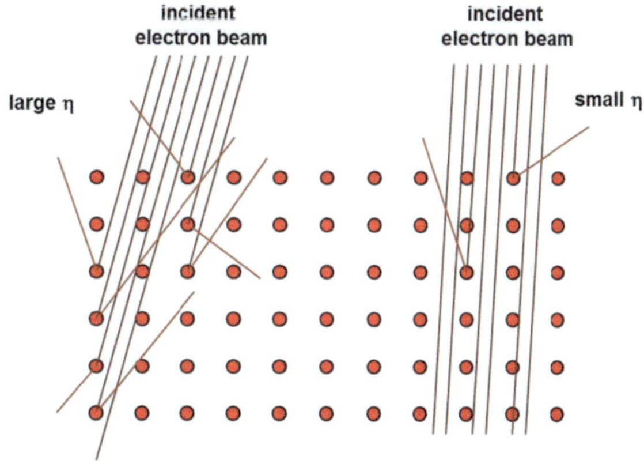

Fig. 1 Schematic representation of the origin of the electron channeling phenomenon, where the intensity of the backscattered electrons, represented by the backscattered electron yield η, depends on the relative angle between the incoming electron beam and the crystal lattice (after Newbury and Yakowtiz [3])

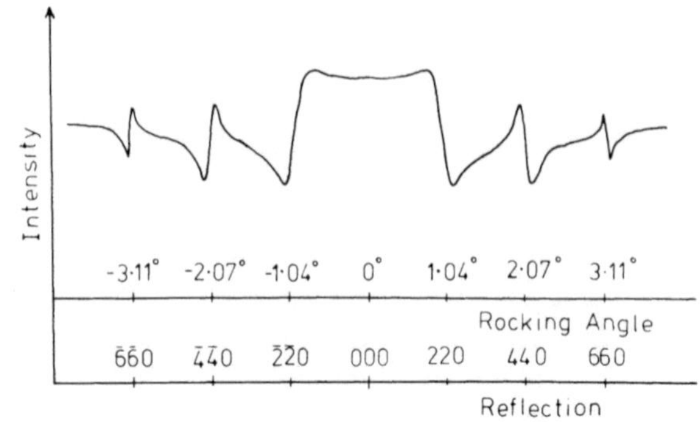

Fig. 2 Variation in the backscattered electron intensity as a function of incident electron angle across a {XX0} Bragg reflection of silicon for an incident electron energy of 30 keV. The Bragg positions and associated Bragg angles are noted. (Reprinted by permission from Newbury and Yakowtiz [3])

electron signal), where the contrast between grains is a consequence of differences in grain orientations, as shown in Fig. 3. This electron channeling phenomenon is manifested as Kikuchi patterns when the electron beam trajectory is systematically varied on a large grain or a single crystal. Indeed, the original observations by Coates [1] were realized as a result of the electron beam trajectory varying during scanning at low magnification.

Fig. 3 Backscattered electron image of polycrystalline commercially pure titanium showing electron channeling contrast resulting from the varying orientations of the grains. Note that within some grains the backscattered electron contrast varies, either from local changes in the grain orientation or as a result of the electron beam trajectory changing slightly during the beam scanning. (Image courtesy of S. Han and M. Crimp)

3 Electron Channeling Patterns and Selected Area Electron Channeling Patterns

The patterns of electron channeling lines that develop because of the change in BSE yield with respect to the crystal orientation are commonly known as electron channeling patterns (ECPs). These patterns are made up of a series of bands (see Fig. 4), where each band represents the electron channeling behavior and orientation of a given set of crystallographic planes. The angular spacing between the two edges of the band represents twice the Bragg angle θ for the given accelerating voltage (and resulting wavelength λ) and the interplanar spacing d for the given (hkl) plane, according to Bragg's Law:

$$\lambda = 2d\sin\theta$$

The center of the pattern represents the orientation of the un-deflected electron beam, and therefore represents the orientation of the crystal with respect to the optic axis of the SEM electron column. Often, higher order channeling bands are observed (as suggested by Fig. 2 and seen in Fig. 4a). The patterns are rotationally correct to the image, meaning the orientations of bands reflect the actual orientations of the {hkl} planes in the material, with the bands being parallel to the lines defined the by the intersection of the projections of the planes with an imaginary plane perpendicular to the optic axis of the microscope.

Such patterns are typically observable on large single crystal specimens imaged using backscattered electrons (silicon wafers are ideal for imaging ECPs), and are sometimes observable on so-called "secondary electron images" (commonly Everhart–Thornley detected images) due to the high energy backscattered electrons contributing to the signal. Nonetheless, these ECPs are rather weak and can be easily

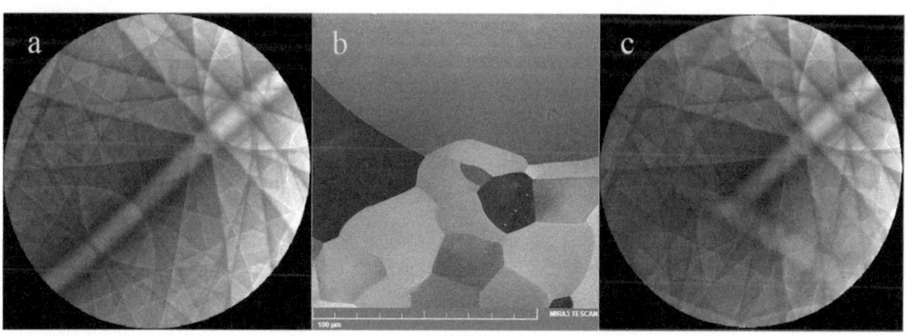

Fig. 4 **a** Selected area electron channeling pattern collected from the large grain in the upper portion of **b**. The pattern in **a** was collected from an area approximately 30 mm in diameter. **c** SACP collected from a somewhat lower position of the top grain in **b**. The lower portion of the pattern in **c** has now come from the not only the upper grains, but also one or more of the grains below the large grain, resulting in the pattern showing channeling characteristics of multiple grains, as evidenced by the broken channeling bands. (Image courtesy of S. Han and M. Crimp)

interfered with by topographical contrast mechanisms. The angular range of the patterns is simply given by the angle over which the indecent electron beam sweeps during scanning. As such, the angular range of the ECPs will decrease as the magnification is increased. Because of this, the patterns are lost when imaging samples even at moderately low magnifications. Regardless, these patterns open a window for examining the crystallography of materials in the SEM.

ECPs were the first tool to allow the morphology of samples imaged in SEM to be directly related to their crystallography. Indeed, a variety of studies of the crystallographic nature of materials were carried out using ECPs in the period of 1970s–1980s (Stickler and Booker [4], Davidson [5], Joy et al. [6]). Some of these studies demonstrated the relationship between electron channeling behavior and the accumulation of dislocations by showing the breakdown of ECPs and channeling contrast with plastic deformation. Nevertheless, with the advent of practical, commercially available electron backscattered patterns/diffraction (EBSP or EBSD) systems, ECPs fell out of favor. The reasons for this are quite clear. First, use of EBSD was facilitated by rapid development of auto-indexing and auto-scanning, making data collection very rapid as a result of a single EBSD pattern being collected in a single scan step on the microscope. This is in contrast to the relatively slow collection of ECPs, which requires scanning over multiple angles to collect a single pattern. But more critically, EBSD offers much higher spatial resolution that ECPs, as the beam is stationary for each EBSD pattern collected. In contrast, ECPs require the scanning over a large area to achieve the angular range needed to make ECPs useful for determining orientation. Regardless, ECPs have the advantage over EBSD in that the orientation determined by the former is directly interpretable with respect to the incident beam/optic axis. EBSD on the other hand typically involves a highly tilted specimen and side mounted detector, requiring indirect determination of the orientations, often based on calibration from a secondary standard. This results in more limited accuracy of EBSD determined orientations with respect to the optic axis (note there has been limited work done on tilt-free EBSD by Vystavěl et al. [7], but the accuracy and precision of this approach are not clear).

Some of the difficulties resulting from the generally large areas from which ECPs are collected are shown in Figs. 4 and 5. When collecting channeling patterns from even large grained polycrystals, abrupt changes in the patterns can occur as a consequence of the tilted beam moving into grains with different orientations, as illustrated by the changes between the ECPs shown in Fig. 4a and c. Similarly, if ECPs are collected from samples with significant amounts of residual dislocation content, the channeling bands become distorted (Fig. 5) as a result of the lattice curvatures induced by the dislocations.

A number of approaches have been used to improve the spatial resolution of electron channeling patterns, collectively known as selected area channeling patterns (SACPs). Generally, these SACPs use a variety of electron optics/scanning coil combinations to rock the beam on an area of interest (Van Essen and Shulson [8], Van Essen et al. [9]), rather than allowing the electron trajectory to vary simply by the normal scanning action of the beam. These approaches allowed SACPs to be collected from regions in the range of 1 micron in diameter (Joy and Newbury [10]).

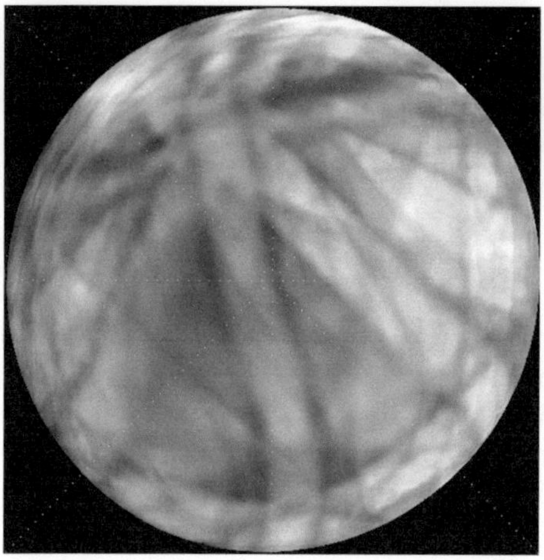

Fig. 5 Selected area channeling pattern collected from deformed (~2% plastic strain) polycrystalline commercially pure titanium. Distortion of the channeling bands results from the variations in crystal orientation within the primary grain. The distortion of the pattern becomes more severe at the outer edges of the pattern, where the beam movement becomes more dramatic during rocking due to greater spherical aberration effects at higher rocking angles. (Image courtesy of S. Han and M. Crimp)

At one time these instrument specific approaches were commercially available on a variety of SEMs. But as electron backscattered diffraction (EBSD) gained in popularity, SACP fell out of favor. Nonetheless, with the recent increase in interest in ECCI and the need for determining orientations precisely with respect to the incoming optic axis/beam direction, a number of approaches for collecting high resolution SACPs (HR-SACPs) from areas as small as 300–500 nm in diameter have been developed (Guyon et al. [11], Kerns et al. [12]). Unfortunately, these approaches are instrument/column specific and are not widely commercially available.

4 Electron Channeling Contrast and Electron Channeling Contrast Imaging (ECCI)

While ECPs and SACPs are manifested by changing the orientation of the incoming electron beam with respect to a fixed crystal orientation, either through scanning or rocking, electron channeling contrast (at least of dislocations) is manifested by having a nominally constant beam orientation scanning across a sample while the orientation of the lattice planes in that sample vary in orientation. In its simplest sense, this electron channeling contrast is seen in the grain-to-grain contrast commonly observed in polycrystalline samples, as illustrated in Figs. 3 and 4b.

Electron channeling contrast imaging (ECCI) is typically meant to be the imaging of sub-grain scale crystallographic defects. As outlined earlier, the ability to image such crystal defects, most commonly dislocations, based on electron channeling was hypothesized by Booker et al. [2]. Nevertheless, the realization of this promise took some time. Based on dynamical theory calculations, Spencer et al. [13] concluded that in order to image dislocations in bulk samples, a field emission gun (FEG) electron source would be required, which were at the time uncommon on SEMs. Thus, early demonstration of dislocation imaging in reflection was carried out using TEM thin foils imaged in scanning electron microscopes [14, 15]. The first successful imaging of dislocations in bulk samples was carried out by Pitaval et al. [16] and Morin et al. [17] who used a cold FEG SEM and a high-energy filter mounted in front of side mounted scintillator detector to increase the signal to noise ratio. This group successfully imaged the dislocations with line resolutions (i.e. dislocation image line widths) of ~10 nm to depths of ~100 nm, and furthermore demonstrated $\mathbf{g}\cdot\mathbf{b} = 0$ invisibility of dislocations. Czernuszka et al. [18] used digital signal processing to overcome the need for an energy filter, and coined the acronym electron channeling contrast imaging (ECCI).

While these studies established the basis for ECCI, the technique was still limited for a number of reasons. To enhance the number of backscattered electrons in the signal (i.e. high η), all these studies were carried out with the sample normal tilted in the range of 60–80° from the optic axis. This configuration was facilitated by a number of custom designed and built detectors. Nevertheless, having the BSE detector at this "high tilt" position was often at odds with the location of the EBSD detector, i.e. the optimum location of both detectors was at the same place. These difficulties left ECCI as a very specialized technique that was only performed in a small number of labs.

Simkin and Crimp [19] were able to take advantage of evolving FEG electron source technology to overcome the limitations of the side detector and demonstrate the ability to carry out ECCI with commercially available microscopes. Taking advantage of a Schottky FEG source, which offers significantly higher beam currents than traditional (at the time) cold FEG sources, this group demonstrated the ability to carry out ECCI with the sample normal parallel (± 30°) to the electron beam trajectory. In this configuration, a standard, four-quadrant Si-diode BSE detector was able to capture the backscattered electrons carrying the channeling contrast signal. Beyond making ECCI available with standard, off-the-shelf components, the low tilt configuration offered a number of other advantages, outlined in Fig. 6 [20]. At high tilts, the resulting SEM image has significant image for-shortening and requires dynamic focusing, both of which complicate the imaging conditions. Furthermore, the beam rocking necessary to collect SACPS compounds the significant variations in the η that occur at high tilts, such that SACPs collected at these high tilts display extreme contrast variations, making the patterns difficult to interpret. High tilts often lead to the sample and or stage components being positioned close to the objective lens pole-piece and BSE detector, which can result in instrument damage or an inability to reach tilt/rotate combinations for optimum imaging conditions. Finally, using a low tilt configuration allows an EBSD detector to be positioned in its normal side-mount position, which allows EBSD to be carried out in conjunction with/complimentary to ECCI. Overall, the low tilt configuration is much more experimentally convenient for most ECCI studies, and is now the most common approach for carrying out ECCI.

5 How to Carry Out Electron Channeling Contrast Imaging

The underlying intensity in ECC images is small and can be easily overwhelmed by topography and debris. Consequently, it is critical to have well prepared surfaces in order to carry out ECCI. Also, because the underlying electron channeling contribution to the BSE intensity is low, ECC imaging requires the enhancement in the contrast from defects to be maximized. Consequently, ECCI is typically carried out under conditions where small changes in orientation associated with the defect will result in a maximum change in the underlying BSE channeling signal. This is accomplished by working close to the Bragg condition, where small changes in lattice orientation will lead to the largest changes in backscattered electron signal (as reflected by the large change in intensity with rocking angle in Fig. 2). In practical terms, this means orienting the crystal near the edge of a specific electron channeling band through a combination of tilts and rotations, so that the optic axis of the microscope (approximately the center of the electron channeling pattern) lies near the edge of the specific band (see Fig. 7). This is equivalent to setting up "2-beam" conditions in TEM, with the ECC image formed primarily due to channeling by the crystallographic planes associated with the channeling band of choice. If possible, one should attempt to avoid affects associated with other channeling bands, but sometimes this can be difficult to achieve. Deviations from the Bragg condition (**s** being the deviation parameter or ω being the dimensionless deviation parameter) were originally examined theoretically by Spencer et al. [13] and experimentally by Simkin and Crimp [19], and are known to result in significant changes in contrast and dislocation line width, as shown in Figs. 8 and 9. More recently these deviation effects have been studied by Zaefferer and Elhami [21] and I. Gutierrez-Urrutia [22], the later showing that optimum line width minimization with acceptable contrast is achieved with significant positive deviations from the Bragg condition. It should be noted that collecting ECC images requires careful focusing and stigmation. Furthermore, because of the low signal to noise of the electron channeling signal, successful imaging requires slow scans, with image collection often requiring scan times in the range of five to ten minutes. Such long imaging times requires good beam stability and minimal stage vibration and drift, and at present precludes the imaging of moving dislocations using in situ deformation. Continuing advances in both electron gun and detector technology are helping to overcome these experimental difficulties.

The approach to carrying out ECCI by setting up specific electron channeling conditions as outlined in the previous paragraph inherently assumes the ability to collect SACPs from the area of interest on the sample. Nevertheless, as outlined earlier, the ability and/or approach to collect SACPs is instrument specific, and many

SEMs do not have the capability capture SACPs. Thus, on many instruments, an indirect approach to setting up the proper channeling conditions is necessary, which entails using EBSD determined orientations. This approach (I. Gurierrez et al. [23]) uses stage tilts and rotations predicted from a high tilt EBSD determined orientation to establish a good "two-beam" channeling condition for imaging at low stage tilts. Once the sample normal orientation is determined using EBSD, either commercial software ([23] used the Tools for Orientation determination and Crystallographic Analysis–TOCA software) or a stereographic projection is used to predict the stage manipulations to achieve the desired channeling conditions. Nonetheless, this approach is subject to potential errors. First, EBSD has inherent errors associated with sample mounting and the calibration approach that leads to an absolute orientation accuracy of ~2° and a precision of ~1° [24]. In contrast, the accuracy and precision of SACPs relative to the optic axis is ~0.1° [11]. Second, this approach is dependent on the mechanical accuracy and precision of the SEM stage, which can vary significantly with instrument. Such stage issues can also be exasperated by the mass of the specimen itself. Because of these limitations, the channeling conditions typically need to be adjusted once the predicted stage manipulations have been carried out in order to optimize the image contrast. This channeling condition modification is facilitated by imaging the region of interest at low magnification, where the edge of the channeling band can often be inferred by the light to dark contrast that occurs due to the small amount of beam trajectory change during beam scanning (i.e. a small portion of the underlying ECP can sometimes be seen). Imaging this edge of a channeling band can also serve to confirm that the intended channeling condition has been reached, as often the orientation of the band edge can be determined and compared with the known band orientation. While this EBSD based approach to ECCI might be less intuitive than setting up electron channeling conditions using SCAP, it opens up the approach to a much larger user base that may not have access to the rather limited number of instruments with SACP capabilities. In fact, a large body of recent studies have been carried out using this approach and it is quite likely that the majority of ECCI studies now being carried out do not use SACPs for setting up channeling conditions.

6 Details of ECCI of Dislocation Structures

As mentioned above, the vast majority of ECCI studies use the technique for imaging and characterizing dislocation structures. The change in contrast associated with dislocations results from the change in local channeling condition that occurs from the local plane bending manifested by a dislocation. Most typically, the dislocation may appear as a bright line (image peak) on a dark background. But depending on the specific channeling condition and deviation parameter, the dislocation can appear as a dark feature on a bright background or even appear as some variation of a combination of a bright/dark line. This variation in contrast conditions was first outlined by Spencer et al. [13]. Because the Bragg/channeling planes are bent in the opposite sense on opposite sides of a dislocation, the image of the dislocation lies to one side of the actual dislocation core. This position shift in the apparent position from the true position is typically in the range of a few tens of nanometers. It is also critical to note that as the dislocation position below the surface increases, the coherency of the incoming beam breaks down due to inelastic scattering event, while the phase shifts associated with the dislocation vary, resulting in variations in the dislocation image position and contrast with depth, as illustrated by Fig. 9 [13]. These effects result in a number of critical issues associated with dislocation images collected by ECCI. First, with all other factors constant, the appearance of a dislocation will vary with depth below the free surface. A dislocation inclined to the sample surface will have variations in contrast and apparent position as the dislocation projects below the surface. This effect is the same as the well-known "dotted" or "zig-zag" contrast typically seen from dislocations observed in diffraction contrast TEM. The second issue that occurs with the coherency breakdown with depth is that the dislocation becomes invisible with depth, typically at the depth of a few extinction distances (note, the depth at which the dislocation disappears will be a function of a number of other variables, including accelerating voltage, electron beam spot size and divergence, channeling condition, and deviation parameter). This effect is illustrated in Fig. 10. At first, this loss of contrast with depth would seem to be a limitation of ECCI; after all, dislocations are observed through the entire depth of a TEM sample. Nevertheless, it should be noted that the depths at which dislocations "disappear" in ECCI is similar to the thickness of TEM foils used for diffraction contrast TEM (and foil thicknesses where chromatic aberration begin to cause blurring of TEM images). Furthermore, if ECC contrast were to develop from dislocations at significantly larger depths, it is likely that the contrast from large numbers of overlapping dislocations at different depths would create confusing images where individual dislocations could not be resolved (and in polycrystalline materials, contrast from grains with different orientations at these depths would confuse the image). At the extreme, if ECCI facilitated imaging of dislocations through the full thickness of a bulk sample (in many cases on the order of a cm), dislocations would never been observed in ECCI, except under exceptional circumstances.

7 Characterization of Dislocations Using ECCI

Because ECC images of dislocations are linked to the underlying crystallography through the associated electron channeling pattern or EBSD pattern, it is possible to characterize the crystallographic nature of these dislocations, in a manner similar to that used in diffraction contrast TEM.

While many studies focus on the Burgers vectors of the dislocations, other crystallographic characteristics, in particular the line directions of dislocations, can be determined. In some cases, line directions can be approximated simply by observing the general direction of a dislocation in relation to a stereographic projection of the region of interest in the given orientation. Because the dislocation contrast decreases as the dislocations project below the sample surface, it is possible to qualitatively determine if a dislocation is highly inclined to the sample surface ("short" dislocations), lies closer to parallel with the surface ("long" dislocations), or at some intermediate condition. Furthermore, because a dislocation will be in sharper contrast near the free surface, it is easy to determine the inclination direction of the dislocation from a single image, something that cannot be easily accomplished from a single TEM image. Additional information, such as preferred dislocation line directions or slip planes, can assist in this stereographic projection-based approach for estimating line directions. An example of such an analysis using stereographic projections, dislocation inclination and projected line length, and underlying information is presented in Fig. 11 [25].

A more rigorous determination of line directions can be carried out using a stereological approach, where multiple images at different tilts and rotations are employed. In this approach, the apparent normal vector {hkl} to the dislocation line direction (i.e. the vector perpendicular to the electron beam and the projected direction of the dislocation) is determined for one imaging condition. This apparent normal vector defines an {hkl} plane that the dislocation line direction lies in (not to be confused with the slip plane). The sample is then tilted to a different imaging condition and the process is repeated, resulting in a different {hkl} plane that the dislocation line lies in. The true < uvw > line direction of the dislocation is simply given by the cross-product of the two apparent line normals (i.e. the dislocation line direction is the intersection of the two planes the dislocation line lies on). A number of issues should be considered when taking this approach. First, maximizing the tilt angle between the two images will enhance accuracy. Second, the apparent dislocation line {hkl} normals may be high index vectors, which can sometimes be difficult to determine solely from an ECP/SACP or EBSD pattern. Determination and visualization of these {hkl} normals is often facilitated using stereographic projections. Using an extension of this approach, it is sometime possible to define slip and fault planes. For example, if a dislocation contains multiple line segments with different orientations, once the line directions of each segment is determined, it is a simple matter to define the plane common to the two line segments by taking their cross-product.

As noted above, the ability to characterize dislocations in terms of their Burgers vectors using ECCI was hypothesized and then demonstrated early in the development of the technique. The approach for this characterization is based on the well-known TEM $\mathbf{g}\cdot\mathbf{b} = 0$ invisibility criteria, with **g** representing the particular diffraction (in TEM) or channeling (in ECCI) condition. In theory, to determine the Burgers vector of a given dislocation, one needs to capture images in a number of different channeling conditions, identify under which channeling conditions the particular dislocation is out of contrast, and then take the cross-product of the {hkl} of two of these out-of-contrast conditions to determine **b**. In practice, it is often somewhat more complicated than this for a number of reasons. First, given the stage tilt and rotation limits of many microscopes, it can be difficult to reach a large number of suitably different channeling conditions. This is compounded by the realization that as the sample is tilted further from having its normal parallel to the electron beam (around 30 degrees), ECC imaging becomes more difficult for a number of reasons, including fewer backscattered electrons reaching a near pole piece detector and difficulty in collecting SACPs at high angles outlined above (see Fig. 6f). Secondly, because ECC imaging is difficult and requires careful attention to focusing, astigmatism, and sample drift and vibration, it is quite easy to get dislocations "out of contrast" for reasons unrelated to the channeling conditions. Consequently, it is very useful to work with images that have some dislocations in contrast and sharp focus while other are out of contrast (i.e. dislocations with different Burgers vectors) in order to ensure the imaging conditions are suitable. Finally, it should be remembered that the $\mathbf{g}\cdot\mathbf{b} = 0$ invisibility criterial is based on isotropic elastic, and dislocations in materials with significant elastic anisotropy may display residual contrast. Given these considerations, care must be taken when drawing conclusions from image contrast. It is not uncommon for users to use not only $\mathbf{g}\cdot\mathbf{b} = 0$ invisibility as a criterion, but the converse $\mathbf{g}\cdot\mathbf{b} \neq 0$ visibility as a secondary criterion to support their conclusions. An example of a Burgers vector analysis is shown in Fig. 11 [25], which shows the relationships between the various channeling patterns, as set up using HR-SACPs, the underlying dislocation line analysis of the (in this case) <111> screw dislocations, and four ECC images that show the variation of contrast of the different dislocations under the different channeling conditions. Studies of the same areas characterized using both ECCI, EBSD, differential image correlation (DIC) approaches show consistent identification of the Burgers vectors [26]. $\mathbf{g}\cdot\mathbf{b} = 0$ can be used to determine the details of complex dislocation configurations, such as tilt and twist subgrain boundaries, as illustrated for a complex boundary in UO_2 shown in Fig. 12 [27].

While the $\mathbf{g}\cdot\mathbf{b} = 0$ invisibility criterion is well known, it is ideally applied to screw dislocations that have all of atomic displacements parallel to their Burgers vectors and assuming isotropic elasticity. However, edge dislocations have atomic displacements not only parallel to **g**, but also normal to their slip planes (defined by $\mathbf{b}\mathrm{x}\mathbf{u} = \mathbf{n}$, where u is the line direction), even in elastically isotropic materials. Thus, edge dislocations are only out of contrast for the combined criteria of $\mathbf{g}\cdot\mathbf{b} = 0$ *and* $\mathbf{g}\cdot\mathbf{b}\mathrm{x}\mathbf{u} = 0$ invisibility. Stated another way, edge dislocations will only be invisible when **g** is parallel to **u**. This combined invisibility criterion is illustrated for ECCI in Fig. 13 [28]. It should be noted that this criterion in it of itself does not lead to the determination of the Burgers vector, as any Burgers vector that is perpendicular to invisibility **g** meets the criterion. Therefore, the criteria must be supplemented by other information, for example multiple line directions for the same dislocation , in order to determine the Burgers vector [28].

Many studies of both structural metals and electronic materials are concerned with the density of dislocations. As dislocation densities are often expressed in terms of number per unit area (length per unit volume), it is generally a simple matter to determine the dislocation density using ECCI by counting the number of times dislocations exit the crystal surface in a given area. (Note that some studies have used an approach that counts the number of dislocation intersections with test lines, but these approaches require the depth of image contrast of the dislocations to be known, complicating the analysis). Nonetheless, a few issues should be considered. First, it is important to determine if any dislocations are out of contrast for the given reflection due to invisibility criteria outlined above. In some cases, it may be necessary to image

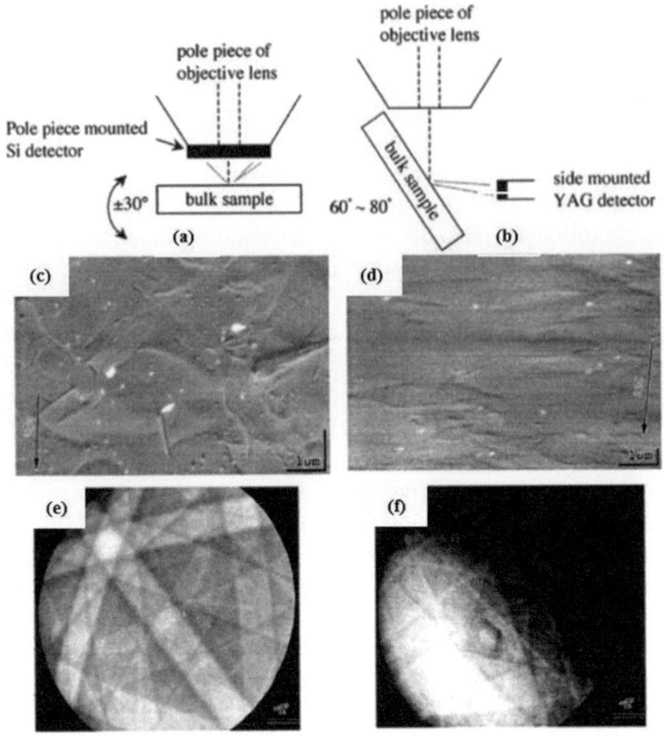

Fig. 6 Comparison of low tilt **a** and high **b** configurations for ECCI. The low tilt configuration makes use of a standard objective lens pole piece-located electron detector (often Si), while the high tilt configuration requires a side mounted detector (often YAG). Compared to the low tilt ECC image of dislocations shown in image **c**, the high tilt configuration results foreshortening, as shown in image **d** of the same area, requires dynamic focusing, and results in large intensity variations across the SACPS, as shown in the comparison of images **e** and **f**.(Reprinted by permission from Crimp [20])

under multiple channeling conditions in order to identify all of the dislocations in a given area and account for these when determining the dislocation density. Secondly, it should be remembered that the dislocation density units of number per unit length is based on a configuration in which the dislocation lines are perpendicular to the area of measurement (in some ways like a flux). The dislocation density determined will be less than the actual dislocation density if the dislocation lines being counted are not perpendicular to the surface of the area being examined. It is a simple matter to correct the dislocation density using a factor of $1/\cos\alpha$, where α is the angle between the (average) dislocation line direction and the surface normal. In a more generalized sense, if random grain or dislocation line directions are being examined, a stereological average of $\alpha = 60°$ can be used [29]. In general, dislocation densities determined by ECCI are limited by the ability to resolve the individual dislocations, which results in a maximum measurable density of ~ $10^{14}/m^2$ [29]. Despite this rather

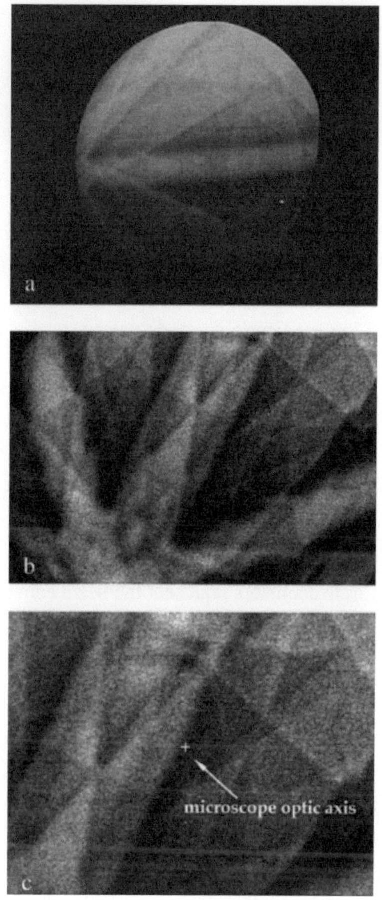

Fig. 7 Sequence of selected area channeling patterns illustrating the steps to setting up the channeling conditions for ECCI. **a** A relatively large angle SACP is obtained and a channeling band of interest identified. **b** The sample is tilted and rotated (most SEMs have tilt-rotate stages) to bring the edge of the desired channeling band near the center of the screen, which corresponds to the optic axis of the microscope. **c** Fine adjustments are made to the specimen orientation to adjust the deviation from the Bragg condition to optimize image contrast. (Reprinted by permission from M. Crimp [20])

low dislocation density limit, it is interesting to note that the dislocation densities determined using ECCI are comparable to those determined from the same areas using high resolution EBSD [30].

8 ECCI of Other Crystal Defects

While the vast majority of ECCI studies have focused on examination and characterization of dislocation structures, it is important to note that ECCI can be used for examining other types of crystalline defects that have historically been characterized using diffraction contrast TEM. These defects are of the planar defect type, and primarily include stacking faults and thin twins, but might also include thin planar precipitates and grain boundaries. While thick twins would be expected to display ECCI contrast from the simple fact that a significant volume of materials is

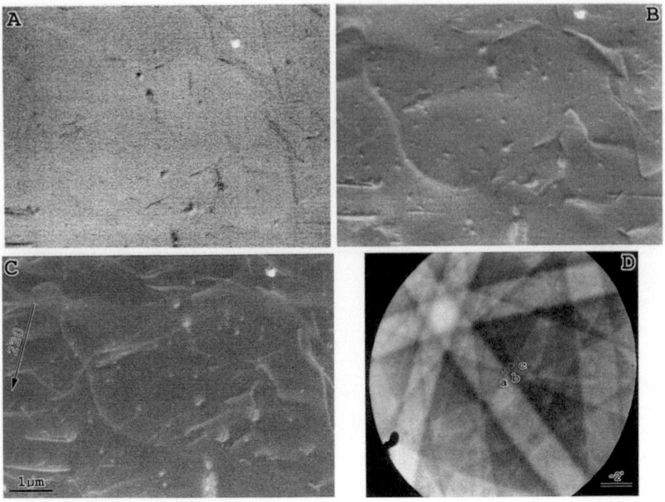

Fig. 8 Series of three ECC images collected with varying deviations from the Bragg condition in Si using g = (220). **a** s < 0, **b** s≈0, and **c** s > 0. **d** SACP illustrating each of the channeling conditions. Circle diameters represent the beam convergence angle, 2a ≈ 5.7 mrad. (Reprinted by permission from Simkin and Crimp [19])

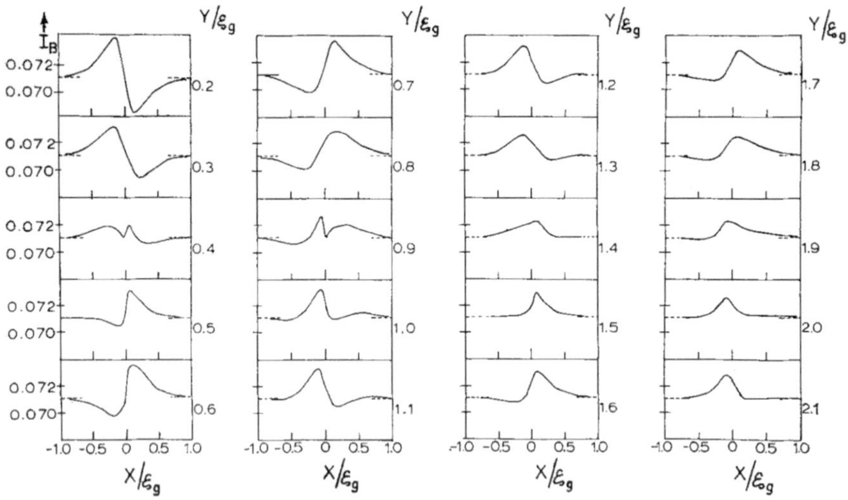

Fig. 9 Dynamical theory computed edge dislocation intensity profiles in Si using a {220} reflection with g·b = 1, with the dislocation line and Burgers vector both parallel to the free surface. x_g represents the extinction distance, while Y represents the depth below the surface. Note how the peak intensity decreases and the sense of the intensity distribution changes with depth below the surface. (Reprinted by permission Spencer et al. [13])

Fig. 10 ECC image of a low angle grain boundary showing dislocations in the boundary as well as in the neighboring subgrains. Note that both the boundary dislocations and those in the matrix show decaying intensity and line broadening as the dislocations project into the material from the surface. "Shorter" dislocations are a result of those dislocations being more highly inclined to the specimen surface, while "longer" dislocations lie more parallel to the surface. (Image courtesy of H. Mansour and N. Maloufi)

at a different orientation [23], which would result in different electron channeling behavior, thin planar defects would not have significant volumes of different orientations and might not be expected to be seen using ECCI based solely on their volumes. Unlike dislocations, for which the local electron channeling behavior is altered due to the relatively long-range strain fields, planer defects generally do not have long range strain fields (except for the bounding dislocations). Thus, ECCI contrast from thin planer defects will result from the phase shift that occurs as a result of incoming electron beam interacting with the atom position shift across the defects (**R**), in a manner consistent with the contrast mechanisms of these defects in diffraction contrast TEM.

A number of studies have imaged stacking faults using ECCI, including Weidner et al. [31], Zarfferer and Elhami [21], Gutierrez-Urrutia [32]. Similar to dislocations inclined to the surface of the sample, ECC contrast from these faults decays with depth below this surface, but like the contrast from faults observed using TEM, bright-dark fringe contrast is observed superimposed on the underlying intensity decay, with two to three fringes observed (Fig. 14). These fringes run approximately parallel to the line defined by the intersection of the fault with the free surface, and each fringe runs along an equi-depth contour in the material. The details of this fringe contrast are dependent on the specific channeling condition in terms of both **g** and **s**, and contrast analysis can be performed according to the $\mathbf{g}{\cdot}\mathbf{R} = 0$ invisibility criteria developed for TEM [32]. Similar decaying fringe contrast is expected from thin twins (where two phase changes would occur, thin precipitates (again with two phase changes), and in some cases grain boundaries of thicker precipitates.

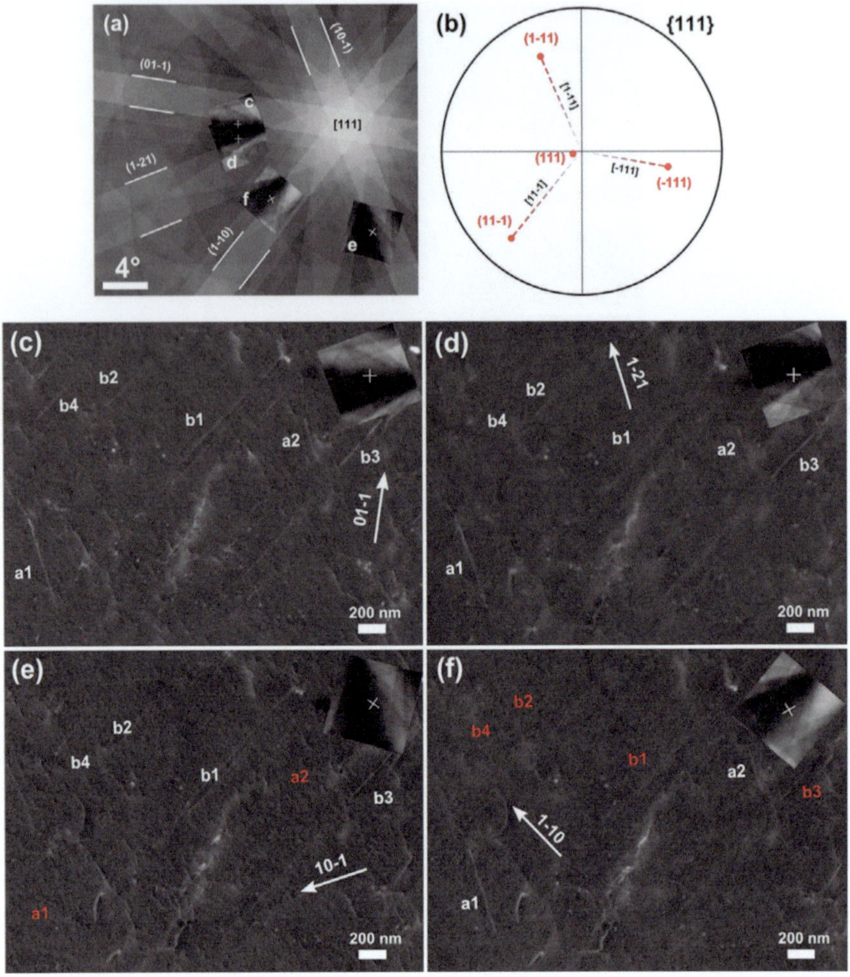

Fig. 11 Characterization of dislocations in a 2%Si interstitial free steel. **a** four superimposed HR-SACPS that indicate the four channeling conditions used in **c–f**. **b** stereographic projection shows the four possible < 111 > directions used to identify the different < 111 > screw dislocation line directions in the bcc steel. $g \cdot b = 0$ contrast analysis confirms the dislocation Burgers vectors, consistent with the line direction analysis. (Reprinted by permission Mansour et al. [25])

9 Sample Preparation for ECCI

While sample preparation for ECCI is much simpler that for TEM, as bulk samples can be used for ECCI and it is not necessary to make thin foils, it is still very critical that the surfaces be high quality and defect free. Indeed, it has been shown that even careful mechanical metallographic preparation can lead to dislocations being generated in the top 100 μm of the material. While the surface may appear defect/

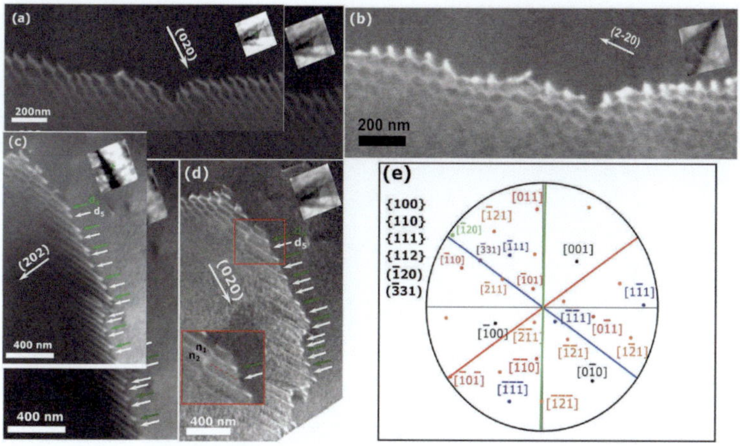

Fig. 12 ECCI analysis of the complex dislocation structure of a subgrain boundary in UO_2. **a** g = (020) and **b** $(2\bar{2}0)$ of the twist portion of the boundary and **c** g = (202) and **d** g = (020) of the tilt portion of the boundary. **e** Stereographic projection showing {100}, {110}, {111}, {112}, $(\bar{1}20)$, $(\bar{3}31)$ poles. The straight lines correspond to the traces of the dislocations: d1 (red); d2 (green); d3, and d4 and d5 (blue). (Reprinted by permission Mansour et al. [27])

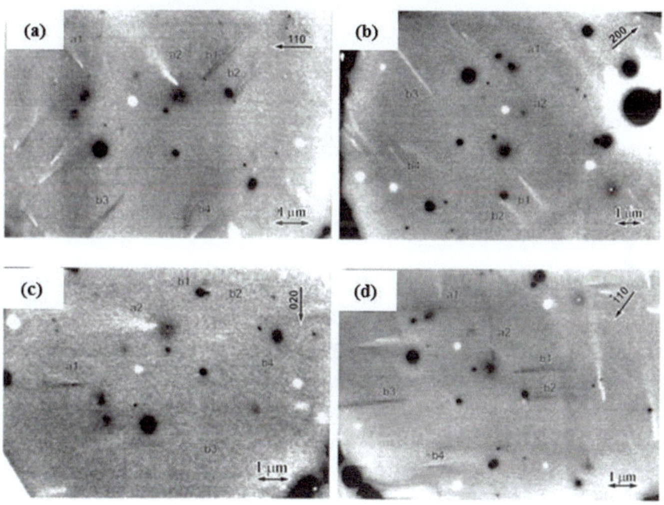

Fig. 13 Demonstration of the g·bxu = 0 invisibility criteria in a B2 FeAl alloy. Dislocations with two line directions have b = < 100 >. Both line directions are in contrast with g = (110) **a** and g = $(\bar{1}10)$ **d**, but each line direction is out of contrast when g is parallel with the line direction at g = (200) **b** and g = (020) **c**. Note significant porosity in this material is revealed as black spots. Images are not rotationally corrected. (Reprinted by permission Crimp et al. [28])

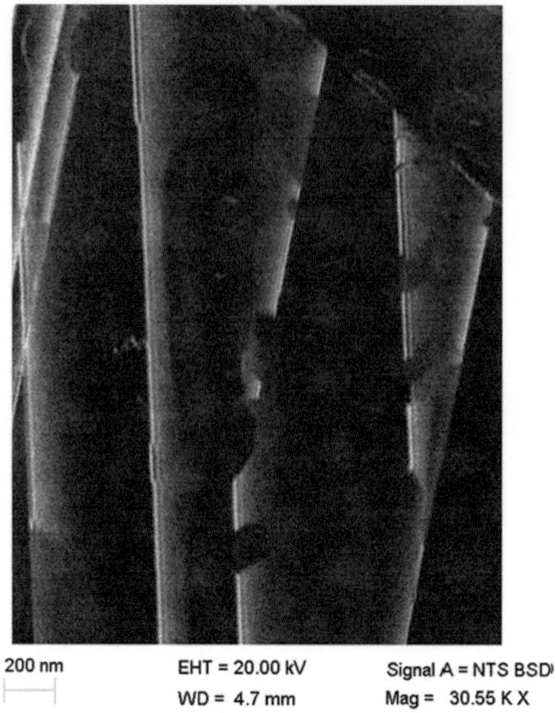

Fig. 14 Images of stacking faults lying on two different sets of planes in an fcc 316L stainless steel. The stacking faults are inclined to the sample surface and decrease in intensity with depth. Fringe contrast is observed superimposed on the decreasing intensity. (Image courtesy of I. Gutierrez-Urrutia.)

topography free using secondary electron (Everheart-Thornley detector) imaging, ECC imaging can reveal scratch-like features that are actually tracks of dislocations that resulted from polishing media moving across the surface, as shown in Fig. 15 [33]. Thus, it is clear that sample preparation must be carried out carefully to eliminate any residual surface damage and to avoid any sample preparation artifacts.

It is not the intention of the section to give extensive details on sample preparation for ECCI, as sample preparation can be very material specific. Instead, a number of general approaches that can be suitable for ECCI sample preparation will be mentioned. Traditional metallographic approaches of cutting, grinding, and polishing are often used in the early stages of ECCI sample preparation. However, both the intermediate and final stages of this preparation must be very carefully carried out. Care must be taken between decreasing polishing size steps to ensure that all residual scratches from previous stages are well eliminated. This often entails spending more time on each stage than one might for more common metallographic applications. The final stages, which may entail a number of different approaches, then must be carefully carried out. Colloidal silica polishing has become common for EBSD sample preparation, and for some materials may be suitable for ECCI. An

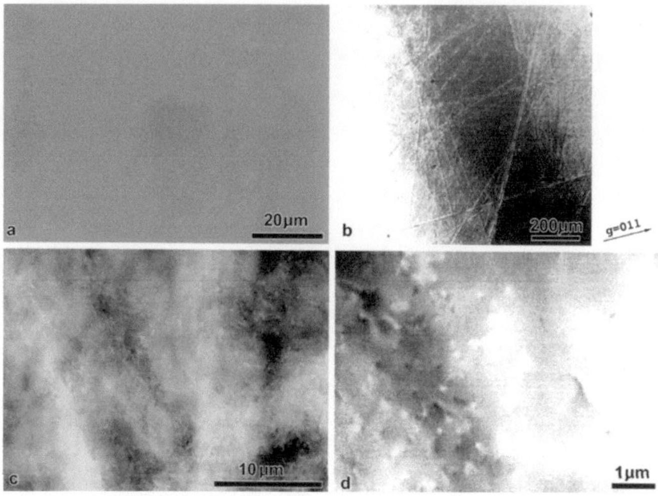

Fig. 15 **a** Secondary electron image indicating no residual scratch topography from sample preparation of a B2 NiAl single crystal. **b–d** ECC images showing evidence of "scratches" that can be attributed to residual dislocations left from the original mechanical sample preparation, with the dislocations being revealed at high magnification **d**. (Reprinted by permission Simkin and Crimp [33])

advantage of colloidal silica is that it typically does not require extensive training or variable control. In other circumstances, electropolishing can be used to eliminate and residual damage layer, but it can be difficult to establish proper conditions and can require significant training or experience to get high quality surfaces (electropolishing also has more safety issues than colloidal silica). Nevertheless, done well, electropolishing can lead to the very best sample preparation for ECCI. Other approaches that are becoming more common include the use of either FIB, PFIB, or broad beam argon-based FIB milling. These techniques can also be applied to ceramics and electronic materials, which may not allow some of the mechanical or electrochemical approaches. Of course, both of these approaches require significant capital costs. Plasma cleaning surface before and even during ECCI sessions can also help clean surfaces for high quality ECCI.

In some extraordinary cases, sample preparation may not even be necessary. Electronic materials and devices are often grown with high quality surfaces suitable for ECCI. Figure 16 shows a developmental device containing multiple layers and materials [34]. In this case ECCI revealed dislocations (and stacking faults nearly parallel to the surface) in the as-grown SiC in a region just outside of the metal contact. It should be noted that because of the heteroepitaxially grown nature of this structure, it was possible to collect selected channeling patterns across a number of the device features, with the various aspects of the devices breaking up the SACP, but not inhibiting the ability to set up channeling conditions. Thus, in practice it is quite easy to carry out ECCI on some structures that at first might appear impossible to image dislocations in.

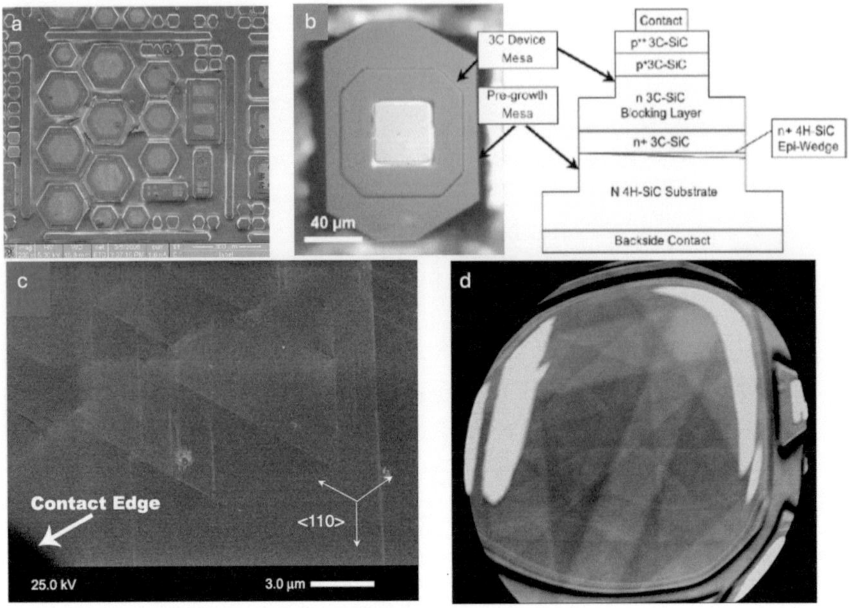

Fig. 16 Example of ECCI carried out without sample preparation on an as-grown electronic devise. **a** Low magnification image of the devise and **b** single feature showing the architecture. **c** ECC image showing dislocations and stacking faults on the SiC device, near the edge of the metal contact. **d** SACP showing the sharp band contrast resulting from the heteroepitaxial structure, with device features interrupting the channeling pattern. (Images **a**, **b**, and **d** courtesy of Speer, Pirouz, and Crimp, **c** Reprinted by permission Speer et al. [34])

Regardless of the sample preparation approach is chosen, it is important to carry out the preparation in a careful, reproducible manner. Doing so will overcome one of the greatest challenges to performing high quality ECCI studies and will allow the user to carry out a wide range of studies in the near surface region of bulk samples.

10 Final Thoughts

ECCI offers the ability to observe crystalline defects in the near surface region of bulk samples, with the advantage of avoiding the requisite thin foils of TEM. This makes ECCI appropriate for a broad range of studies that would be limited by TEM sample preparation and minimal thin areas, as well as in-situ and sequential studies where examination of the same area after different treatments is desired, such as mechanical deformation studies.

Since ECCI was first envisioned and carried out, the quality of images has clearly improved, as a result of advances in scanning electron microscopes and detector technology. Improved electron probe quality in terms of smaller probe sizes, decreased beam convergence angles, and lower energy spreads have led to sharper, higher contrast images. Higher efficiency detectors backscattered electron detectors have resulted in decreased image collection times, which in turn makes sample drift issues less of a concern. Likewise, more stable microscope stages and better vibration isolation improves ECC imaging. As outlined earlier, the ability to collect SACP in various instruments continues to be developed, and it is likely that as more researchers use ECCI for their studies, microscope manufactures will respond to this increasing demand with greater SACP capabilities. Overall, it is expected that the ability to carry out high quality ECCI will continue to increase over the coming years.

References

1. Coates DG (1967) Kikuchi-like reflection patterns obtained with the scanning electron microscope. Philos Magaz 16:1179–1184
2. Booker GR, Shaw AMB, Whelan MJ, Hirsch PB (1967) Some comments on the interpretation of the kikuchi-like reflection patterns observed by scanning electron microscopy. Philos Magaz 16:1185–1191
3. Newbury DE, Yakowitz H (1975) Contrast mechanisms of special interest in materials science. In: Goldstein JI, Yakowitz H (eds) Practical scanning electron microscopy; electron and ion microprobe analysis. Plenum Press
4. Stickler R, Booker GR (1972) Electron microscopy and the structure of materials. In: Thomas G, Fulrath RM, Fisher RM (eds). University of California Press, pp 301–312
5. Davidson DL (1984) Uses of electron channelling in studying material deformation. Int Metal Rev 29:75–95
6. Joy DC, Newbury DE, Davidson DL (1982) Electron channeling patterns in the scanning electron microscope. J Appl Phys 53(8):R81–R122
7. Vystavěl T, Stejskal P, Unčovský M, Stephens C (2018) Tilt-fee EBSD. Microsc Microanal 24:1126–1127
8. van Essen CG, Shulson EM (1969) Selected area channeling patterns in the scanning electron microscope. J Mater Sci 4:336–339
9. van Essen CG, Shulson EM, Donaghay RH (1970) Electron channelling patterns from small (10μm) selected areas in the scanning electron microscope. Nature 225:847–848
10. Joy DC, Newbury DE (1972) Scanning electron microscope selected area channeling patterns from 1 micron specimen areas. J Mater Sci 7:714–716
11. Guyon J, Mansour H, Gey N, Crimp MA, Chalal S, Maloufi N (2015) Sub-micron resolution selected area electron channeling patterns. Ultramicroscopy 149:34–44
12. Kerns RD, Balachandran S, Hunter AH, Crimp MA (2020) Ultra-high spatial resolution selected area electron channeling patterns. Ultramicroscopy 210:112915
13. Spencer JP, Humphreys CJ, Hirsch PB (1972) A dynamical theory for the contrast of perfect and imperfect crystals in the scanning electron microscope. Philos Magaz 26:193–213
14. Clark DR (1971) Observation of crystal defects using the scanning electron microscopy. Philos Magaz 24:973–979
15. Stern RM, Ichinokawa T, Takashima S, Hashimoto H, Kimoto S (1972) Dislocation images in the high resolution scanning electron microscope. Philos Magaz 26:1495–1499
16. Pitaval M, Morin P, Baudry J, Viaria E, Fontaine G (1977) Advances in crystalline contrast from defects. Scan Electron Microsc 1:439–444
17. Morin P, Pitaval M, Besnard M, Fontaine G (1979) Electron-channeling imaging in scanning electron microscopy. Philos Magaz A 40:511–524
18. Czernuszka J, Long JT, Hirsch PB (1991) Electron channeling contrast imaging (ECCI) of dislocations in bulk samples. MRS Proc 209:289–291

19. Simkin BA, Crimp MA (1999) An experimentally convenient configuration for electron channeling contrast imaging. Ultramicroscopy 77:65–75
20. Crimp M (2006) Scanning electron microscopy imaging of dislocations in bulk materials, using electron channeling contrast. Microsc Res Tech 69:374–381
21. Zaefferer S, Elhami N-N (2014) Overview no. 154, theory and application of electron channelling contrast imaging under controlled diffraction conditions. Acta Mater 75:20–50
22. Gurierrez-Urrutia I (2019) Quantitative analysis of electron channeling contrast of dislocations. Ultramicroscopy 206:112826
23. Gurierrez-Urrutia I, Zaefferer S, Rabbe D (2009) Electron channeling contrast imaging of twins and dislocations in twinning-induced plasticity steels under controlled diffraction conditions in a scanning electron microscope. Scr Mater 61(2009):737–740
24. Humphreys FJ (2001) Review: grain and subgrain characterisation by electron backscatter diffraction. J Mater Sci 36:3833–3854
25. Mansour H, Guyon J, Crimp MA, Gey N, Beausir B, Maloufi N (2014) Accurate electron channeling contrast analysis of dislocations in fine grained bulk materials. Scr Mater 84–85:11–14
26. Sperry R, Han S, Chen Z, Daly SH, Crimp MA, Fullwood DT (2021) Comparison of EBSD, DIC, AFM, and ECCI for active slip system identification in deformed Ti-7Al. Mater Charac 173:110941
27. Mansour H, Crimp MA, Gey N, Iltis X, Maloufi N (2019) Dislocation analysis of a complex sub-grain boundary in UO_2 ceramic using accurate electron channeling contrast imaging in a scanning electron microscope. Ceram Inter 45:18666–18671
28. Crimp MA, Simkin BA, Ng BC (2001) Demonstration of the g·b x u = 0 edge dislocation invisibility criterion for electron channelling contrast imaging. Philos Magaz Lett 81:833–837
29. Crimp MA, Hile JT, Bieler TR, Glavicic MG (2004) Dislocation density measurements in commercially pure titanium using electron channeling contrast imaging. TMS Lett 1:101–102
30. Dunlap BE, Ruggles TJ, Fullwood DT, Jackson B, Crimp MA (2020) Comparison of dislocation characterization by electron channeling contrast imaging and cross-correlation electron backscattered diffraction. Ultramicroscopy 184:125–133
31. Weidner A, Martin S, Klemn V, Martin U, Biermann H (2011) Stacking faults in high-alloyed metastable austenitic cast steel overserved by electron channelling contrast imaging. Scr Mater 64:513–516
32. Gurierrez-Urrutia I (2021) Analysis of electron channeling contrast of stacking faults in fcc materials. Microsc Microana 27:318–325
33. Simkin BA, Crimp MA (2000) Use of electron channelling contrast imaging to assess near-surface mechanical damage. Philos Magaz Lett 80:395–400
34. Speer KM, Heudeck PG, Crimp MA, Burda C, Pirouz P (2007) Possible formation mechanisms for surface defects observed in heteroepitaxially grown 3C-SiC. Phys Stat Solid A 204:2216–2221

Part II
Application of Advances in Electron Backscattered Diffraction

Chapter 6
Mesotexture: A Key to Understanding Microstructural Evolution During Grain Boundary Engineering

R. Jeyaraam, Srikanth Vedantam, and V. Subramanya Sarma

1 Introduction

In the paradigm of materials science and engineering, continuous efforts are being made to enhance the properties of materials to improve efficiency, safety and reliability. The correlations between processing-microstructure-properties and performance is often referred to as materials tetrahedron and has been the cornerstone of materials science and engineering over the last few decades. It is well known that the properties of materials are controlled by their microstructure. In the industrial manufacture of metallic components, thermomechanical processing (TMP) is widely employed to tailor the microstructure and achieve the desired properties. The TMP usually involves single or multiple iterations of deformation (cold/ hot) and/ or annealing. The parameters of TMP i.e., stress/strain state, extent of deformation/ strain, annealing temperature and time have significant influence on the resulting microstructure.

Microstructure is often described by parameters such as grain size, grain shape, crystallographic texture, character and distribution of grain boundaries (GBs)/ interfaces [1]. Among the many microstructural parameters mentioned above, grain boundary character distribution (GBCD) (also referred to as mesotexture) has considerable influence on the microstructure evolution during processing as well as on the final properties [2]. Therefore, the study of mesotexture in materials during

R. Jeyaraam
NAND Technology Integration, Micron Semiconductor Asia Pte Ltd, Singapore, Singapore

S. Vedantam
Department of Engineering Design, Indian Institute of Technology Madras, Chennai, India

V. S. Sarma (✉)
Department of Metallurgical and Materials Engineering, Indian Institute of Technology Madras, Chennai, India
e-mail: vsarma@iitm.ac.in

S. Suwas and D. P. Field (eds.), *Advances in Texture, Microtexture, and Allied Techniques*, Materials Horizons: From Nature to Nanomaterials,
https://doi.org/10.1007/978-981-96-5346-1_6

processing is very important for understanding not only the microstructural evolution during TMP but also the final properties of materials. Electron back scatter diffraction (EBSD) is an indispensable tool for the study of micro and mesotexture of materials [3].

Grain Boundary Engineering (GBE) is a well-established microstructural engineering approach for modifying the properties of materials by manipulating the GBCD, the GB network topology and connectivity in the microstructure [4]. The fundamental aim of GBE is to increase the fraction of coincident site lattice (CSL) (often referred to as Σ boundaries) boundaries (Σ3 to Σ29) and to modify the GB network [5]. The low Σ GBs,[1] are considered special because of their supposedly lower energy, mobility and diffusivity in comparison to RHAGBs [5]. A higher fraction/density of these boundaries have been shown to increase resistance to corrosion [6, 7], creep [8, 9], fatigue [10, 11], liquation cracking [12, 13] and fracture [14, 15]. Some of the methods that are followed to accomplish this are TMP, rapid solidification, magnetic annealing, magnetic crystallisation, to name a few [16]. TMP is the most commonly employed method to engineer microstructures in face centred cubic (FCC) materials with low and medium stacking fault energy (SFE).

Among the special boundaries (SBs), annealing twin boundaries (ATBs), referred to as Σ3 in the CSL notation, occur in higher frequency in low/medium SFE FCC materials. The formation and evolution of ATBs can be engineered by varying the TMP parameters. Following their formation, ATBs may interact with one another and other CSL boundaries leading to a generation of higher order ATBs. Under suitable processing conditions, these newly generated ATBs could become part of the GB network and disrupt random high angle grain boundary (RHAGB) connectivity resulting in a GBE microstructure. The optimum TMP parameters to achieve the GBE microstructure are observed to be material dependent. Majority of TMP schedules reported in the literature for achieving the GBE microstructure are identified by trial and error based on the semi-empirical knowledge on the role of material and TMP parameters on ATB formation and their evolution [17–20]. A clear understanding of the mechanisms of formation, evolution and interactions of ATBs as a function of material properties and TMP parameters is essential for reducing the developmental cycle in identifying the suitable TMP parameters through computational models. The recent advances in experimental (3D-EBSD, EBSD during in-situ deformation/heating etc.,) and multiscale computational modeling tools have contributed to an improved understanding of the formation and evolution of ATBs. Among the computational tools, full-field[2] (phase field model [21], Monte Carlo [22], Cellular Automata [23]) models with appropriate inputs from experimental results have the capability to simulate microstructure development during TMP and to identify the optimum conditions to obtain the desired microstructure.

[1] The Σ3 - Σ29 GBs are often referred to as special boundaries (SBs). The physical relevance of these notations are insignificant for high Σ boundaries because of their less periodic boundary structures.

[2] Full-field models refer to numerical methods where the local behaviour of the GBs and their network (poly- crystal topology) are incorporated in the simulations.

In this chapter, we highlight the significance of micro and mesotexture in understanding the mechanisms of microstructural evolution and the development of suitable computational models to simulate microstructural evolution in the context of GBE. The chapter has been organised as follows: In Sects. 2 and 3, the geometry of grain boundaries and the fundamentals of microstructure evolution during annealing of deformed materials are briefly presented. The important aspects of GBE viz., mechanisms of GBE, annealing twin boundaries, quantification of a GBE microstructure, grain boundary networks and twin related domains, significance of grain boundary planes, modelling and simulation of GBE microstructure evolution during TMP are presented in Sects. 4, 5, 6, 7, 8 and 9. Conclusions and possible future directions of research in the area of GBE are presented in Sect. 10.

2 Mesotexture and Geometry of Grain Boundaries

The microtexture measurements obtained from EBSD provide the orientation as well their spatial distribution in a given microstructure [3]. The mesotexture, i.e., GBCD on the other hand, is evaluated from the microstructure data of neighboring grain orientations. Some important details pertaining to mesotexture and GB character relevant to the scope of this chapter are given below. A more detailed description can be found in standard textbooks [3, 24].

The interface (GB) between two differently oriented grains is a region of a disturbed lattice of a few atomic dimensions width. The GBs have five macroscopic degrees of freedom (DOF), viz., the angle θ and the axis of misorientation $\langle uvw\rangle$ (3 DOFs), the position of the GB plane (2 DOFs). GBs can also be classified based on the axis of misorientation between the grains as tilt and twist boundaries. For a tilt boundary, the axis is parallel to the GB plane and perpendicular in the case of a twist boundary. For a curved boundary, the tilt/ twist nature varies based on the position of the GB plan normal with respect to the axis of misorientation. If the misorientation angle is less than 15°, the GBs are referred to as low angle grain boundaries (LAGBs), and GBs with misorientation angles > 15° are classified as high angle grain boundaries (HAGBs). The misorientation of GBs are calculated from the orientations of the abutting grains. Orientations of individual grains are usually represented using a set of three Euler angles (ϕ_1, ϕ and ϕ_2) [3]. From the orientations, misorientation matrix of the interface is calculated. Based on the symmetry matrices[3] used in the calculation, 1152 misorientation matrices are obtained. The angle-axis pair for each of the 1152 misorientation matrices are determined, and the smallest angle of misorientation and the corresponding axis is then assigned to the GB by convention. For example, a GB with a misorientation angle-axis pair of 60° $\langle 111\rangle$ can also be represented through other pairs like 70.53° $\langle 110\rangle$. The smallest angle, which is 60° and its corresponding axis $\langle 111\rangle$ is commonly used to denote this GB. The procedure of determining the character of a GB is described in Appendix A.

[3] 24 matrices in the case of cubic materials.

For misorientations < 15° (LAGBs), the tilt and twist boundaries can be considered as arrangements of arrays of edge and screw dislocations respectively. The energy of the GB has a linear relationship with the misorientation angle up to 15∘ [25]. With an increase in the misorientation angle beyond this, the dislocation cores overlap, and the GBs can no longer be described with the dislocation model, and the GB energy does not vary linearly with the misorientation angle. However, certain HAGBs with particular angle and axis of misorientations are observed to have periodic boundary structures with lower energy and have a fraction of lattice points from both the lattices being in coincidence. They are termed as the coincidence site lattice (CSL) boundaries.

The concept of CSL boundaries was first put forward in 1949 to explain the recrystallisation texture of Cu [5, 26]. When the lattices of the abutting grains are notionally superimposed and rotated about the axis of misorientation $\langle uvw \rangle$ through the angle, a number of lattice points common to both the lattices will be in coincidence. These GBs are denoted using the term Σ, which is the reciprocal density of the coinciding lattice points.

For certain Σ boundaries, more than one misorientation angle- axis combination will give rise to the same number of coincidence sites. For example, two misorientation angle-axis exist for a Σ17 boundary, viz., 28.1° about $\langle 100 \rangle$ and 61.9° about $\langle 221 \rangle$, and they are designated as Σ17a and Σ17b. Following the experimental studies of GBs through field ion microscopy in 1964 [27], the concept of deviations in CSL boundaries was introduced. In real polycrystalline materials, GBs are observed to deviate from the ideal misorientation of coincidence (say 60° $\langle 111 \rangle$ in the case of Σ3 boundaries). For small angular deviations, the coincidence structure of the boundary is conserved by the presence of arrays of dislocations namely, secondary intrinsic grain boundary dislocations (SIGBDs). The maximum angular deviation $\Delta\theta_m$ that can be accommodated in a CSL boundary Σx is given by,

$$\Delta\theta_m = \frac{v_0}{x^c} \tag{1}$$

where, v_0 is a constant which depends on the angular limit set for an LAGB and this is usually 15°. Therefore, an LAGB can be expressed in CSL notation as a Σ1 boundary. The value of $\Delta\theta_m$ for a given Σ boundary depends on c. Brandon's criterion with the value of c as 1/2 [28] is the most widely used deviation criterion in the literature. Gleiter and Pumphrey proposed a value of 2/3 [29]. Based on experimental observations, Palumbo and Aust suggested a more restrictive criterion using a value of 5/6 for c [30]. During deformation, dislocation accumulation near the CSL boundaries leads to a deviation in the misorientation angle. However, during annealing, which causes recovery and annihilation of dislocations, the deviations in CSL boundaries are observed to reduce. The algorithm for determining the deviation angle of a GB from the ideal coincidence of a given Σ boundary is explained in Appendix B. High angle grain boundaries which are not CSL boundaries (particularly SBs) are termed as random high angle grain boundaries (RHAGBs).

3 Microstructural Evolution During Annealing of Deformed Materials

Strain-annealing and strain-recrystallisation based TMP schedules have been proposed to develop a GBE microstructure [31, 32]. A critical analysis of mesotexture in materials subjected to strain-annealing and strain-recrystallisation TMPs have shown that the RHAGB network is significantly disrupted during strain-annealng processing when compared strain-recrystallisation. This suggests that strain-annealing based TMP is more suitable to achieve the GBE microstructure [33]. In the strain-annealing type processing, the material is subjected single or iterative type processing involving cycles of deformation (typically by rolling to strains <15%) and annealing sequences. The extent of deformation, temperature, time and number of cycles of processing are material dependent parameters and often optimised by trial and error [18, 19, 34, 35]. To understand the microstructure evolution during strain-annealing processing, the GBCD and stored energy distribution in the deformed state needs to be evaluated. When a material is subjected to strain, the defect density (predominantly dislocations at low strains) increases with strain which in turn increases the stored energy. It may be noted that stored energy is the driving force for the strain induced boundary migration (SIBM) as well as recrystallisation, the major mechanisms responsible for the evolution of the mesotexture and microstructure during the subsequent annealing. At low strains, strong stored energy gradients develop due to the anisotropy i.e., orientation dependency of the strain accumulation.

The deformed microstructures are unstable and during annealing, changes in microstructure occur through static recovery, SIBM/recrystallisation and grain growth leading to a reduction in the stored energy. Materials which are subjected to deformation at elevated temperatures may undergo dynamic recovery and dynamic recrystallisation and these topics are beyond the scope of this chapter. At a given temperature, all the above mechanisms may be operating concurrently in different regions of the sample while one of them could be dominant. Recovery is a softening process that is predominant when the deformed material is heated to relatively low temperatures ($<0.3T_m$, where T_m denotes the homologous melting temperature) at which recrystallization or GB migration cannot occur due to low thermal energy. If recovery is predominant, it could lead to a reduction in the overall stored energy and could affect the SIBM/ recrystallisation kinetics during prolonged annealing. A physically based model was developed in our group to understand the role of recovery on recrystallisation kinetics [36].

During the continued annealing of deformed materials, microstructure may evolve through SIBM and/or recrystallisation depending on the extent of strain, the annealing temperature and time. Recrystallization involves formation of strain free grains (nucleation) from the deformed matrix through migration of HAGBs followed by growth of the nuclei into the deformed matrix. It has been well established that a critical amount of strain (≈15–20%) is required for static recrystallization to occur. It is also well accepted that there are very small volumes of relatively lower stored energy in the deformed materials which could be considered as potential nuclei and these could grow during annealing consuming the deformed matrix [37]. On the contrary, during annealing of materials subjected to a strain <10%, SIBM is found to be predominant. The driving force for SIBM is the heterogeneous distribution of stored energy [37]. The GBs surrounding the grains with lower stored energy migrate faster and consume the grains with relatively higher stored energy. The quantification of stored energy and its heterogeneity can be obtained from the microtexture data using the parameters such as grain orientation spread (GOS), kernel average misorientation (KAM), grain average misorientation (GAM), kernel average neighbor misorientation (KANM) and grain reference orientation deviation (GROD). The details about each of these parameters are discussed in detail in [38].

4 Mechanisms of Grain Boundary Engineering

The mechanisms that contribute to the generation of a GBE microstructure are,

1. Formation of ATBs.
2. Interactions of ATBs with one another and other CSL GBs leading to the introduction of more CSL boundaries in the GB network.

During SIBM or recrystallisation, ATBs form behind GBs and TJs at suitable TMP parameters for a given material. During the evolution of the microstructure, the formed ATBs could interact with one another and other CSL boundaries resulting in the formation of more CSL boundaries, particularly of $\Sigma 3^n$ type ($\Sigma 3$, $\Sigma 9$, $\Sigma 27$, etc.). The newly generated CSL boundaries could become part of the GB network, and disrupt the connectivity of RHAGBs, thereby resulting in a GBE microstructure. The interactions of CSL boundaries, in general, and their outcomes are governed by Σ-combination rule[39].[4] In a triple junction (TJ), if two of the GBs are of $\Sigma 3$, the third GB could either be a $\Sigma 1$ or $\Sigma 9$. Similarly, an interaction of $\Sigma 3$ and $\Sigma 9$ in a TJ could result in $\Sigma 27$ or $\Sigma 3$. Randle postulated that during the interaction of $\Sigma 3$ and $\Sigma 9$ boundaries, the probability is higher for the resultant GB to be $\Sigma 3$ [17, 40]. The mechanism by which $\Sigma 3$ boundaries are said to proliferate during microstructural evolution is referred to as the "$\Sigma 3$ regeneration model" [17]. In general, the interaction reaction can be stated as,

$$\Sigma 3^{n+1} + \Sigma 3^n \rightarrow \Sigma 3 \tag{2}$$

where $n = 1, 2, 3...$

However, in our recent work [41], we have shown that this reaction has relatively lower probability among all the probable ones. The interactions of $\Sigma 3^n$ boundaries in the context of GB networks are discussed in Sect. 7. ATBs are the primary source for the formation of higher order twin boundaries, which eventually disrupt the

[4] If x, y and z denote the Σ values of the three CSL boundaries in a TJ, $x.y = k^2.\ z$, where, k is the common divisor of x and y.

RHAGB network. In view of this, a thorough understanding of the conditions for their formation and evolution is crucial. Recently Bozzolo and Bernacki [42] have reviewed the current status of the knowledge on the ATB formation and evolution. In the following section, a brief summary of the current understanding of ATBs is provided.

5 Annealing Twins

Twins are produced either during deformation or on annealing following deformation. The former is called deformation/mechanical twins while the latter is termed as annealing twins. Annealing and mechanical twins can be distinguished from their distinct morphologies. In 2D, annealing twins appear as broad bands with straight sides, while the deformation twins are largely lens-shaped. Crystallographically, annealing, and mechanical twins are identical. Annealing twins are commonly observed in FCC materials with low to medium SFE. They consist of stacking faults on consecutive $\{111\}$ planes and are produced by the glide of Shockley partials. The grain boundary between the annealing twin and its parent grain is referred to as the annealing twin boundary (ATB).

Based on the CSL model, ATBs are also referred to as $\Sigma 3$ boundaries. ATBs in FCC materials have a misorientation of 60° about $\langle 111 \rangle$ with the orientation of the parent grain in which they form. ATBs are generally of two types, viz., (a) coherent, in which the boundary plane is of $\{111\}$ type and (b) incoherent, where the plane has indices other than $\{111\}$. The coherent ATBs have a planar morphology, while the incoherent ones are mostly curved. The coherent and incoherent ATBs have the same misorientation angle while they differ by the interface plane inclination. It is important to note that the coherent and incoherent ATBs differ in properties such as energy and mobility to a considerable extent [43].

From observations of 2-D microstructures, ATBs are classified into four different morphologies (Fig. 1). Type-A twins are those which form near the TJs and are present as single twin boundaries without a pair. Type-B twins are an extended form of type-A and exist as pairs, and the twin runs through the grain. Type-C twins terminate inside a grain while type-D are island grains with twin misorientation with the matrix.

Liu et al. [44] have performed 3D EBSD studies by serial sectioning on grain boundary engineered 316 stainless steel and have reported the observation of four categories of 3D morphologies of annealing twins, viz., single-face ATBs (type-A), tunnel ATBs (type- B), Occluded ATBs (type-C) and parallel-multiplanar ATBs (type-D) (Fig. 2). Though many ATBs in the 3D GBE microstructure could be classified into the above mentioned types, the authors have reported that some ATBs have complex morphologies. They have concluded that these morphologically complex twins do not belong to a single grain but an assembly of many small grains of similar orientations merged into a single grain. They also suggested that these twins could also arise due to the standard cleanup procedure that is employed before analysing

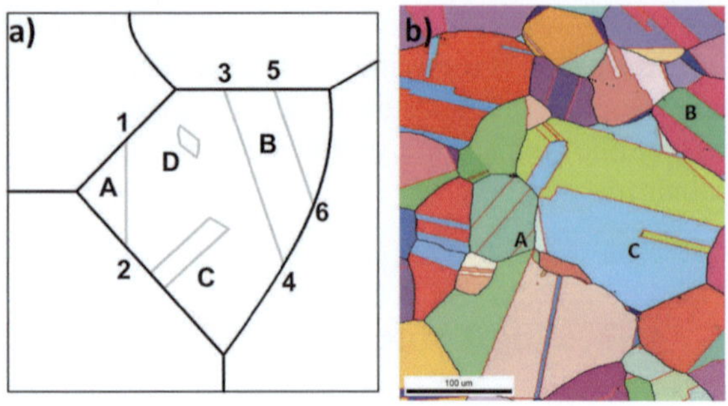

Fig. 1 **a** Schematic of morphologies of annealing twins in a 2-D microstructure **b** a grain boundary engineered microstructure of Alloy-59 showing different morphologies of twins

the data. It should also be noted that several twins of the same variant with respect to the parent grain impinging with one another could result in complicated twin morphologies.

The formation of ATBs have been explained by number of models such as grain encounter model [45, 46], stacking fault packets model [47], growth accident model [48–50] and grain boundary dissociation model [51, 52]. In the grain encounter model, it is proposed that the impingement of grains having twin orientation relationship results in the formation of an ATB [46]. This mechanism does not account for the high fraction of ATBs obtained during TMP as the probability for such occurrences is very low ($\approx 2\%$) [53]. Dash [47] presented the stacking fault packet generation model, in which ATBs nucleate due to accumulation of packets of stacking faults during GB migration. The twins formed during this process have coherent and incoherent segments. The growth accident theory was originally proposed by Carpenter and Tamura [48] and developed by Gleiter [49] and Pande [50]. According to this model, ATBs form on migrating RHAGBs when stacking fault accidents occur. Goodhew

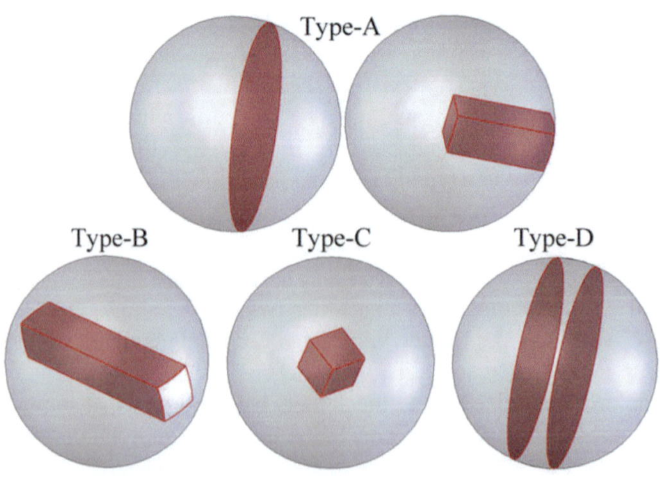

Fig. 2 Schematic showing the 3D morphologies of annealing twins [44]

[52] proposed a mechanism based on the observation of ⟨110⟩ twist boundaries of near coincidence with Σ boundaries that readily dissociate into twins. All these models except the grain encounter model indicate that twin formation is associated to 111 facets of the parent GB.

Many experimental observations show that ATBs form behind migrating GBs and triple junctions (TJs) due to growth accidents [54–57]. It has also been proposed that growth accidents are more probable at higher GB migration velocities [58–60]. Among the several factors responsible for twin formation, the SFE is seen to control the twin boundary density in a microstructure [61–63], as materials with high SFE tend to result in lower twin content following TMP. For a given SFE, the parameters of TMP i.e., annealing temperature [50], time, prior strain[64] are also observed to play an important role in the formation and evolution of ATBs.

Jin et al. [65] observed the formation of ATBs during recrystallisation of 304 stainless steel using in-situ EBSD. It was observed that the twin boundary density increased with an increase in the percentage of recrystallisation. High stored energy in the microstructure affected the ATB density as it influenced the GB migration velocity, consistent with the growth accident theory of [49]. They also reported that the rate of twin generation was more at the beginning of recrystallisation and decreased with the progress of recrystallisation as the average GB migration velocity reduced with the progress of recrystallisation due to the decrease in driving force i.e., stored energy. It was also speculated that, during grain growth, the evolution of ATBs was mainly governed by interactions of the boundaries and disappearance of twins by the migration of HAGBs. Jin et al. [55] also studied twin generation in Ni following straining to 25%. They observed that the ATB density increased during recrystallisation while there was a significant drop during grain growth due to annihilation. In addition, it was reported that the migration of convex GBs nucleated coherent ATBs (energetically favorable) unlike the concave GBs, where incoherent ATBs nucleated. Convex boundaries were seen to be abundant during recrystallisation as there was sufficient driving force to assist in the negative curvature migration. This indicated that both the GB tortuosity and its migration velocity contribute significantly to twinning. Bozzolo and Bernacki have pointed out that, during recrystallization, the ATBs formed are predominantly incoherent and the fraction of coherency significantly improves during the subsequent grain growth [42]. Straight (coherent) ATBs are seen to be more stable during grain growth due to their lower energy and mobility while the incoherent ATBs tend to get annihilated. Coherent ATBs tend to disappear only when the TRD that they are a part of shrink and disappear. However, there is still no clear understanding of the conditions that lead to the formation of coherent and incoherent ATBs [42].

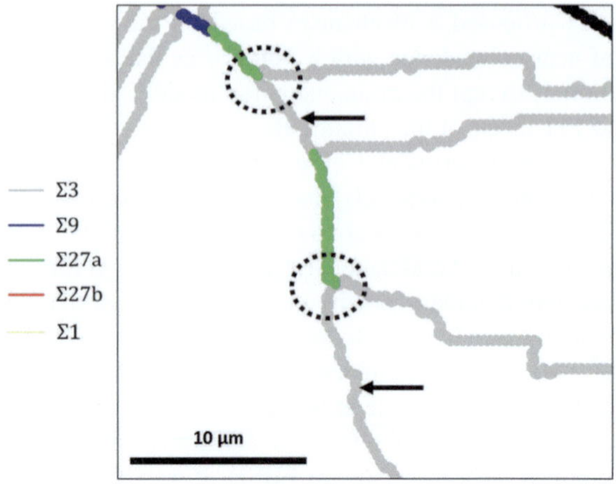

Fig. 3 A snippet of a microstructure from strain annealed Ni showing inconsistent triple junctions

6 Quantification of a GBE Microstructure

A GBE microstructure is generally characterized by a high length/number fraction (approx. > 70%) of SBs (Σ3-Σ29) with a significant disruption of the RHAGB network. To quantify the extent of GBE in a given microstructure, CSL boundaries are identified based on the angle-axis pair determined from the orientation data[5] obtained by EBSD using the deviation criterion discussed in Sect. 2.2).[6] The length/number fraction, density and distribution of $\Sigma 3^n$ (Σ3, Σ9 and Σ27)[7] are then evaluated. It may be noted that the statistics of CSL boundaries and the TJ character distribution are seen to vary significantly based on the deviation criterion employed. Figure 3 shows a snippet of a microstructure of GBE processed commercial-pure Nickel. It can be observed that two TJs (denoted by dashed circles) of the type Σ3-Σ27a-Σ3 are found. These TJs do not satisfy the Σ combination rule and are crystallographically not permissible. This implies that the GBs in the TJ could have been wrongly identified. Further analysis revealed that Σ81d GB which has an angle-axis pair of 60.4° ⟨443⟩ is incorrectly identified as a Σ3 as this is within the tolerance limit of the Σ3 misorientation of 60° ⟨111⟩. The two GBs denoted by the solid arrows have a deviation of about 7.2° from the ideal Σ3 misorientation, which is also within limit as per Brandon's criterion. As the misorientation deviation criterion is satisfied, the GBs have been assigned as Σ3. Similarly, other inconsistent junctions that have been identified by TSL-OIM™ software are, Σ13-Σ3-Σ7, Σ25-Σ9-Σ5, and Σ13-Σ13-Σ3, to name a few. This issue has been highlighted earlier in the literature, and the importance of more stringent deviation criteria has been emphasised [66–69].

The length fraction of an SB is given by the ratio of the length of the boundary to the overall length of GBs in the microstructure. Similarly, number fraction gives the

[5] Refer to Appendix A.

[6] Refer to Appendix B.

[7] These are the predominant SBs observed in GBE microstructures.

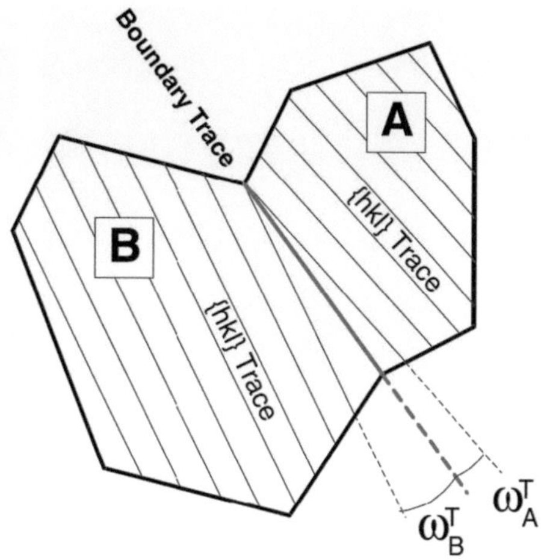

Fig. 4 Schematic showing the boundary trace and trace deviation angle [70]

ratio of number of pixels/data points pertaining to an SB to the overall number of pixels/data points of GBs in the microstructure. But as the overall length of the GB dynamically changes through the duration of annealing, an appropriate representation would be the twin boundary density L_A given by,

$$L_A = \frac{l_{ATB}}{A} \tag{3}$$

where l_{ATB} and A are the length of the ATBs and the area of the scan, respectively.

As discussed in the previous section, a coherent ATB has a boundary plane of {111}. In 2D EBSD measurements, the inclination of the boundary plane cannot be measured. Wright and Larsen [70] have proposed a method of measuring the angle between the GB trace on the surface and the trace of {111} plane on both the sides of the GB. Incoherent ATBs are distinguished from the coherent ones by applying a tolerance of trace deviation. Figure 4 shows a schematic of the reconstructed twin boundary and the trace deviation angle [70]. The procedure of trace deviation determination can be found in [70]. Usually, ATBs with trace deviation less than an angle say 8°[8] are considered as coherent.

The extent of coherency of ATBs is represented as a fraction of coherency f_c which is given by,

$$f_c = \frac{l_{cATB}}{l_{ATB}} \tag{4}$$

where l_{cATB} represents the length of the coherent ATBs in the microstructure determined by employing a trace deviation tolerance.

[8] This is the default criterion in the TSL OIM™ software. A larger trace deviation tolerance angle would be required to identify coherent ATBs if the EBSD step size is coarser (≥ 1).

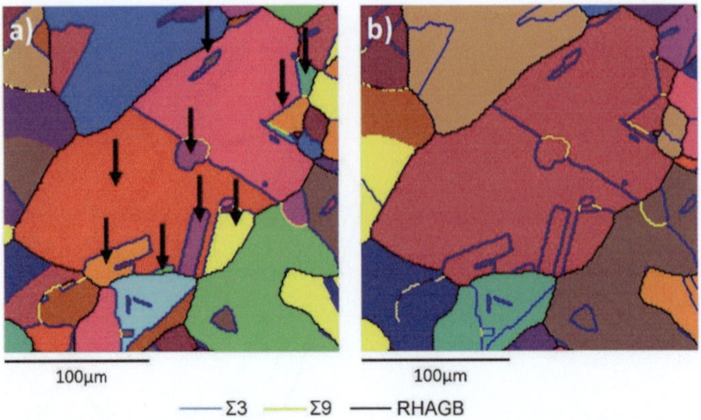

Fig. 5 **a** Unique grain color map, and **b** Unique twin related domain (TRD) color map of the same region of interest (ROI) in a grain boundary engineered microstructure

The extent of twinning in a microstructure is quantified by the average number of twins per grain, *Navg*. If N_1 and N_2 are the number of grains in the microstructure with and without considering twins as grains, the average number of twins per grain is given by,

$$N_{avg} = \frac{N_1 - N_2}{N_2} \tag{5}$$

A GBE microstructure is characterized by a relatively higher value of N_{avg}. The network of SBs can be anlysed through the triple junction (TJ) character distribution. A TJ is referred to as J_n ($n = 0,1,2,3$), where n denotes the number of SBs (Σ3-Σ29) in it. Figure 6 shows the schematic of the different types of TJs based on the number of SBs present in them. A higher fraction of J_2 and J_3 junctions in a microstructure will result in the disruption of the RHAGB network. A $(J_2 + J_3)$ fraction of 0.4–0.5 is considered to result in a GBE microstructure. The evolution of a GBE microstructure can also be described using twin related domains (TRDs). Twin related domains (TRD) are clusters or groups of grains that are related to one another through $\Sigma 3^n$ misorientations and are separated from one another by RHAGBs. Annealing twins that form from a single grain or nucleus[9] interact with one another and other GBs resulting in more boundaries with $\Sigma 3^n$ character. All the grains in a TRD are related by $\Sigma 3^n$ misorientation regardless of whether they are adjacent are not. Figure 5a shows the unique grain color map of a GBE microstructure. In Fig. 5b, each TRD is assigned a unique color. It can be noticed that the grains indicated by the arrows in Fig. 5a are all shown to be a part of a single TRD in Fig. 5b.

TRDs are described by the average size (analogous to grain size), length of the longest chain (LLC), the number of grains per TRD and polysynthetism (p) [71]. LLC is the highest $\Sigma 3^n$ misorientation that exists in a TRD. Polysynthetism is the

[9] Twins that form from the same parent grain.

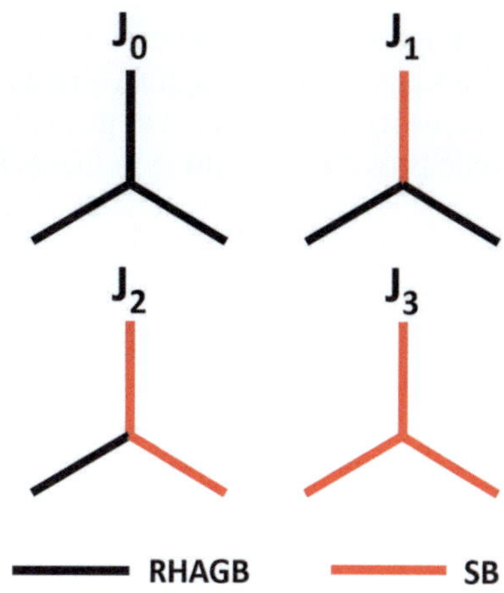

Fig. 6 Character of a triple junction based on the number of special boundaries in it

ratio of the number of grains to the number of unique orientations in a TRD. Through these parameters, the extent and order of multiple twinning during microstructural evolution can be described. The TRD analysis is often carried out using the software ARPGE [72]. Fractal analysis is usually being done on GBE microstructures to quantify the extent of RHAGB connectivity [33]. This analysis involves determination of the longest uninterrupted RHAGB network. The fractal dimension *Df* is calculated using the formula [33],

$$N(x) = x^{-Df} \tag{6}$$

where, $N(x)$ denotes the number of boxes of length x that covers the longest uninterrupted RHAGB.

Yuan Jin [73] compared the ATB density and average number of twins per grain between 2 and 3D microstructures. A 3D GBE microstructure has been reconstructed from a number of 2D sections using Dream 3D software [74]. In a 3D microstructure, the density of ATB L_V is given as,

$$\mathrm{L_V} = \frac{\mathrm{A_{ATB}}}{\mathrm{V}} \tag{3}$$

where A_{ATB} and V are the area of the twin boundaries and the volume of the microstructure, respectively. It has been observed that, the ATB density calculated in a number of 2D sections (L_A) are in the same range with the value determined for the entire 3D microstructure (L_V). However, due to the complex morphologies of the twins in 3D, the average number of twins per grain (N_{avg}) in the 2D and 3D data do not match and the 2D value could be under-estimating the real quantity of twins in a 3D microstructure.

The GB maps of the as received (AR) and GBE processed commercial-pure (CP) Nickel, Inconel 617 and Cantor alloy are shown in Fig. 7. It can be observed that, following GBE, a dense network of $\Sigma 3^n$ boundaries is formed. In Inconel 617 and Cantor alloy, though a considerably high density of ATBs are seen in the AR microstructure compared to that in CP-Nickel, GBE has enhanced the mutual interactions of ATBs resulting in the increase in density of higher order ATBs, thereby disrupting the RHAGB network. The GBE quantifying parameters discussed above are listed in Table 1. It can be observed that, following GBE processing, the grain size, fraction and densities of $\Sigma 3^n$ boundaries, average number of twins per grain, TRD parameters increased in all the three material systems. In this analysis, Palumbo Aust's deviation criterion has been used to characterize the CSL boundaries. A trace deviation tolerance of 8° has been used to identify coherent ATBs. It can also be seen that while the increase in ATB density following GBE processing is significant in CP Nickel compared to the alloys, the absolute ATB density is the highest in the Inconel 617. The extent of coherency is also seen to decrease in CP Ni, while in Inconel 617 and Cantor alloy, the fraction of coherency is seen to increase following GBE processing. This indicates that majority of ATBs formed in CP Ni following GBE processing are of incoherent type. While it is well known that the SFE of the material influences the extent of twinning, the factors affecting the GB plane preference are not yet well established.

7 Grain Boundary Networks and Twin Related Domains

The GB network and RHAGB connectivity is determined by the interactions of CSL boundaries during the microstructure evolution. Among the interactions of CSL boundaries, those involving Σ3 boundaries are of particular interest. A Σ3 interacting with $\Sigma 3^n$ boundaries (n is an integer and describes the order of twinning) generates more such boundaries. These interactions are usually referred to as multiple twinning. Over the last decade, the formation of variants of $\Sigma 3^n$ boundaries through multiple twinning has been mathematically analysed using rotation groups of CSL misorientations and possible outcomes of interactions have been calculated [72, 77, 78]. Though the outcomes of interactions of CSL boundaries are known, the exact crystallographic conditions under which these interactions result in a boundary of a particular character are not yet clearly elucidated.

Our group carried out a systematic study investigating the relationship between the misorientation axes of the three CSL boundaries in a TJ, in addition to the combination rule [41]. This resulted in determining the probabilities of interactions of a selected $\Sigma 3^n$ boundaries as shown in Fig. 8.

Experimental observations indicate the presence of a higher fraction of Σ3-Σ9-Σ3 TJs relative to Σ3 -Σ9-Σ27 TJs [17, 40, 57]. It has been proposed that the interaction of Σ3 and Σ9 boundaries results in a Σ3 and has a higher preference over Σ27. However, from Fig. 8b, it is evident that 50% of the combinations of Σ3 and Σ9 result in the third GB being a Σ27b, while the chances of it being Σ27a and Σ3 are 25% each. When Σ3 boundaries impinge, the probability of the resulting GB being Σ9 is 0.75 (Fig. 8a). The higher fraction of J_3 junctions in GBE microstructures being Σ3-Σ9-Σ3

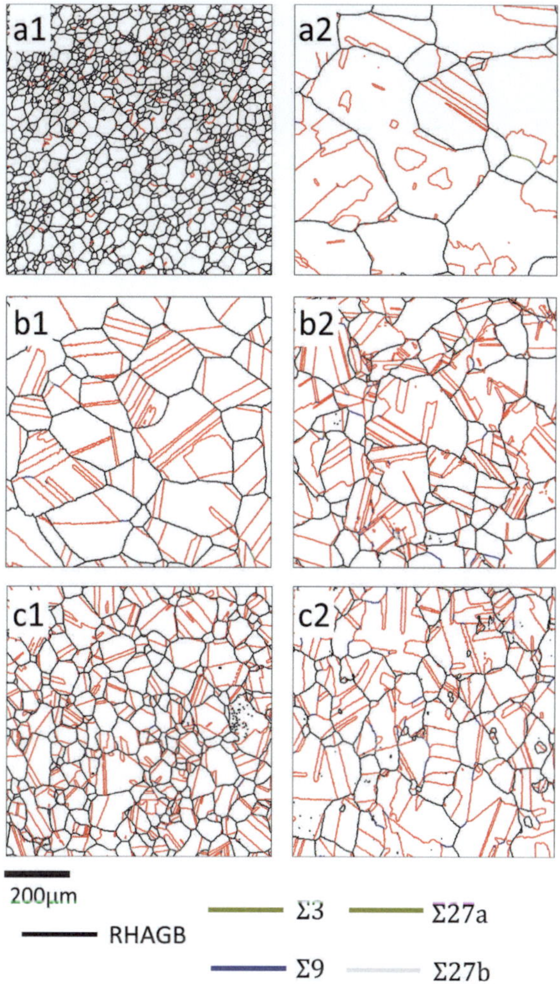

Fig. 7 Grain boundary maps of the as-received and grain boundary engineered microstructures of **a1**, **a2** Commercial-pure Nickel; **b1**, **b2** Inconel 617 and **c1**, **c2** Cantor alloy

are plausibly due to twin impingement ($\Sigma3 + \Sigma3 \rightarrow \Sigma9$) rather than Σ3 regeneration ($\Sigma3 + \Sigma9 \rightarrow \Sigma3$). Similarly, when two GBs of a J_3 junction are Σ9 and Σ27b or Σ27a, the probability that the third GB will be Σ9 is 0.083 [79] (Fig. 8f and g), and this is much less than the probability for generation of Σ27 during $\Sigma3 + \Sigma9$ reaction, 0.75 (Fig. 8b). However, we see that the interaction $\Sigma3 + \Sigma27$ yields a probability of 0.25 for Σ9 (Fig. 8d and e). This interaction is less likely to contribute to the fraction of Σ3-Σ9-Σ27 TJs, owing to the relatively low density of Σ27 boundaries compared to Σ3 boundaries. Hence it is reasonable to conclude that Σ3-Σ9- Σ27 TJs are mostly a result of $\Sigma3 + \Sigma9 \rightarrow \Sigma27$ interaction, rather than $\Sigma3 + \Sigma27 \rightarrow \Sigma9$ or $\Sigma9 + \Sigma27 \rightarrow \Sigma3$. The above analysis contribute to a better understanding of the mechanism of GBE. In addition, these interactions play a crucial role in the disruption of the RHAGB

Table 1 Various parameters of the as-received and grain boundary engineered microstructures of commercial-pure Nickel

Microstructural parameter	CP-Nickel (SFE: 128 mJm^{-2})[43]		IN617 (SFE: 45 mJm^{-2}) [75]		Cantor alloy (SFE: 25–30 mJm^{-2})[76]	
	AR	GBE	AR	GBE	AR	GBE
Grain size (with twins), μm	15 ± 2	40 ± 3	9 ± 1.2	14 ± 2	10 ± 1	18 ± 1
Grain size (without twins), μm	22 ± 3	90 ± 8	20 ± 3.2	40 ± 2.5	19 ± 3	56 ± 7
Σ3 length %	8 ± 6	51 ± 1	38 ± 1.5	53 ± 3	45 ± 0.6	62 ± 1
Σ9 length %	1 ± 0.1	2.5 ± 0.5	0.5 ± 0.04	2.8 ± 0.25	1.2 ± 0.1	4 ± 0.5
Σ27 length %	0.1 ± 0.01	1 ± 0.2	0.1 ± 0.001	0.5 ± 0.02	0.4 ± 0.1	2 ± 0.3
$J_2 + J_3\%$	9 ± 1	38 ± 2	19 ± 2.5	41 ± 2.5	20 ± 1	40 ± 2
$Navg$	0.5 ± 0.09	5.5 ± 0.25	1 ± 0.1	5.25 ± 0.3	2 ± 0.6	7 ± 1.6
Σ3 density mm^{-1}	9 ± 1	26 ± 2	40 ± 2.8	75 ± 4.2	44 ± 2.1	45 ± 2.2
Σ9 density mm^{-1}	0.1 ± 0.01	2 ± 0.2	1.2 ± 0.21	3.9 ± 2.2	0.5 ± 0.1	3.2 ± 0.2
Σ27 density mm^{-1}	0.1 ± 0.001	0.5 ± 0.01	0.1 ± 0.005	0.89 ± 0.02	0.1 ± 0.05	1 ± 0.5
Fraction of coherency	0.8 ± 0.01	0.56 ± 0.05	0.6 ± 0.035	0.7 ± 0.028	0.54 ± 0.03	0.6 ± 0.07
TRD size, μm	20 ± 3	75 ± 5	19 ± 4	35 ± 3	21 ± 2	40 ± 6
Avg. no. of grains per TRD	1 ± 0.4	12.5 ± 0.4	2.1 ± 0.1	10 ± 2.5	3.4 ± 0.4	13 ± 3
LLC	0.1 ± 0.06	1.5 ± 0.3	0.9 ± 0.01	1.3 ± 0.09	0.9 ± 0.05	1 ± 0.3
Polysynthetism	1 ± 0.4	4.5 ± 1	1.1 ± 0.05	4.2 ± 0.28	1.5 ± 0.08	4.6 ± 0.2

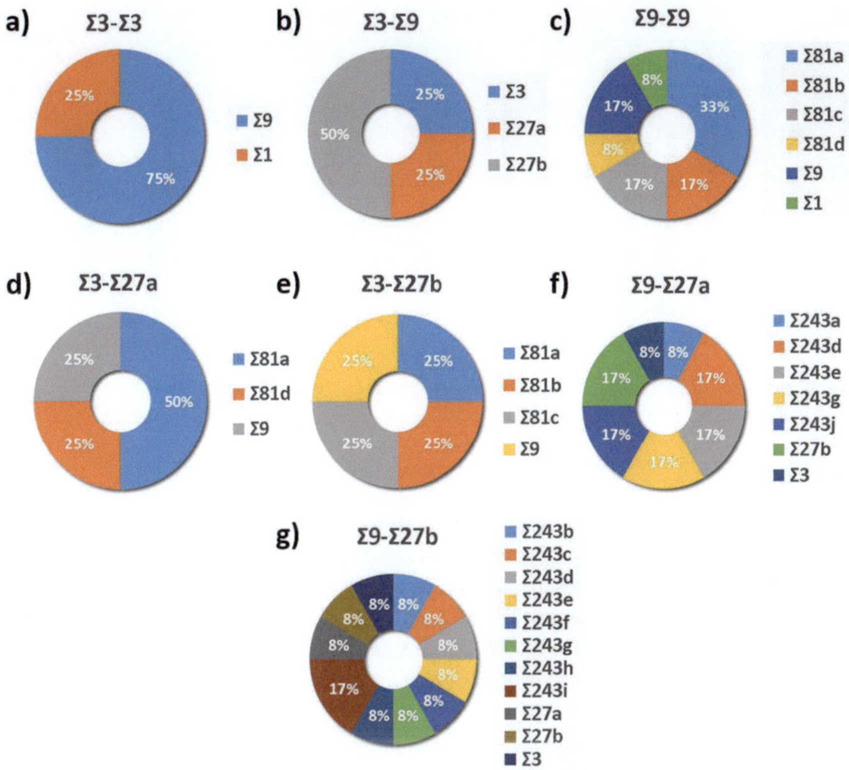

Fig. 8 Pie charts showing the possible outcomes and their probabilities during interactions of $\Sigma 3^n$ boundaries in triple junctions

network. Multiple twin related interactions resulting in the formation of higher order ATBs forms the basis of the formation and evolution of TRDs.

The J_3 junction analyses have also been performed on as-received and GBE microstructures of a few FCC materials, viz., CP Ni, Ni-based superalloys like Hastelloy (C276), Alloy 59, and Inconel 617 [12] to understand the mechanism of GBE across material systems. From Fig. 9a, it is seen that the fraction of Σ3-Σ9-Σ3 TJs is higher than Σ3-Σ9-Σ27 in GBE and non-GBE conditions in all the materials evaluated. In CP Ni, the increase in Σ27 fraction following GBE is lower compared to that of Σ9 (Fig. 9b). This is the reason for the occurrence of more number of Σ3-Σ9-Σ3 TJs are present compared to Σ3-Σ9-Σ27 TJs (Fig. 9a). On the other hand, in the Ni-based superalloys, there is a significant increase in the fraction of Σ3 as well as Σ9 and Σ27 boundaries following GBE (Fig. 9b). This is because more Σ9 and Σ27 are seen to have been generated by interactions of Σ3 boundaries. This has led to an increase in the fraction of Σ3-Σ9-Σ27 TJs in Ni alloys. It can be concluded that the evolution of $\Sigma 3^n$ boundaries in CP Ni and Ni based alloys are considerably different. Generation of Σ27 from Σ3 + Σ9 interaction is relatively higher in the

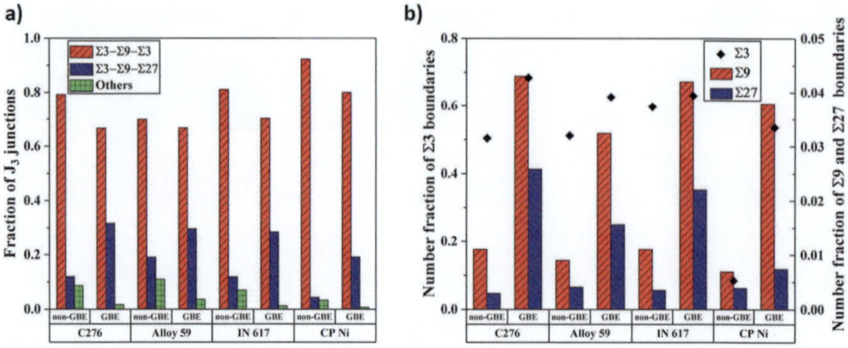

Fig. 9 **a** Fractions of Σ3-Σ9-Σ3, Σ3-Σ9-Σ27 and other J_3 triple junctions, **b** number fractions of Σ3, Σ9 and Σ27 boundaries in Ni-based superalloys and CP Ni in non-GBE and GBE conditions [41].

Ni-based superalloys as shown by the fraction of Σ3-Σ9-Σ27 TJs (Fig. 9a). While primary twinning is dominant during GBE processing of CP Ni, the formation of higher-order twin boundaries is significant in the Ni alloys. This could be due to the relatively higher fraction of pre-existing Σ3 boundaries in Ni-based superalloys in the non-GBE condition ($\approx$ 40–50%) in comparison to CP Ni ($\approx$ 15%). This, in turn, is due to the lower stacking fault energies of the Ni-based superalloys, which promotes the formation of ATBs.

It has been experimentally observed that large TRDs have fewer unique orientations than smaller ones [80]. Higher polysynthetism is an indication of large TRD size [81]. Recrystallised microstructures are characterised by small TRDs. The growth of TRDs is found to depend on the microstructure evolution mechanism, i.e., SIBM or recrystallisation. Prithiv et al. [33] investigated the role of SIBM and recrystallization on GB networks. It has been found that, during SIBM, the ATBs that form due to growth accidents have a higher probability to interact with one another and other CSL boundaries resulting in the proliferation of SBs and break the RHAGB connectivity. On the contrary, the ATBs and higher order variants that form during recrystallization and grain growth do not form a part of the GB network resulting in an uninterrupted RHAGB network. This also results in smaller TRDs compared to large TRD sizes seen in SIBM microstructures [33, 80]. Figure 10 shows a schematic representation of formation, evolution and interactions of ATBs during recrystallization (a1–a3) and during recrystallization followed by SIBM (b1–b3). It can be seen that recrystallization results in smaller TRDs due to impingement with less interactions between ATBs while SIBM provides more scope for ATB interactions resulting in large TRDs.

The morphologies of annealing twins in 3D are known to be more complicated than in the 2D microstructures and the extent of RHAGB network disruption also varies in 3D microstructures. Liu et al. [44] have studied the morphologies of twins in 3D microstructures and reported that the connectivity of RHAGBs have

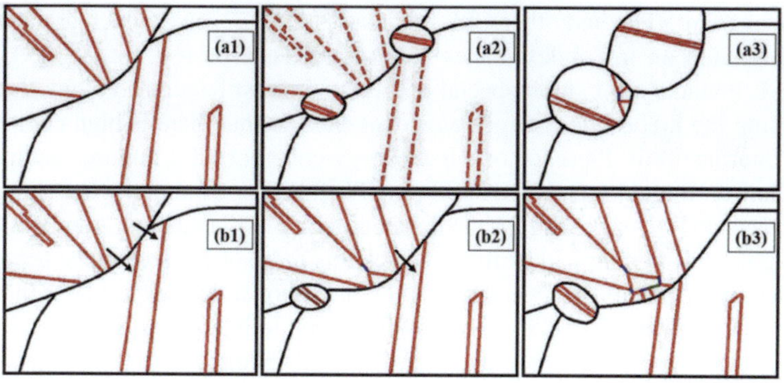

Fig. 10 **a1–a3** Schematic showing twinning and their interactions during recrystallization; **b1–b3** during recrystallization and SIBM. The black boundaries are RHAGBs while the red, blue and green colors indicate Σ3, Σ9 and Σ27 boundaries respectively [33].

not been disrupted significantly following GBE. This clearly indicates the significant difference in the network topology in a 3D GBE microstructure compared to 2D counterparts. This underscoes the necessity for 3D characterization of GBE microstructures.

8 The Significance of the Grain Boundary Plane

It is generally believed that CSL boundaries with low Σ values (particularly with Σ value < 29) have a higher density of coincidence sites and have relatively lower energy [5] in comparison to RHAGBs. However, there are a number of examples which contradict the direct correlation of the Σ value of a GB and its energy. Some researchers showed that low Σ GBs need not be associated with special properties based on Σ value alone as the CSL model does not account for the GB plane character [82, 83].

For example, a Σ9 {221} GB has a higher energy compared to a Σ11 {113} [83]. It is also well known that a Σ3 {111} (a coherent ATB) has a lower energy compared to a Σ3 which has planes other than {111} (incoherent ATB). It has been shown that the mobility of an ATB is closely related to the boundary inclination [84]. Rahimi et al. [85] has reported that some incoherent ATBs are susceptible to intergranular cracking. Gertsman et al. [66] show that most of the CSL boundaries that exhibited cracking had very large deviations (permissible under Brandon's criterion) from ideal CSL misorientation. It is generally agreed that CSL boundaries closer to ideal misorientation are more likely to have low index boundary planes [86, 87]. It is now known that the higher order twin boundaries are mainly formed out of geometrical necessity during multiple twinning [41]. These boundaries will exhibit special properties only if at least one of them has a low index plane. It is observed that incoherent ATBs and other CSL boundaries ($\Sigma5$, $\Sigma7$, etc.) which are generally not terminated on low index planes behave as RHAGBs [66, 82, 88, 89]. For a low CSL boundary to exhibit special properties it must be aligned along the plane containing the highest density of coincident sites so that there is high coincidence at the interface [86]. Presence of a low energy coherent ATB imposes geometrical constraint on the adjoining RHAGBs in a TJ [90]. For example, when a coherent ATB meets an RHAGB, then for the third boundary in the junction, four among the five degrees of freedom will be fixed. Hence it is more often likely to be terminated by a low index plane [82]. King et al. [91] have shown that several GBs that were neither low angle nor low CSL to be quite resistant to intergranular stress corrosion cracking. Detailed analysis revealed that these GB planes were close to low index planes. This brings into focus the need for redefining the term 'Special Boundaries'. An SB can now be defined as the one that contains at least one low index plane [40].

From the above discussion, it is clear that for the identification of a special GB in a microstructure, it is necessary to determine the crystallographic orientation of the GB plane [40]. Research over the last two decades highlighted the importance of GB plane character and structure and how they alter the traditional idea of GBE based on CSL model. Development of 3D EBSD techniques [92] using dual beam microscopes has contributed to the evaluation of a large number of GB planes and their distributions in polycrystalline materials. However, since 3D EBSD techniques are not still routinely available, GB plane information is still obtained from 2D EBSD data using five-parameter grain boundary analysis based on stereological methods [3, 93].

Grain boundary plane distributions determined through five-parameter grain boundary analysis in GBE Brass revealed that most of the GBs were terminated on low index planes [94]. Tilt and twist boundaries are often terminated by at least one low index boundary plane and have lower free volume and low energy compared to other general boundaries [82]. A similar analysis on GBE Cu and GBE Ni also revealed several boundaries vicinal to [110] and [011] planes (excluding $\Sigma3$ data) [82, 88]. The population distribution of planes also revealed several asymmetrical tilt grain boundaries (ATGBs) with at least one low index plane.

In the recent times, the concept of grain boundary plane engineering (GBPE) has been proposed, in which, the GB planes in the microstructure are tailored to reorient to have low energy orientations by processing such as low temperature annealing and doping [82, 94]. In case there is insufficient energy for boundary migration, reorientation is observed to occur through boundary rotation [86]. Thus, the occurrence of boundary reorientation in a material could lead to a grain boundary plane engineered microstructure. Another method to achieve GBPE is through the introduction of dopants or impurities in the material as the impurity additions has been shown to influence the type of GB planes in the material [82].

Sharma and Shekhar [95] reported that, during annealing of Cu–Zn alloy and 316 Stainless steel, a majority of the formed ATBs deviate by $< 1°$ from the ideal misorientation. It was emphasised that the angular deviation ($\Delta\theta$) was more significant than the deviation of the GB plane. The interactions of ATBs resulting in $\Sigma3^n$ GBs were not affected by the coherency of ATBs as long as the deviations in angle were permissible as per the deviation criterion employed. Sharma and Shekhar [96] also observed that the majority of the ATBs that formed during recrystallisation (in-situ studies on Cu–Zn alloy and ex-situ studies of 316 stainless steel) were incoherent and had low angular deviations. During the grain growth, the deviation of ATBs increased from the ideal misorientation. The five-parameter grain boundary analysis discussed above is a statistical method and largely depends on the 2D EBSD data to get the GB plane information. The reconstruction of the GBs also depend on the step size of the EBSD scan. It may also be noted that, this analysis does not account for the curvatures of GB planes in 3D.

An et al. [97] carried out a study of GB plane distribution and corrosion behavior of the SBs in a GBE microstructure of 304L stainless steel through serial sectioning based 3D EBSD. They have reported that, a majority of $\Sigma 3$, $\Sigma 9$ and $\Sigma 27a$ boundaries terminated on {111} planes, while $\Sigma 27b$ showed a higher preference for {112} planes. Coherent ATBs were seen to be resistant to intergranular corrosion compared to the incoherent ones. In addition, irrespective of the character of the GB, their plane is seen to play a dominant role. GBs with higher atomic packing density showed better corrosion resistance in both RHAGBs and CSL boundaries.

9 Modelling and Simulation of GBE Microstructure Evolution During TMP

The generation of a GBE microstructure during TMP is governed by (a) the rate of formation of twins, (b) the twin morphology, and (c) their interactions during GB migration. In order to understand the complex interplay of these factors in a systematic manner, an appropriate modeling approach is necessary ([42]). In the context of GBE, modelling and simulation tools should account for the formation of twins, their evolution and interaction. The length scales of the available models generally range from an atomistic-level which focus on property determination of GBs to mesoscale like full-field simulations involving microstructure evolution.

The models proposed by Gleiter [49] and Pande [50] (briefly discussed in Sect. 4) are based on the growth accident theory and predict the evolution of ATB density. The experimental results [73] are not consistent with the Gleiter's model which is based on classical nucleation theory wherein the twin formation process has been associated with the activation enthalpy of GB migration. On the other hand, Pande's model has been shown to be consistent with many experimental findings [65, 98]. The model which was formulated to model the ATB evolution during curvature driven grain growth has been modified to incorporate stored energy as well [54]. The change in the number of twins per grain as a function of the GB migration driving force and grain size from the above models are not consistent with the experimental results [98].

Yuan Jin [73] presented a mean field model to account for the reduction in ATB density during the grain growth regime. The model is described as,

$$\frac{L_i(t+\Delta t)-L_i(t)}{L_i(t)} = b\frac{D_i(t+\Delta t)-D_i(t)}{D_i(t)}, i = 1,2,\ldots n \tag{8}$$

where L_i and D_i denote the density of ATBs and the corresponding average grain size of a grain i. The value of the constant b is influenced by two factors. If the ATB is coherent, the value of $b > 1$ and for an incoherent segment, $b < 1$. The constitutive input to the model was the area distribution of the grain size and the ATB length of each category. The value of b was determined by fitting the ATB density evolution data from experiment results. This model has an advantage over Pande's model as this requires only one model parameter b.

Jin [73] developed a level set model for grain growth in the presence of ATBs. Before the simulation, ATBs (coherent and incoherent) were inserted at certain locations in the microstructure. The overall twin boundary density was found to decrease gradually with time. During the simulation, the coherent twin boundaries were observed to be immobile, and the change in ATB density was due to the shrinking of the incoherent twins. The simulation results were seen to agree with experimental observations. It was also observed that, while the coherent ATBs extend, the new segments of the boundary deviated significantly from coherency and became incoherent.

Tang et al. [99] simulated the grain growth in twinned microstructures using a multi- phase field model (MPFM) in which the influence of network topology on the evolution was studied. The initial microstructures with a varying fraction of SBs were synthesised using Monte Carlo Potts model. The ratio of mobilities of RHAGBs to $\Sigma 3^n$ GBs ($\Sigma 3$, $\Sigma 9$ and $\Sigma 27$) was taken to be 100:1. In the microstructures where the constraints were applied, the SB densities continuously decreased with simulation time while those densities increased in the other microstructure. Similarly, the evolution of TJs was found to be sensitive to the topology of the GBs.

Our group has developed a MPFM which incorporated the formation of ATBs of different morphologies as well as their interactions with other GBs during grain growth (strain induced boundary migration) [100]. This model has been developed only to simulate the process of GBE during strain annealing. The growth accident theory of twin formation has been introduced in a probabilistic manner. The morphologies of the annealing twins arising from growth accidents in 2D microstructures are referred to as types A, B and C (as shown in Fig. 1a) ([58]). It was suggested that sufficiently high GB velocity promotes the occurrence of growth accidents leading to formation of twins. The GB migration rate is in-turn influenced by the stored energy as well as its mobility. The stored energy is incorporated by varying the free energy parameter and the mobilities of the GBs are assigned based on their character. The decay of stored energy due to recovery is also incorporated in the model. The model is capable of capturing the twin formation, saturation and annihilation regimes during GBE (Fig. 11).

Recrystallization involves formation of strain free grains or nuclei in the deformed matrix which grow by consuming the deformed matrix [101]. The primary mechanisms involved are the nucleation of the grains followed by the migration of high angle grain boundaries (HAGBs) due to the significant stored energy difference

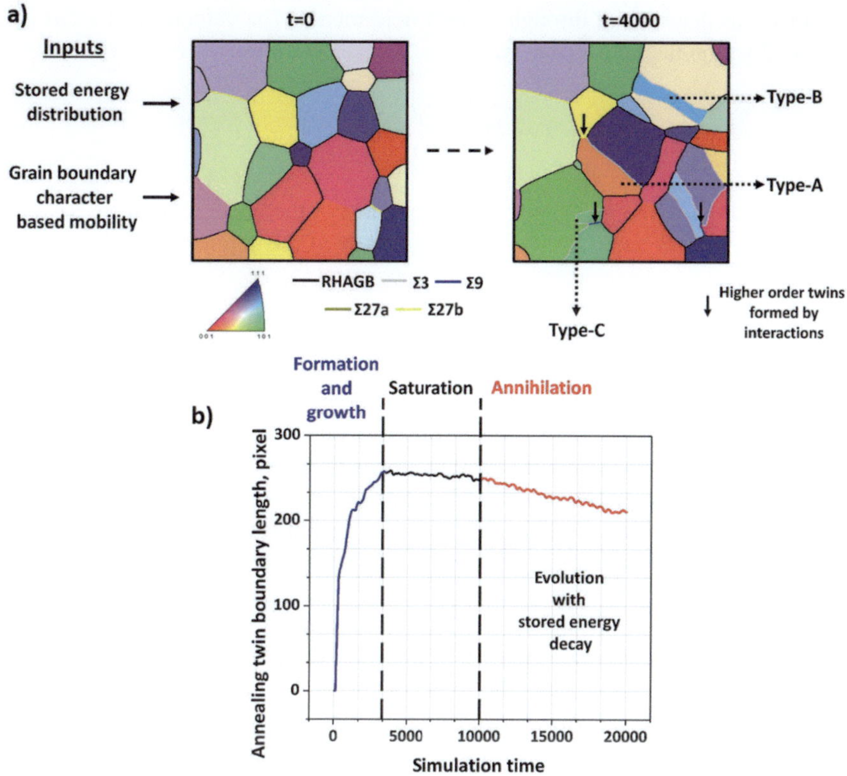

Fig. 11 **a** Inverse pole figure maps of a polycrystalline microstructure showing formation of different morphologies of annealing twins during strain induced boundary migration. The inputs for the model are also indicated. **b** The plot shows the evolution of annealing twins with annealing time showing regimes of formation, saturation and annihilation

between the nuclei and the deformed grains. ATBs form during the process of migration of the HAGBs due to growth accidents [49]. To simulate recrystallization, various approaches such as cellular automata [102], Monte-carlo [103] and phase field modelling [104, 105] have been developed. The migration of recrystallization front has been modelled through a phase field model by Moelans et al. [104] by using separate order parameters for the deformed and strain free grains. Our group had developed a phase field model incorporating the stored energy of deformation determined through kernel average misorientation (KAM) to modify the free energy functional of the model [106]. The primary goal of the study was to compare the role of mode of deformation on recrystallization kinetics in commercial pure Nickel and Titanium. To simulate recrystallization of a deformed microstructure during annealing the stored energy distribution and mobility of the boundaries were taken as the constitutive inputs. There have been a lot of experimental and simulation works determining the mobilities of grain boundaries [107, 108]. In this work, the mobility was determined through a mean field model. The velocity v of a GB of a recrystallization nuclei is given by,

$$v = M\left(\frac{-2\gamma}{R(t)} + E_{st}(t)\right) \quad (9)$$

where γ is the GB energy in Jm^{-2}, R is the radius of the nucleus at time t, M is the mobility of the GB and $E_{st}(t)$ is the stored energy of the deformed matrix at time t given by,

$$E_{st}(t) = E_0 + E_1 e^{-\nu t} \quad (10)$$

where, E_0, E_1 and ν are the fitting parameters. Mobility M is obtained from the best fit of the experimental data of (t). Using the different values of M for the different modes of deformation, the recrystallized area fraction is obtained as a function of time and annealing temperature. The inverse pole figures of the simulated microstructures at various duration during annealing of rolled and torsion tested samples at 848 K are shown in Fig. 12a–d and e–h respectively. The corresponding changes in fraction of recrystallization with time from the simulated and experimental microstructures are shown in Fig. 12i. To model the GBE microstructure evolution during strain-recrystallization processing formation of twins during recrystallisation needs to be developed and incorporated into the above model.

10 Conclusions and Future Directions

The concept of GBE has evolved significantly over the past 35 years due to the advancements in characterization of micro and mesotexture as well as computational techniques for studying microstructural evolution. It is now widely accepted that classification of low Σ boundaries as SBs based on the CSL model is not correct and the GB plane character determines the interfacial properties viz., energy, mobility, etc. Five-parameter analysis is currently widely used to study the GB plane orientations from 2D EBSD data. However, this approach is statistical in nature and it is necessary to evaluate the 3D microstructure to identify the GB planes and their distributions.

Though annealing twins have been studied for many decades, there are still many unanswered questions related to the formation and evolution of coherent and incoherent twin boundaries [42]. In the last few years, the development of in-situ EBSD [56, 59, 109], Synchrotron X-ray based near field high-energy diffraction microscopy (nf-HEDM) [73, 110] studies during annealing and 3D EBSD based on serial sectioning have provided a better understanding of the formation and evolution of ATBs during SIBM and recrystallization. Characterization of 3D microstructures began with serial sectioning followed by EBSD [44]. Ion-milling and plasma FIB

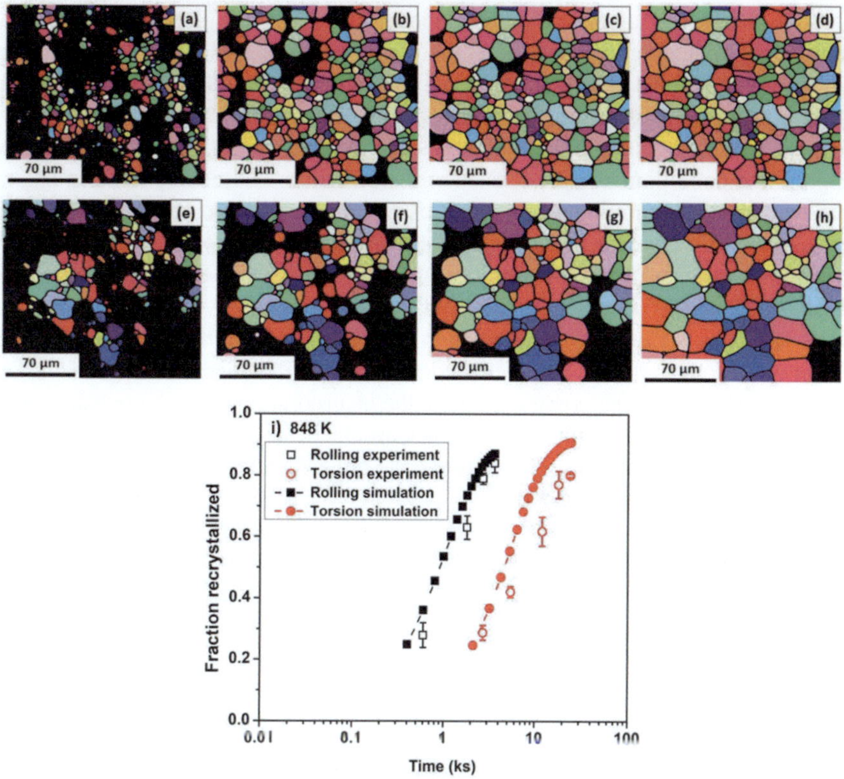

Fig. 12 **a–d** Phase field simulations of recrystallization of rolled microstructures subjected to equivalent strain of $\in = 0.5$ annealed at 848 K for different durations. **e–h** show such simulations of torsion samples subjected to the same equivalent strain. **i** shows the variation in fraction of recrystallization with time in the experimental and simulated microstructures [106]

technology were then used to carry out serial sectioning with precise control of material thickness removed [111]. Recently, a commercial version of Tribeam microscope developed at University of South Carolina Beaufort (USCB) has been made available for 3D microstructure studies [42, 112, 113]. With advancements in 3D orientation microscopy [97], we are in a position to characterize the GB plane precisely and get insights into the formation and interactions of the SBs. It is believed that significant improvements in the spatial and angular resolutions of the 3D microstructure characterisation techniques [42, 44, 73, 97, 110, 111, 111–113] can be expected in the coming years and this will certainly help in identifying the mechanisms of twin formation and evolution during annealing.

Many physical based models have been developed for the formation and interactions of twins during the annealing. Among these, the growth accident model is believed to be consistent with the experimental results. There have been many attempts to develop full field simulation models to predict the microstructural evolution during annealing. Among these, the phase field models looks very promising because they facilitates following the microstructure evolution in 2D and 3D with precise measurements of grain size distribution, GBCD and GB network topology. In this context, crystallographic conditions like the character of the parent GB plane, appropriate conditions for the formation of incoherent and coherent ATBs has to be incorporated to mimic the formation of ATBs in real microstructures for closer predictions with experiments. 3D phase field models are already in use to study grain coarsening under a variety of conditions [114, 115]. A number of tools for generation of 3D synthetic microstructures have been developed over the last few years [74]. These tools facilitate the simulation of grain growth, formation of ATBs and their interactions in 3-D. A complete understanding of the microstructural evolution during GBE will only be possible through the incorporation of the knowledge gathered from the above mentioned state-of-the art characterization techniques into 3D full field simulation models. That will enable better prediction of the evolution of GBE microstructures during TMP.

Acknowledgements The authors acknowledge the financial support of DST-SERB, Govt. of India (Grant no: EMR/2017/004548). They also wish to thank Dr. Stuart Wright for the valuable discussions regarding the TSL-OIM^{TM} EBSD analysis software. The authors extend their gratitude to Ms. Shakthipriya, Mr. Hemanth Thota and Mr. Vijay Bharadwaj for their help in the experiments.

Appendix 1 Determination of Misorientation of an Interface

The elements of the orientation matrix from the Euler angles (ϕ_1, ϕ_2) are given by ([3]),

$$\begin{aligned}
g(1,1) &= \cos\phi_1 \cos\phi_2 - \sin\phi_1 \sin\phi_2 \cos\phi \\
g(1,2) &= \sin\phi_1 \cos\phi_2 - \cos\phi_1 \sin\phi_2 \cos\phi \\
g(1,3) &= \sin\phi_2 \sin\phi \\
g(2,1) &= -\cos\phi_1 \sin\phi_2 - \sin\phi_1 \cos\phi_2 \cos\phi \\
g(2,2) &= -\sin\phi_1 \sin\phi_2 + \cos\phi_1 \cos\phi_2 \cos\phi \\
g(2,3) &= \cos\phi_2 \sin\phi \\
g(3,1) &= \sin\phi_1 \sin\phi \\
g(3,2) &= -\cos\phi_1 \sin\phi \\
g(3,3) &= \cos\phi
\end{aligned} \tag{11}$$

Crystallographically related solutions for the orientation matrices shall be obtained by pre-multiplying with the symmetry matrices S. In cubic systems, 24 crystallographically related solutions are available due to the sample symmetry [3]. The 24 symmetry matrices are,

$$\begin{bmatrix}1&0&0\\0&1&0\\0&0&1\end{bmatrix}\begin{bmatrix}0&0&\bar{1}\\0&\bar{1}&0\\\bar{1}&0&0\end{bmatrix}\begin{bmatrix}0&0&\bar{1}\\0&1&0\\1&0&0\end{bmatrix}\begin{bmatrix}\bar{1}&0&0\\0&1&0\\0&0&\bar{1}\end{bmatrix}$$
$$\begin{bmatrix}0&0&1\\0&1&0\\\bar{1}&0&0\end{bmatrix}\begin{bmatrix}1&0&0\\0&0&\bar{1}\\0&1&0\end{bmatrix}\begin{bmatrix}1&0&0\\0&\bar{1}&0\\0&0&\bar{1}\end{bmatrix}\begin{bmatrix}1&0&0\\0&0&1\\0&\bar{1}&0\end{bmatrix}$$
$$\begin{bmatrix}0&\bar{1}&0\\1&0&0\\0&0&1\end{bmatrix}\begin{bmatrix}\bar{1}&0&0\\0&\bar{1}&0\\0&0&1\end{bmatrix}\begin{bmatrix}0&1&0\\\bar{1}&0&0\\0&0&1\end{bmatrix}\begin{bmatrix}0&0&1\\1&0&0\\0&1&0\end{bmatrix}$$
$$\begin{bmatrix}0&1&0\\0&0&1\\1&0&0\end{bmatrix}\begin{bmatrix}0&0&\bar{1}\\\bar{1}&0&0\\0&1&0\end{bmatrix}\begin{bmatrix}0&\bar{1}&0\\0&0&1\\\bar{1}&0&0\end{bmatrix}\begin{bmatrix}0&1&0\\0&0&\bar{1}\\\bar{1}&0&0\end{bmatrix}$$
$$\begin{bmatrix}0&0&\bar{1}\\1&0&0\\0&\bar{1}&0\end{bmatrix}\begin{bmatrix}0&0&1\\\bar{1}&0&0\\0&\bar{1}&0\end{bmatrix}\begin{bmatrix}0&1&0\\1&0&0\\0&0&\bar{1}\end{bmatrix}\begin{bmatrix}0&\bar{1}&0\\0&0&\bar{1}\\1&0&0\end{bmatrix}$$
$$\begin{bmatrix}\bar{1}&0&0\\0&0&1\\0&1&0\end{bmatrix}\begin{bmatrix}0&0&1\\0&\bar{1}&0\\1&0&0\end{bmatrix}\begin{bmatrix}0&\bar{1}&0\\\bar{1}&0&0\\0&0&\bar{1}\end{bmatrix}\begin{bmatrix}\bar{1}&0&0\\0&0&\bar{1}\\0&\bar{1}&0\end{bmatrix}$$

For the purpose of misorientation calculation, the Euler angles of the grains are converted into orientation matrix g. The misorientation matrix M_{12} of the GB between the grains 1 and 2 is calculated through,

$$M_{12} = (S_i g_1)^{-1}(S_i g_1),\ i, j = 1, 2, \ldots, 24 \tag{12}$$

where g_1 and g_2 are the orientation matrices of the grains 1 and 2 respectively. S_i and S_j are two of the 24 symmetry matrices. A total of 1152 values are obtained for the misorientation matrix M_{12} through Eq. 12. From the misorientation matrix M, the angle θ and the axis $\langle uvw$ are determined using,

$$\theta = \cos^{-1}((\mathrm{tr}(M) - 1)/2) \tag{13}$$

$$u = \frac{M(2,3) - M(3,2)}{2\sin\theta},\ v = \frac{M(3,1) - M(1,3)}{2\sin\theta},\ \text{and} w = \frac{M(1,2) - M(2,1)}{2\sin\theta} \tag{14}$$

Algorithm 1 Determining the angle and axis of misorientation of an interface.

Input: Euler angles ϕ_1, ϕ and ϕ_2 of the two grains
Output: Angle, ° and axis of misorientation
Compute: Orientation matrices g_1 and g_2 from the respective Euler angles using Eq. 11
for $i = 1$ to 24 **do**
for $j = 1$ to 24 **do**

6: **Compute**: Misorientation matrix M using Eq. 12
7: **Compute**: Angle using Eq. 13 and store in an array theta(i, j)
8: **end for**
9: **end for**
10: **Assign**: The minimum value of array theta(i, j) to θ.
11: **Compute**: The misorientation matrix M_{min} corresponding to θ using Eq. 12 inputting appropriate values of i and j
12: **Compute**: The axis of misorientation corresponding to M_{min} using Eq. 14.
13: **Return**: The angle θ and axis $\langle uvw \rangle$ of misorientation.

It has to be noted that, instead of determining $24 * 24 = 576$ misorientation matrices, it is enough to calculate only 24 matrices. The minimum misorientation angle can be obtained with only these matrices. In that case, the for-loop in step 5 of Algorithm 1 can be avoided by using a constant value of j.

Appendix 2 A Calculation of deviation of a CSL boundary

The elements of the misorientation matrix from the angle-axis pair θ <*uvw*> are given by,

$$\begin{aligned} M(1,\ 1) &= (1 - u^2)\cos\theta + u^2 \\ M(1, 2) &= uv(1 - \cos\theta) + w\sin\theta \\ M(1, 3) &= uw(1 - \cos\theta) - v\sin\theta \\ M(2, 1) &= uv(1 - \cos\theta) - w\sin\theta \\ M(2, 2) &= (1 - v^2)\cos\theta + v^2 \\ M(2, 3) &= vw(1 - \cos\theta) + u\sin\theta \\ M(3, 1) &= uw(1 - \cos\theta) + v\sin\theta \\ M(3, 2) &= vw(1 - \cos\theta) - u\sin\theta \\ M(3, 3) &= (1 - w^2)\cos\theta + w^2 \end{aligned} \tag{15}$$

A deviation matrix M_{dev} of an interface is determined from its misorientation matrix M_{real} and that of the ideal CSL boundary M_{CSL}. The deviation matrix captures both the angle and axis deviation. From the deviation matrix, deviation angle Δθ is calculated.

$$M_{dev} = M_{CSL}^{-1} M_{real} \tag{16}$$

$$\Delta\theta = \cos^{-1}((\text{tr}(M_{dev} - 1))/2) \tag{17}$$

Using the angle-axes of CSL boundaries, misorientation matrices M_{CSL} are determined. The misorientation matrices change with the variant of the axis $\langle uvw \rangle$. Hence the matrix for each variant of the axis have to be calculated.

Table 2 An example look-up table showing the minimum and maximum accepted angle of misorientation of $\Sigma 3^n$ boundaries

GB	Min., °	Max., °
Σ3	58	62
Σ9	37	42
Σ27a	29	34
Σ27b	34	37

Algorithm 2 Determining the deviation of an interface from an ideal $\Sigma 3^n$ misorientation.

Input : Misorientation of the interface M_{min}, misorientation matrices of $\Sigma 3^n$ boundaries.
Output : Deviation from coincidence, °
Store : The misorientation matrix of Σ3, Σ9, Σ27a and Σ27b pertaining to all the variants of axis of misorientation in separate multi-dimensional arrays $cslmatrix_c$ of size $3 \times 3 \times v$. The suffix c denotes 3, 9, 27a or 27b, while v denotes the no. of axis variants. For example, the misorientation matrices of Σ3 boundary are stored in the array $cslmatrix_3$ of size $3 \times 3 \times 8$.
Generate : A look-up table *cslangle* containing the $\Sigma 3^n$ boundaries with corresponding minimum and maximum accepted angles of misorientation. An example is shown in Table 2
Assign : The matrix M_{min} computed as described in Algorithm 1 to M_{real}
Determine : The possible character of the interface from the value of θ using the look- up table *cslangle*. The corresponding array of misorientation matrices, $cslmatrix_c$ will be used. For example, if the value of θ is 58.5°, it may have a character of Σ3. $cslmatrix_3$ will be employed in the calculation. The variable v takes the value of 8.
for $i = 1$ to v **do**
Assign: The value of $cslmatrix_c$ (:,:, v) to M_{CSL}
Compute : Deviation matrix M_{dev} using Eq. 16
Compute: Angle of deviation using Eq. 17 and store in an array dev(v)
end for
Assign : The minimum value of array dev(v) to Δθ
Return : The angular deviation of the interface from ideal CSL misorientation, Δθ.

References

1. Verlinden B, Driver J, Samajdar I, Doherty RD (2007) Thermo-mechanical processing of metallic materials, 1st edn. Elsevier, ISBN 9780080444970
2. Flewitt PEJ, Wild RK (2001) Grain boundaries: their microstructure and chemistry. John Wiley & Sons Ltd., England, ISBN 0-471-97951-1

3. Engler O, Randle V (2008) Introduction to texture analysis. Taylor & Francis, ISBN 978-1-84816-115-3
4. Watanabe T (1984) An approach to grain boundary design for strong and ductile polycrystals. Res Mech 11:47–84
5. Randle V (1996) The role of coincidence site lattice in grain boundary engineering. The University Press, Cambridge
6. Palumbo G, King PJ, Aust KT, Erb U, Lichtenberger PC (1991) Grain boundary design and control for intergranular stress-corrosion resistance. Scr Metal Mat 25:1775–1780
7. Lehockey EM, Limoges D, Palumbo G, Sklarchuk J, Tomantschger K, Vincze A (1999) On improving the corrosion and growth resistance of positive Pb-acid battery grids by grain boundary engineering. J Power Sour 78:79–83
8. Thaveeprungsriporn V, Was GS (1997) The role of coincidence-site-lattice boundaries in creep of Ni-16Cr-9Fe at 360 °C. Metal Mater Trans A 28:2101–2112
9. Lehockey EM, Palumbo G (1997) On the creep behaviour of grain boundary engineered nickel. Mater Sci Eng A 237:168–172
10. Kobayashi S, Hirata M, Tsurekawa S, Watanabe T (2011) Grain boundary engineering for control of fatigue crack propagation in austenitic stainless steel. In Procedia Eng 10:112–117
11. Gao Y, Kumar M, Nalla RK, Ritchie RO (2005) High-cycle fatigue of nickel-based superalloy ME3 at ambient and elevated temperatures, Role of grain-boundary engineering. Metal Mater Trans A 36A:3325–3333
12. RamReddy K, Kumar EN, Jeyaraam R, Ram DJ, Sarma VS, Vedantam S (2019) Effect of grain boundary character distribution on weld heat-affected zone liquation cracking behavior of AISI 316Ti austenitic stainless steel. Mater Char 152:115–123
13. Qian M, Lippold JC (2003) The effect of annealing twin-generated special grain boundaries on HAZ liquation cracking of nickel-base superalloys. Acta Mater 51:3351–3361
14. Sinha S, Kim DI, Fleury ESSE, Suwas S (2014) Effect of grain boundary engineering on the microstructure and mechanical properties of copper containing austenitic stainless steel. Mater Sci Eng A 626:175–185
15. Lim LC, Watanabe T (1990) Fracture toughness and brittle-ductile transition controlled by grain boundary character distribution (GBCD) in polycrystals. Acta Metal Mater 38:2507–2516
16. Watanabe T (2011) Grain boundary engineering: historical perspective and future prospects. J Mater Sci 46:4095–4115
17. Randle V (1999) The effects of thermomechanical processing on interfacial crystallography in metals. Mater Sci For 294–296:51–58
18. Guyot B, Lee SL, Richards NL (2005) Effect of small strain levels on special boundary distribution in commercially pure nickel. J Mater Eng Perfor 14:85–90
19. Guyot BM, Richards NL (2005) A study on the effect of cold rolling and annealing on special grain boundary fractions in commercial-purity nickel. Mater Sci Eng A 395:87–97
20. Owen G, Randle V (2006) On the role of iterative processing in grain boundary engineering. Scr Mater 55:959–962
21. Moelans N, Blanpain B, Wollants P (2008) An introduction to phase- field modeling of microstructure evolution. CALPHAD Comput Coupling Phase Diagr Thermochem 32:268–294
22. Anderson MP, Srolovitz DJ, Grest GS, Sahni PS (1984) Computer simulation of grain growth-I Kinetics. Acta Metal 32:783–791
23. Liu Y, Baudin T, Penelle R (1996) Simulation of normal grain growth by cellular automata. Scr Mater 34:1679–1683
24. Suwas S, Ray RK (2014) Deformation textures. ISBN 9781447163138. https://doi.org/10.1007/978-1-4471-6314-5
25. Read WT, Shockley W (1950) Dislocation models of crystal grain boundaries. Phy Rev 78:275–289
26. Kronberg ML, Wilson FH (1949) Secondary recrystallization in copper. Met Trans 185:501–514

27. Brandon DG, Ralph B, Ranganathan S, Wald MS (1964) A field ion microscope study of atomic configuration at grain boundaries. Acta Metal 12:813–821
28. Brandon DG (1966) The structure of high-angle grain boundaries. Acta Metal 14:1479–1484
29. Gleiter H, Pumphrey P (1976) Grain boundary structure. Mat Sci Eng 25:159–164
30. Palumbo G, Aust KT (1990) Structure-dependence of inter granular corrosion in high purity nickel. Acta Metal Mat 38:2343–2352
31. Thomson CB, Randle V (1996) A study of twinning in nickel. Scr Mat 35:385–390
32. Randle V, Owen G (2006) Mechanisms of grain boundary engineering. Acta Mater 54:1777–1783
33. Prithiv TS, Bhuyan P, Pradhan SK, Sarma VS, Mandal S (2018) A critical evaluation on efficacy of recrystallization vs. strain induced boundary migration in achieving grain boundary engineered microstructure in a Ni-base superalloy. Acta Mater 146:187–201
34. Lee SL, Richards NL (2005) The effect of single-step low strain and annealing of nickel on grain boundary character. Mater Sci Eng A 390:81–87
35. Lee SL, Richards NL (2005) Influence of long term annealing on grain boundary character distributions in nickel. Mater Sci Eng A 405:74–85
36. De PS, Vuppuluri A, Vadlamani SS, Vedantam S (2021) A physically based model of the effect of recovery and clustering on recrystallization kinetics. J Mater Sci 56:7082–7093
37. Humphreys F, Hatherly M (2004) Recrystallization and related annealing phenomena, 2nd edn. Elsevier, ISBN 9780080441641
38. Wright SI, Nowell MM, Field DP (2011) A review of strain analysis using electron backscatter diffraction. Microsco Microanal 17:316–329
39. Miyazawa K, Iwasaki Y, Ito K, Ishida Y (1996) Combination rule of Σ values at triple junctions in cubic polycrystals. Acta Cryst 52:787–796
40. Randle V (2004) Twinning-related grain boundary engineering. Acta Mater 52:4067–4081
41. Jeyaraam R, Vedantam S, Sarma VS (2019) On the significance of misorientation axes of CSL boundaries in triple junctions in cubic materials. Mater Charac 152:276–281
42. Bozzolo N, Bernacki M (2020) Viewpoint on the formation and evolution of annealing twins during thermomechanical processing of FCC metals and alloys. Metal Mat Trans A 51:2665–2684
43. Porter DA, Easterling KE (1992) Phase transformations in metals and alloys, 2nd edn. Springer, ISBN 9780442316389
44. Liu T, Xia S, Zhou B, Bai Q, Rohrer GS (2018) Three-dimensional study of twin boundaries in conventional and grain boundary- engineered 316L stainless steels. J Mater Res 33:1742–1754. ISSN 20445326.
45. Burgers WG (1946) Stimulation crystals and twin-formation in recrystallized aluminium. Nature 157:76–77
46. Nielsen JP (1967) The origin of annealing twins. Acta Metal 15:1083–1085
47. Dash S, Brown N (1963) An investigation of the origin and growth of annealing twins. Acta Metal 11:1067–1075
48. Carpenter HCH, Tamura S (1926) Experiments on the production of large copper crystals. Proc R Soc A Math Phys Eng Sci 113
49. Gleiter H (1969) The formation of annealing twins. Acta Metal 17:1421–1428
50. Pande CS, Imam MA, Rath BB (1990) Study of annealing twins in fcc metals and alloys. Metal Trans A 21:2891–2896
51. Meyers MA, Murr L (1978) A model for the formation of annealing twins in FCC metals and alloys. Acta Metal 26:951–962
52. Goodhew PJ (1979) Annealing twin formation by boundary dissociation. Met Sci 13:108–112
53. Lin B (2015) Investigating annealing twin formation mechanisms in face-centered cubic nickel. PhD thesis, Carnegie Mellon University, Pittsburgh, PA, USA
54. Cahoon JR, Li Q, Richards NL (2009) Microstructural and processing factors influencing the formation of annealing twins. Mater Sci Eng A 526:56–61
55. Jin Y, Lin B, Bernacki M, Rohrer GS, Rollett AD, Bozzolo N (2014) Annealing twin development during recrystallization and grain growth in pure nickel. Mater Sci Eng A 597:295–303

56. Song KH, Chun YB, Hwang SK (2007) Direct observation of annealing twin formation in a Pb-base alloy. Mater Sci Eng A 454–455:629–636
57. Lin B, Rohrer GS, Rollett AD, Jin Y, Bozzolo N, Bernacki M (2013) Evolution of microstructure in pure nickel during processing for grain boundary engineering. Mater Sci For 753:97–100
58. Mahajan S, Pande CS, Imam MA, Rath BB (1997) Formation of annealing twins in FCC crystals. Acta Mater 45:2633–2638
59. Bozzolo N, Jacomet S, Loge RE (2012) Fast in-situ annealing stage coupled with EBSD, a suitable tool to observe quick recrystallization mechanisms. Mater Char 70:28–32
60. Li Q, Cahoon JR, Richards NL (2009) Effects of thermo-mechanical processing parameters on the special boundary configurations of commercially pure nickel. Mater Sci Eng A 527:263–271
61. Charnock W, Nutting J (1967) The effect of carbon and nickel upon the stacking-fault energy of iron. Met Sci J 1:123–127
62. Rath BB, Imam MA, Pande CS (2000) Nucleation and growth of twin interfaces in Fcc metals and alloys. Mater Phys Mech 1:61–66
63. Pande CS, Ashraf Imam M (2009) Grain growth and twin formation in boron-doped nickel polycrystals. Mater Sci Eng A 512:82–86
64. Wang X, Dallemagne A, Hou Y, Yang S (2016) Effect of thermomechanical processing on grain boundary character distribution of hastelloy X alloy. Mater Sci Eng A 669:95–102
65. Jin Y, Bernacki M, Roher GS, Rollett AD, Lin B, Bozzolo N (2013) Formation of annealing twins during recrystallization and grain growth in 304L austenitic stainless steel, vol 753. In: 5th International conference on recrystallization and grain growth, pp 113–116
66. Gertsman VY (2001) Geometrical theory of triple junctions of CSL boundaries. Acta Cryst 57:369–377
67. Gertsman VY, Tangri K (1995) On the relationship between $\Sigma 3^n$ boundaries meeting at a triple junction. Scr Met Mat 32:1649–1652
68. Padmanabhan KA, Gleiter H (2014) On the structure of grain/interphase boundaries and interfaces. Beilstein J NANO 5:1603–1615
69. Gertsman VY, Tangri K (1991) A study of grain boundary statistics in 304 and 316L stainless steels. Philos Magaz A Phys Conden Matter Struct Defects Mech Properti 64:1319–1330
70. Wright SI, Larsen RJ (2002) Extracting twins from orientation imaging microscopy scan data. J Micro 205:245–252
71. Cayron C (2011) Quantification of multiple twinning in face centred cubic materials. Acta Mater 59:252–262
72. Cayron C (2007) Multiple twinning in cubic crystals, Geometric/algebraic study and its application for the identification of the $\Sigma 3^n$ grain boundaries. Acta Cryst 63:11–29
73. Jin Y (2014) Annealing twin formation mechanism. PhD thesis, Carnegie Mellon University
74. Groeber MA, Jackson MA (2014) DREAM.3D: A digital representation environment for the analysis of microstructure in 3D. Integr Mater Manufact Innov 3:5
75. Emiliani M, Richman M, Brown R (1990) Characterization of sputtered inconel 617. J Mater Sci 25:144–148. ISSN 00222461
76. Thota H, Jeyaraam R, Bairi LR, Tirunilai AS, Kauffmann A, Freudenberger J, Heilmaier M, Mandal S, Vadlamani SS (2021) Grain boundary engineering and its implications on corrosion behavior of equiatomic CoCrFeMnNi high entropy alloy. J Alloy Compd 888:161500
77. Reed BW, Minich RW, Rudd RE, Kumar M (2004) The structure of the cubic coincident site lattice rotation group. Acta Cryst 60:263–277
78. Reed BW, Kumar M (2006) Mathematical methods for analyzing highly- twinned grain boundary networks. Scr Mater 54:1029–1033
79. Kumar M, Schuh CA (2006) Preface to the viewpoint set on grain boundary engineering. Scr Mater 54:961–962
80. Bober DB, Lind J, Mulay RP, Rupert TJ, Kumar M (2017) The formation and characterization of large twin related domains. Acta Mater 129:500–509

81. Barr CM, Thomas S, Hart JL, Harlow W, Anber E, Taheri ML (2018) Tracking the evolution of intergranular corrosion through twin-related domains in grain boundary networks. Mater Degrad 2:14
82. Randle V (2010) Role of grain boundary plane in grain boundary engineering. Mater Sci Technol 26:774–780
83. Priester L (1989) Geometrical speciality and special properties of grain boundaries. Sci-New York 24:419–438
84. Osher S, Sethian JA (1988) Fronts propagating with curvature- dependent speed: algorithms based on Hamilton-Jacobi formulations. J Comput Phys 79:12–49
85. Rahimi S, Engelberg DL, Duff JA, Marrow TJ (2009) In situ observation of intergranular crack nucleation in a grain boundary controlled austenitic stainless steel. J Microsco 233:423–431
86. Randle V (1998) The role of the grain boundary plane in cubic polycrystals. Acta Mater 46:1459–1480
87. Thomson CB, Randle V (1997) Fine tuning at Σ3n boundaries in nickel. Acta Mater 45:4909–4916
88. Randle V, Rohrer GS, Miller HM, Coleman M, Owen GT (2008) Five-parameter grain boundary distribution of commercially grain boundary engineered nickel and copper. Acta Mater 56:2363–2373
89. Xia S, Li H, Liu TG, Zhou BX (2011) Appling grain boundary engineering to Alloy 690 tube for enhancing intergranular corrosion resistance. J Nucl Mater 416:303–310
90. Randle V (1999) Mechanism of twinning induced grain boundary engineering in low stacking fault energy materials. Acta Mater 47:4187–4196
91. King A, Johnson G, Engelberg D, Ludwig W, Marrow J (2008) Observations of intergranular stress corrosion cracking in a grain-mapped polycrystal. Science 321:382–385
92. Bobrowski P, Faryna M, Glowinski K (2016) Evaluation of grain boundary plane distribution in yttria stabilized polycrystalline zirconia based on 3D EBSD analysis. Mater Charac 122:137–141
93. Saylor DM, El-Dasher BS, Adams BL, Rohrer GS (2004) Measuring the five-parameter grain-boundary distribution from observations of planar sections. Metal Mater Trans A 35:1981–1989
94. Randle V (2006) Special boundaries and grain boundary plane engineering. Sci Mater 54:1011–1015
95. Sharma NK, Shekhar S (2017) Cut-off deviation for CSL boundaries in recrystallized face-centered cubic materials. Philos Magaz 6435:1–14
96. Sharma NK, Shekhar S (2020) New insights into the evolution of twin boundaries during recrystallization and grain growth of low-SFE FCC alloys. Mater Charac 159:110015
97. An D, Griffiths TA, Konijnenberg P, Mandal S, Wang Z, Zaefferer S (2018) Correlating the five parameter grain boundary character distribution and the intergranular corrosion behaviour of a stainless steel using 3D orientation microscopy based on mechanical polishing serial sectioning. Acta Mater 156:297–309
98. Jin Y, Bernacki M, Agnoli A, Lin B, Rohrer G, Rollett A, Bozzolo N (2015) Evolution of the annealing twin density during δ-Supersolvus grain growth in the nickel-based superalloy Inconel™ 718. Metals 6:5
99. Tang M, Reed BW, Kumar M (2012) Coarsening kinetics of topologically highly correlated grain boundary networks. J Appl Phys 112:043505
100. Jeyaraam R, Sarma VS, Vedantam S (2020) Phase field modelling of annealing twin formation, evolution and interactions during grain growth. Comput Mater Sci 182:109787
101. Humphreys FJ (2001) Grain and subgrains characterisation by electron backscatter diffraction. J Mater Sci 36:3833–3854
102. Raabe D (2002) Cellular automata in materials science with particular reference to recrystallization simulation. Ann Rev Mater Sci 32:53–76
103. Rollett AD, Srolovitz DJ, Doherty RD, Anderson MP (1989) Computer simulation of recrystallization in non-uniformly deformed metals. Acta Metal 37:627–639

104. Moelans N, Godfrey A, Zhang Y, Jensen DJ (2013) Phase-field simulation study of the migration of recrystallization boundaries. Phys Rev B 88:1–10
105. Moelans N, Zhang YB, Godfrey A, Jensen DJ (2015) A phase-field simulation study of irregular grain boundary migration during recrystallization. IOP Conf Ser Mat Sci Eng 89. ISSN 1757899X
106. Athreya CN, Mukilventhan A, Suwas S, Vedantam S, Sarma VS (2018) Influence of the mode of deformation on recrystallisation behaviour of titanium through experiments, mean field theory and phase field model. Model Simul Mater Sci Eng 26:035004
107. Lens A, Maurice C, Driver JH (2005) Grain boundary mobilities during recrystallization of Al-Mn alloys as measured by in situ annealing experiments. Mater Sci Eng A 403:144–153
108. Olmsted DL, Holm EA, Foiles SM (2009) Survey of computed grain boundary properties in face-centered cubic metals-II: Grain boundary mobility. Acta Mater 57:3704–3713
109. Jin Y, Lin B, Rollett AD, Rohrer GS, Bernacki M, Bozzolo N (2015) Thermo- mechanical factors influencing annealing twin development in nickel during recrystallization. J Mater Sci 50:5191–5203
110. Bhattacharya A, Shen YF, Hefferan CM, Li SF, Lind J, Suter RM, Rohrer GS (2019) Three-dimensional observations of grain volume changes during annealing of polycrystalline Ni. Acta Mater 167:40–50
111. Kelly MN, Glowinski K, Nuhfer NT, Rohrer GS (2016) The five parameter grain boundary character distribution of α-Ti determined from three-dimensional orientation data. Acta Mater 111:22–30
112. Echlin MP, Straw M, Randolph S, Filevich J, Pollock TM (2015) The TriBeam system: femtosecond laser ablation in situ SEM. Mater Charac 100:1–12
113. Randolph SJ, Filevich J, Botman A, Gannon R, Rue C, Straw M (2018) In situ femtosecond pulse laser ablation for large volume 3D analysis in scanning electron microscope systems. J Vac Sci Tech B 36:06JB01
114. Yang M, Wang L, Yan W (2021) Phase-field modeling of grain evolutions in additive manufacturing from nucleation, growth, to coarsening. Comput Mater 7:56
115. Kim HK, Kim SG, Dong W, Steinbach I, Lee BJ (2014) Phase-field modeling for 3D grain growth based on a grain boundary energy database, Model. Simul Mater Sci Eng 22:034004

Chapter 7
Controlling the Twinning-Induced Grain Boundary Character Distribution

Shuang Xia and Qin Bai

1 Introduction

Grain boundaries (GBs) are recognized as weak positions that are susceptible to fast diffusion of elements, segregation of solutes and impurities, precipitation of secondary phase particles, intergranular corrosion and cracking. The existence of GBs in polycrystal make it behave very much differently than single crystal, even though they have the same composition and crystal structure. Not only GB density (inversely proportional to grain size) but also GB structure have very big influence on properties of polycrystalline materials.

At least five parameters are needed to define the geometry of a GB structure: the relative misorientation (3 parameters) between the two grains adjoining the boundary and the GB plane normal directions with respect to the two adjoining grains (2 parameters). In most cases, GBs are not straight neither in two-dimensional (2D) plane sections nor in three-dimensional (3D) spaces. An identical misorientation can be used to describe a GB regardless of the position and shape of the GB plane. Misorientation is often defined by the rotation from the coordinate of the first grain to the second grain, by using three Euler angles or by using a pair of rotation axis and angle, by which the first lattice rotates to coincident with the second one. While

S. Xia (✉) · Q. Bai
School of Materials Science and Engineering, Shanghai University, Shanghai, PR China
e-mail: xs@shu.edu.cn

Q. Bai
e-mail: baiqin31@shu.edu.cn

Q. Bai
Key Laboratory for Microstructures, Shanghai University, Shanghai, PR China

S. Xia
State Key Laboratory of Materials for Advanced Nuclear Energy, Shanghai University, Shanghai, PR China

S. Suwas and D. P. Field (eds.), *Advances in Texture, Microtexture, and Allied Techniques*, Materials Horizons: From Nature to Nanomaterials,
https://doi.org/10.1007/978-981-96-5346-1_7

no identical crystallographic plane can be used to define a curved GB. Therefore, the full five-parameter description is only applicable to straight GBs or straight segments of a GB. On the other hand, in 2D characterizations, inclination angle between the GB plane with respect to the surface is always unknown, resulting in the unknown GB plane orientations even if it is a straight GB. Therefore, GBs in research works are usually described by some of the five parameters. In most cases, only the misorientation (3 parameters) is used to describe GB structure by ignoring the GB plane orientations.

With the help of EBSD, misorientations across GBs can be directly obtained. A character can be assigned to a GB if the deviation between the measured misorientation and a misorientation defined in an already established model is smaller than a predefined tolerance (maximum allowed deviation). The coincidence site lattice (CSL) model is widely used in describing GB misorientations with Σ value representing the reciprocal density of coincidence sites [1]. The tolerances can be defined according to the wildly-used Brandon criterion [2], Palumbo-Aust criterion [3] or any other criterions that can be selected or self-defined by users in the EBSD analysis systems. A misorientation is designated with a Σ value if its deviation from the ideal CSL misorientation stays within the accepted tolerance. Automatic indexing EBSD systems dramatically facilitate the quantitative analyzing of relatively large amounts of GBs on 2D surfaces of polycrystalline materials. The GBCD describes the statistical proportions of different grain boundary types, represented either by length fraction or number fraction.

Different character GBs can have very big variation in proportion. This is due to not only the nature of the material but also the processing history. For example, in fcc materials with low stacking fault energy, such as brass and austenitic stainless steels, annealing twins are readily to appear in a deformed and subsequently annealed state resulting in relatively high proportion of Σ3 boundaries. While for the same materials, if in an as-cast state, annealing twins are less frequently observed. Aluminum or its alloy rarely have annealing twins neither in cast state nor in thermal-mechanically processed state, although they also have fcc structure. This is because of the much higher stacking fault energy of aluminum or its alloy.

Controlling the GBCD will dramatically alter the GB associated properties of materials [4]. Since Watanabe [5] introduced the approach of "GB design," further developed and recognized as grain boundary engineering (GBE) [6], GBCD has been proven to be a very important material parameter in deciding GB related properties. Twinning-induced GBE can be effectively utilized in FCC materials with low to medium stacking fault energies. The fraction of $\Sigma3^n$ (n = 1, 2, 3) type low ΣCSL ($\Sigma \leq 29$) GBs can be increased through multiple twinning via thermal–mechanical processing (TMP), resulting in great enhancement of GB related properties. Therefore, the twinning induced GBE is the most successful and promising one to be applied in industry.

There are two fundamental problems that attracting many attentions in the researches of twinning induced GBE: (1) the mechanism of $\Sigma3^n$ GBs enhancement through TMPs and (2) how the resulting gain boundary network affect material properties. The formation of large size twin-related domain [7] or highly-twinned grain-cluster [8] is now commonly recognized as the typical microstructure of twinning-induced GBE materials. Multiple twinning from a single nucleus gives rise to an extensively twinned grain cluster during recrystallization. The large size and complex shape of heavily twinned grain-clusters played crucial role in enhancement of GB related properties.

In this chapter, the author's and co-workers' practices of applying EBSD in charactering GBs during the investigation of problems of twinning-induced GBE are briefly reviewed. We will discuss the features of the microstructures produced by twinning induced GBE, the GBCD evolution during TMP, the connectivity of random GBN, the highly twinned microstructure as characterized with 3D EBSD, and finally the effects of highly twinned microstructure on some GB related properties. However, our work is based on many excellence results of other researchers. Although they are not included here due to the chapter length limitation, they paved the way for our work and gave us inspirations.

2 Microstructure Features of Twinning Induced GBE

2.1 Differences Between GBE and Non-GBE (Conventional) Microstructure

The GBE microstructure can be easily recognized from an optical micrograph which has abundant annealing twins that are connected with each other. Figure 1a, e shows optical micrographs of GBE and Non-GBE (conventional) Alloy 690 tube cross-sections [9]. Nickel-based Alloy 690 (Ni–30Cr–10Fe, wt.%) is employed as a steam generator tube material in pressurized water reactors, primarily for its superior resistance to intergranular stress corrosion cracking (IGSCC). Many band-morphology annealing twins appear in the GBE sample (Fig. 1a). These annealing twins are parallel to or have certain angles with each other, indicating specific crystallographic relationships among them. This microstructure contrasts with the Non-GBE sample (Fig. 1e), which consists solely of discrete intragranular twins.

Figure 1b and 1f illustrate the grain boundary network (GBN) with various GB types, as determined by EBSD for both specimens. The GBE-treated specimen shows clusters of Σ3 (red), Σ9 (blue), and Σ27 (green) grain boundaries. These $\Sigma 3^n$ (n = 1, 2, 3, …) boundaries link together, forming triple junctions like Σ3–Σ3–Σ9 and Σ3–Σ9–Σ27, while being enclosed by random grain boundaries (black). The GBE tube specimen consists of approximately 75% low Σ CSL grain boundaries, predominantly $\Sigma 3^n$ grain boundaries. Figure 1f illustrates the GBN of the Non-GBE specimen, which contains a smaller number of higher-order twin boundaries (Σ9 and Σ27), having approximately 50% of low Σ CSL grain boundaries.

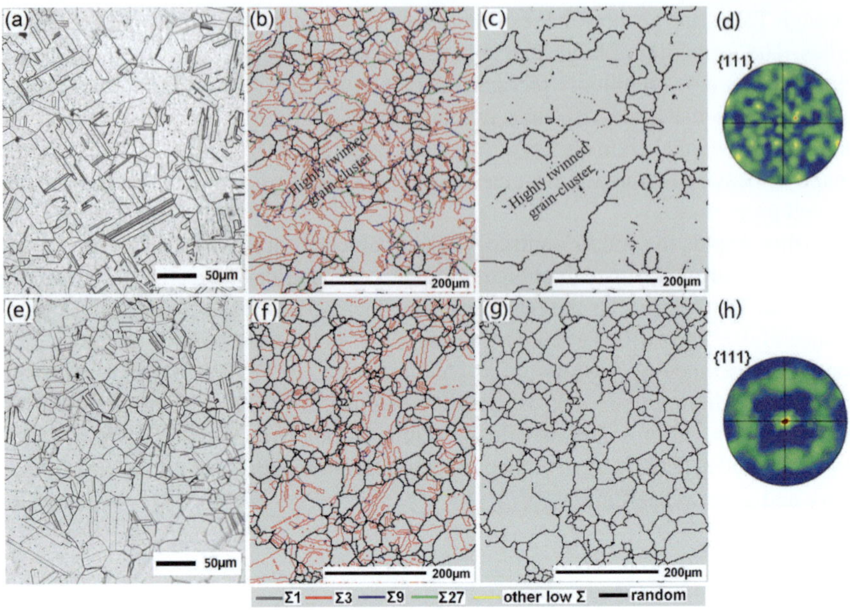

Fig. 1 The optical micrographs, GBN maps, random GB maps and {111} pole figures of GBE (**a**, **b**, **c** and **d**) and Non-GBE (**e**, **f**, **g** and **h**) tube specimens of Alloy 690 [9]

The random GBs are extracted from GBN and shown in Fig. 1c and g. The Non-GBE specimen exhibits a significantly higher density of random GBs compared to the GBE specimen. The grain assembly encircled by random GBs was termed a grain-cluster. Highly-twinned grain-clusters, which is also termed as twin related domain (TRD), can be observed in GBE specimen, and one of them is labeled in Fig. 1b and c. The grain-cluster features grains with $\Sigma 3^n$ mutual misorientations, while the outer boundaries are crystallographically random in relation to neighboring grains [10]. This feature will be discussed in a further detail in the following sections.

Figure 1d and h shows the {111} pole figures of the GBE and NonGBE tube specimens. It is obviously that the NonGBE have strong fiber textures with {111} and {001} components parallel to the tube axis. This is due to the processing history that comprised several cycles of heavy cold rolling and intermediate heat treatment, which normally produces strong fiber texture of tubes. While for the GBE tube, the formation of quite large population of annealing twins resulted in a quite random texture [11].

Therefore, the twinning induced GBE microstructure differs very much from that of convectional materials at least in four prospects: (1) the appearance of large amount interconnected annealing twins; (2) high proportion of Σ3 boundaries, normally higher than 60% in length, and relatively high proportion of twin variants, with the summation of Σ9 and Σ27 around 10%; (3) the development of big, heavily twinned grain-clusters (or TRD); (4) randomized texture.

2.2 Method for Determining High Order Twin Relationships

A microstructure composed of large, highly twinned grain-clusters is a typical characteristic of GBE materials. In this type of grain-cluster, every grain exhibits $\Sigma 3^n$ reciprocal misorientations, whether or not they are direct neighbors. This section briefly describes the method to identify high order twin relationships through twin chain analysis [8].

A grain-cluster comprising 91 grains in a GBE Alloy 690 specimen is shown in Fig. 2. Within these grains, a fully connected line can be constructed using only Σ3 boundaries. For instance, from grains 1 to 91, one possible path is: 1 → 2 → 3 → 23 → 28 → 36 → 53 → 62 → 76 → 78 → 80 → 88 → 89 → 91, which crosses 13 Σ3 boundaries. Likewise, all grains in this cluster create a twin chain, as illustrated in Fig. 2b. In this sequence, short red lines connect adjacent grains that have a Σ3 twin misorientation.

However, in Fig. 2b, higher order twin relationships ($\Sigma 3^n$) between any two non-adjacent grains cannot be immediately determined. For instance, the path between grains 1 and 91 crosses 13 Σ3 boundaries; nevertheless, the resulting orientation relationship between these grains corresponds to $\Sigma 3^5$ rather than $\Sigma 3^{13}$, with a calculated misorientation of 36.3°/[245], classified as Σ 243d, as per the references [11, 12]. This occurs because some grains along the route share the same orientation. That is, after several twinning operations, some grains revert to their original orientation, for example grains 1 and 53, 2 and 23, and 88 and 91. Therefore, grain 1 can directly jump to grain 53 by skipping the in-between grains on the route, and then misorientation between grain 1 and 62 is equivalent to that between grain 53 and 62. In the same way, the misorientation between grain 80 and 91 is equivalent to that between grain 80 and 88. Thus, from grains 1 to 91, the chain is simplified to (1 → 2 → 3 → 23 → 28 → 36 → 53) → 62 → 76 → 78 → 80 → (88 → 89 → 91), and in the final form: 1 → 62 → 76 → 78 → 80 → 91, including 5 independent twinning operations. Therefore, the misorientation between grain 1 and 91 is $\Sigma 3^5$.

The grains in this cluster exhibit 28 distinct orientations, as illustrated in Fig. 2c. A simplified twin chain comprising these 28 orientations is shown in Fig. 2d. An orientation can have more than one grains that locate at different places and adjoin to different grains. Red lines in this sequence link adjacent orientations that maintain Σ3 twin relationships. The hierarchical structure of $\Sigma 3^n$ relationships is clearly discernible within this chain. Figure 2c provides the necessary reference to determine the misorientation between grains 1 and 91, which correspond to orientations S1 and S21, respectively. The path from orientation S1 to S21 is exclusively recognized as S1 → S5 → S8 → S20 → S22 → S21, as shown in Fig. 2d. Hence, the $\Sigma 3^n$ relationship between grains 1 and 91 is determined to be n = 5. This technique can similarly be applied to identify the $\Sigma 3^n$ relationships among other grains in the cluster.

There are 378 mutual misorientations between the 28 orientations. As shown in Fig. 2d, each of these 378 misorientations corresponds to a $\Sigma 3^n$ relationship. The twin relationship order distribution is shown in Fig. 2e, with the maximum twin order being 9 in this cluster. All 378 misorientations are of the $\Sigma 3^n$ type.

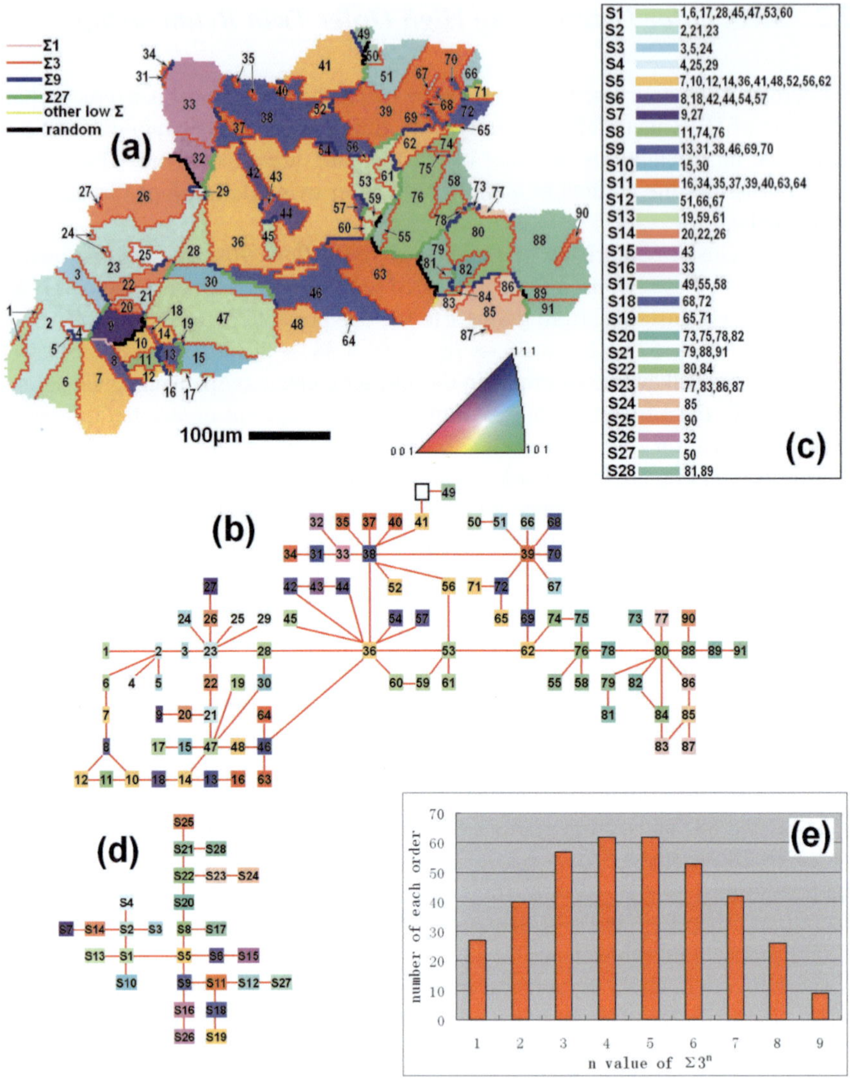

Fig. 2 Twin-chain analysis of a big highly-twinned grain-cluster in a GBE Alloy 690 specimen: **a** Orientation map of the grain cluster containing 91 grains; **b** Twin chain representation of these 91 grains, with red lines indicating Σ3 twin relationships; **c** Classification of the 91 grains into 28 distinct orientations; **d** Twin chain representation of the 28 orientations, where red lines denote Σ3 twin relationships; **e** Distribution of different orders of twin relationships [8]

2.3 High Order Twin Relationships Overlapping with Other Character Boundaries

Last section briefly describes the method for analyzing higher order twin relationships ($\Sigma 3^n$). This method is quite useful to analyze pseudo random GBs inside grain-clusters. These "random" GBs are not actually random, but higher order twin boundaries ($\Sigma 3^n$). Generally, commercial EBSD analysis software determine a GB character according to its misorientation (in the form of three Euler angles or rotation axis and angle pairs) [13, 14]. If a measured misorientation deviates less than the predefined tolerance of a misorientation of a GB character, the measured misorientation was then designated as this character. However, this method may cause mistakes, if applying to analyzing higher order twin relationships ($\Sigma 3^n$, $n > 3$). The following paragraphs will discuss this topic.

Inside the grain cluster in Fig. 2, some GBs are automatically identified as random by the EBSD system and presented in Table 1. In fact, these are higher-order $\Sigma 3^\mathrm{n}$ ($\mathrm{n} > 3$) grain boundaries, as identified through the orientation twin chain analysis in Fig. 2. According to the EBSD system, the boundary between grains 9 and 10 is classified as a random grain boundary. However, the higher order twin relationship between them is $\Sigma 3^4$. Since grains 9 and 10 are linked to orientations S7 and S5, correspondingly, as depicted in Fig. 2c, and the path from S7 to S5 is exclusively recognized as S7 $\rightarrow$ S14 $\rightarrow$ S2 $\rightarrow$ S1 $\rightarrow$ S5 as stated by Fig. 2d, which includes 4 independent twinning operations and hence a $\Sigma 3^4$ association among grain 9 and 10 is determined. Using the orientation twin chain in Fig. 2d, high-order $\Sigma 3^\mathrm{n}$ ($\mathrm{n} > 3$) misorientations can be quickly identified. Consequently, these "random" GBs within the cluster are fundamentally different from those formed randomly due to the encounter of grains with different orientations during recrystallization.

In addition to these pseudo-random GBs within the grain cluster, some high-order $\Sigma 3^\mathrm{n}$ ($\mathrm{n} > 3$) boundaries may be mistakenly identified as low Σ CSL GBs by the EBSD analysis system. As presented in Table 1 and illustrated in Fig. 2, the misorientation between grains 63 and 83 closely aligns with a $\Sigma 7$ CSL relationship, deviating by 2.6°. Likewise, the misorientation between grains 65 and 74 closely aligns with a $\Sigma 21\mathrm{b}$ CSL relationship, deviating by 2.4°. According to the Brandon criterion, both boundaries meet the criteria for low-Σ CSL grain boundaries. However, when the Palumbo–Aust criterion, with its tighter tolerances compared to the Brandon criterion, is applied, the boundary between grains 63 and 83 is the only one classified as a low-Σ CSL, whereas the boundary between grains 65 and 74 is not included in the low-Σ CSL classification. Although their classification as low Σ CSL boundaries, these grain boundaries actually are high-order $\Sigma 3^\mathrm{n}$ grain boundars, as demonstrated through the orientation twin chain analysis. This is due to the fact that certain high-order twin misorientations naturally coincide with low-Σ CSL misorientations within specific tolerances. Thus, the random GBs and low-Σ CSL GBs within the cluster are both products of multiple twinning events.

Table 1 Misorientations of 'random' grain boundaries and 'non-$\Sigma 3^n$ type' low-Σ CSL boundaries within the grain-cluster depicted in Fig. 2. Actually they are higher order $\Sigma 3^n$ type GBs [8]

Grain pair-(orientation pair)	Boundary character		Actual $\Sigma 3^n$ relationship	Measured misorientation
	Brandon criterion	Palumbo–Aust criterion		
9/10-(S7/S5)	Random		$\Sigma 3^4$	54.0°[2 −2 3]/ Σ81b
26/32-(S14/S26)	Random		$\Sigma 3^6$	53.7°[−3 −5 8]/ Σ729c
49/50-(S17/S27)	Random		$\Sigma 3^6$	28.9°[−18 5 −15]/Σ729r
41/50-(S5/S27)	Random		$\Sigma 3^4$	60.9°[4 4 −3]/ Σ81d
55/59-(S17/S13)	Random		$\Sigma 3^4$	38.6°[−1 5 3]/ Σ81c
55/63-(S17/S11)	Random		$\Sigma 3^4$	53.9°[3 2 −2]/ Σ81b
63/79-(S11/S21)	Random		$\Sigma 3^6$	54.3°[−7 13 20]/ Σ729c
85/89-(S24/S28)	Random		$\Sigma 3^4$	55.0°[22 15 15]/ Σ81b
8/9-(S6/S7)	Σ1		$\Sigma 3^5$	11.8°[10 −25 −7] /Σ243c
63/83-(S11/S23)	Σ7/2.6 °		$\Sigma 3^6$	39.1°[7 7 −8]/ Σ729a
65/74-(S19/S8)	Σ21b/2.4°	Random	$\Sigma 3^5$	43°[11 11 −20]/ Σ243a

Note The EBSD analysis system automatically determines the "Boundary character" column, based on the Brandon criterion [2] and the Palumbo–Aust criterion [3]. Orientation twin chain analysis is used to define the "Actual $\Sigma 3^n$ relationship" column. The "Measured misorientation" column is presented in the format: Misorientation/Closest higher-order twin relationship, following the designations of higher-order twin relationships by Gottstein [11] and Randle [12]

2.4 Dominant Area Orientations in Highly-Twinned Grain-Cluster

As shown in the previous sections, in the highly-twinned grain-cluster, some orientations have many grains, while some other orientations only have one grain. A face-centered cubic (fcc) grain has at most four first-order twinning orientations. However, it is important to note that the probabilities of forming these four orientations are not equal.

Figure 3 displays the distributions of the average area fraction and the grain count per orientation, including the number fraction, for 30 large, highly twinned grain clusters in GBE-treated Alloy 690 [17]. The five largest-area orientations,

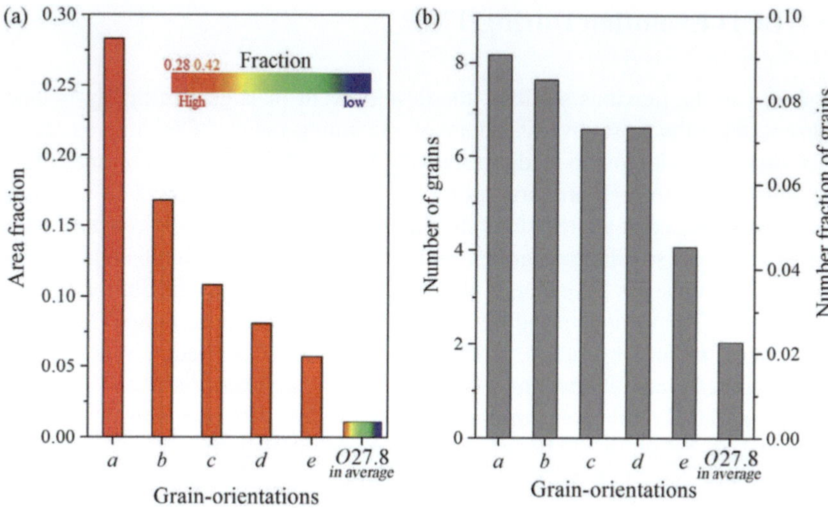

Fig. 3 Distribution of the average area fraction and the average number (number fraction) of grains per orientation in 30 highly twinned grain-clusters, where orientations a–e refer to the five largest-area orientations for each of the 30 grain-clusters; "O27.8" represents the mean value of the remaining 27.8 orientations, excluding orientations a–e, given that the mean orientation count per grain-cluster is 32.8 across the 30 grain-clusters

labelled "a" through "e," are individually presented, in contrast, the remaining orientations, which on average number 27.8, are grouped together into a single category. Figure 3a reveals that the area fractions of orientations within a grain cluster differ significantly. The five most dominant orientations collectively occupy an average of 69.9% of the grain-cluster area. The largest-area orientation, labeled "a," occupies an average of 28.3% of the grain-cluster area, while the remaining 27.8 orientations each cover significantly smaller fractions. Additionally, the five largest-area orientations account for 36.7% of the total number of grains in a grain-cluster, which is greater than the number contributed by any of the other 27.8 orientations. Thus, a grain-cluster has leading orientations that occupy the majority of both the area and the grains within the cluster. It is essential to recognize that the dominant orientations are linked to the grain-cluster rather than the entire specimen, and distinct grain-clusters within a specimen can possess unique dominant orientations. This observation indicates that preferential orientations emerge during multiple twinning, and twins with these orientations exhibit a greater tendency for growth. Additionally, most misorientations among these dominant orientations are of low-order $\Sigma 3^n$-type ($n = 1, 2$), while the four sub-dominant orientations in the grain-cluster predominantly arise as twinning variants of the primary dominant orientation. Twinning in crystalline materials produces twin orientations in both forward (higher) and backward (lower) directions, with a greater likelihood of backward twinning. Multiple twinning encourages the selection of a specific range of preferential orientations, contributing to the dominance of certain orientations in the grain-cluster.

3 GBCD Evolution During TMP

As shown in the previous sections, the development of large size highly-twinned grain-cluster is the most obvious feature of a twinning-induced GBE microstructure in fcc materials with low to medium stacking fault energy. This kind of microstructure corresponds to high proportion of $\Sigma 3^n$ (n = 1, 2, 3) type low Σ CSL GBs. It can be interpreted that a microstructure characterized by large grain clusters, which enclose relatively smaller twin grains within them, is likely to exhibit a high proportion of $\Sigma 3^n$ boundaries [18]. One of the fundamental challenges in Grain Boundary Engineering (GBE) is understanding how thermomechanical processing (TMP) influences microstructural evolution, ultimately leading to the formation of large-sized, highly twinned grain clusters and a correspondingly high fraction of $\Sigma 3^n$ (n = 1, 2, 3) type low-Σ CSL grain boundaries (GBs). Two widely discussed viewpoints in the literature [19–23] on the TMPs of twinning-induced GBE are as follows: (1) The fraction of $\Sigma 3^n$ (n = 1, 2, 3…) GBs can be altered in the GBCD, while other low-Σ CSL GBs remain rare, no matter the TMPs applied; and (2) the degree of deformation applied in TMPs is relatively low, typically ranging from 3 to 30%. Therefore, the focus shifts to understanding how a slight deformation followed by heat treatment leads to the development of a highly twinned grain-cluster microstructure and the corresponding high fraction of $\Sigma 3^n$ GBs.

3.1 *Multiple Twinning During Recrystallization*

The twin-chain, formed through continuous twinning (multiple twinning) originating from a single nucleus during recrystallization in FCC materials with low stacking fault energy, was extensively investigated during the 1980s and 1990s. With the help of high voltage electron microscopy (HVEM), Haasen, Kopezky, Berger, Wilbrandt and et al. [24–26] performed many excellent works on in-situ recrystallizing single crystals and growth competition experiments in order to better understand how new orientations develop during primary recrystallization. Through the process of twinning, new crystallographic orientations are introduced relative to the original recrystallized grain, facilitating the formation of higher mobility boundaries at the recrystallization front. These experiments revealed that Σ3 boundaries (first order annealing twins) and higher order twin boundaries $\Sigma 3^n$ are produced during primary recrystallization starting from a single nucleus. It is plausible that the microstructural evolution during annealing after strain plays a critical role in driving the changes observed in both the microstructure and the grain boundary character distribution (GBCD).

With the help of automatic indexing EBSD, the characterization of partially recrystallized microstructure may facilitate the understanding of the formation of highly-twinned grain-cluster. With controlled annealing time of deformed specimens, partially recrystallized microstructures can be obtained. The Alloy 690 specimens, which were annealed at 1100 °C for 3 min (specimen A*) and 1 min (specimen C*) after undergoing 5 and 20% cold rolling, respectively, were examined using EBSD and the resulting maps of average misorientation for individual grains are presented in Fig. 4 [27]. Based on the color code, it can be observed that the blue regions correspond to areas with smaller average misorientation within the grains, indicating recrystallized regions. In contrast, the un-recrystallized areas, which exhibit larger average misorientation within the grains, are represented by green or even warmer hues on the map. The recrystallized and un-recrystallized regions are delineated by random boundaries. In the un-recrystallized areas, the grain boundaries, including Σ1 boundaries (with misorientations ranging from 2 to 15°) and Σ3 boundaries, will be eliminated as the recrystallizing grains grow, leading to the consumption of these boundaries during the recrystallization process.

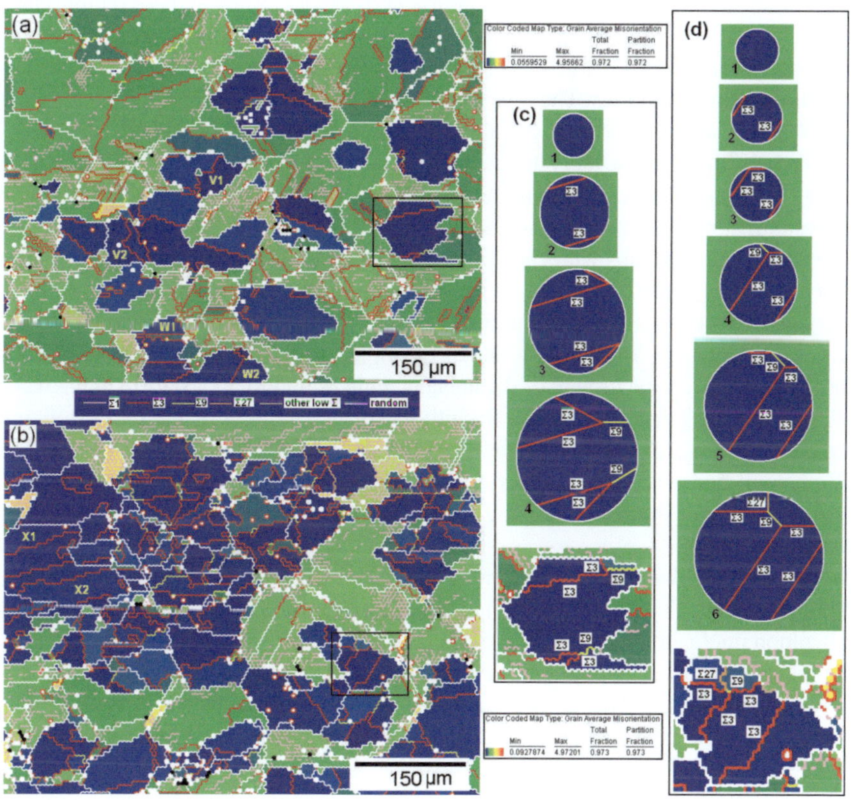

Fig. 4 The maps illustrating multiple twinning in the partially recrystallized specimens A* (5% cold rolled + 1,100 °C/3 min) and C* (20% cold rolled + 1,100 °C/1 min) are presented as follows: **a** and **b** show the grain average misorientation maps of specimens A* and C*, respectively. Meanwhile, **c** and **d** present schematic diagrams illustrating the potential multiple twinning processes within the two recrystallized regions marked by the black framework in **a** and **b** [27]

In specimen A*, most of the recrystallized grains are surrounded by non-recrystallized regions, as illustrated in Fig. 4a. It is evident that some of the recrystallized areas contain straight-sided Σ3 boundaries, which correspond to annealing twins. Furthermore, smaller recrystallized areas do not exhibit Σ3 boundaries, whereas larger recrystallized regions tend to have a higher concentration of Σ3 boundaries. This suggests that as recrystallization progresses, the growing grains consume the remaining un-recrystallized material, leaving behind a continuous sequence of twin boundaries following the movement of random high-angle boundaries that initially delineated the two regions. In the partially recrystallized microstructure, the twin chains are clearly distinguishable. In Fig. 4a, the twin chain from V1 to V2 crosses four twin boundaries, while the chain from W1 to W2 crosses five. In specimen C* (Fig. 4b), most of the areas appear blue, indicating a higher proportion of recrystallized regions. A significant number of grain clusters in these regions have already come into contact and interacted. In the recrystallized areas, the twin chains are readily discernible. For instance, the twin chain joining grain X1 and X2 encounters three twin boundaries. The misorientation characteristics of the grain pairs (V1/V2, W1/W2, and X1/X2) indicate alignment with $\Sigma3^n$ type boundaries.

The twin chains in the recrystallized regions are formed through multiple twinning events originating from each single orientation, or in other words, from a single recrystallization nucleus. As a result, all the grains within these chains exhibit $\Sigma3^n$ orientation relationships with each other. To further elucidate the multiple twinning processes, the potential formation mechanisms of the two recrystallized regions, marked by the black framework in Fig. 4a and b, are schematically shown in Fig. 4c and d, respectively. Take Fig. 4c as an example, step 1: recrystallization nucleus forms; step 2: with the growth of the nucleus, twin boundaries (Σ3) form in the wake of recrystallizing front GBs; step 3: with the further growth of the recrystallizing region, Σ3 boundaries form on other {111} planes; step 4: upon further growth, the Σ3 encountered with the previous Σ3 boundaries to generated Σ9 GBs. With the further growth of the recrystallizing region, large size highly-twinned grain-cluster will be produced.

The recrystallization regime during Thermomechanical Processing (TMP) for Grain Boundary Engineering (GBE) is schematically depicted in Fig. 5 [16]. Recrystallization nucleation is initiated by Strain induced Boundary Migration (SIBM). As the nucleus grows, it sweeps away the deformed matrix, and simultaneously, grain-clusters begin to form with the initial occurrence of twinning events. Following this, the grain-cluster continues to grow, with ongoing twinning events occurring in the wake of the migrating recrystallization front grain boundaries, which progressively sweep away the deformed matrix. This process continues until the entire volume is fully recrystallized. The resulting microstructure is characterized by large, highly twinned grain-clusters, each containing abundant internal $\Sigma3^n$ boundaries.

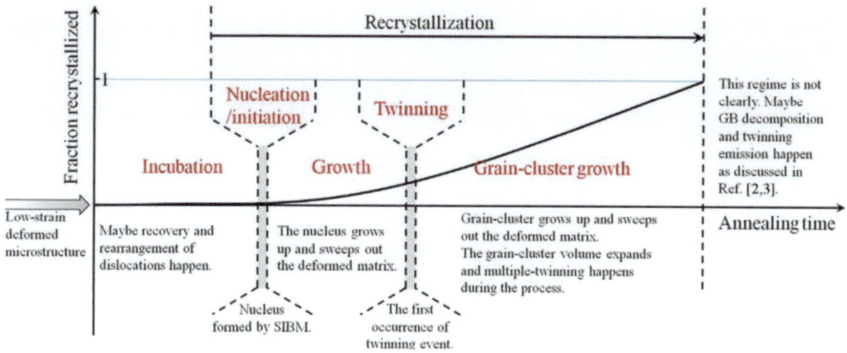

Fig. 5 Illustrative depiction of recrystallization stages during GBE-processing [16]

3.2 *Effects of Pre-Strain*

The influence of pre-straining, succeeded by brief high-temperature annealing, on the Grain Boundary Character Distribution (GBCD) and the resulting microstructure of Alloy 690 has been examined in the literature [27]. Plate samples were solution-annealed and processed by cold rolling with reductions between 5 and 50%, and then annealed at 1100 °C for 5 min before undergoing water quenching (WQ). The Grain Boundary Character Distribution (GBCD) of these samples is shown in Table 2. In the sample subjected to annealing at 1100 °C for 5 min following 5% cold rolling, fraction of low-Σ CSL grain boundaries increased to over 70%, consisting of approximately 60% Σ3 boundaries and 9% from the combined Σ9 and Σ27 boundaries. This suggests that low-level strain followed by high-temperature annealing can significantly enhance the creation of annealing twins in Alloy 690. As the cold rolling reduction increased to 35 and 50%, the fraction of low-Σ CSL grain boundaries declined considerably, with Σ3 boundaries accounting for approximately 40% and the combined Σ9 and Σ27 boundaries around 1%. This suggests that the likelihood of multiple twinning events decreases with higher pre-strain. The fraction of $\Sigma3^n$ boundaries diminished as the cold rolling reduction increased, causing a reduction in the proportion of low-Σ CSL boundaries. Therefore, the fraction of $\Sigma3^n$ boundaries is highly influenced by the level of pre-strain.

It is evident that small amount of pre-stain and subsequent annealing is beneficial to the formation of high proportion of $\Sigma3^n$ boundaries. Numerous studies have also employed small amounts of deformation, typically ranging from 3 to 30%, for the thermomechanical processing (TMP) of grain boundary engineering (GBE) [15, 20, 21, 23, 28]. The reason can be understood following Fig. 6. It is well known that lower pre-strain leads to a reduced nucleation density during recrystallization [29]. As a result, recrystallization nuclei in a specimen with lower pre-strain have a greater area to consume compared to those in a specimen with higher pre-strain. As the nucleus expands within this larger area, multiple twinning events can continuously form, leading to much larger grain clusters and a higher density of $\Sigma3^n$ boundaries

Table 2 The effect of pre-strain on the GBCD of Alloy 690 (the GB characters are defined according to Palumbo–Aust criterion) [27]

Cold rolling (%)	Annealing time (min) at 1100 °C	Σ 1(%)	Σ 3(%)	Σ9 + Σ27 (%)	Other low Σ CSL(%)	Total low ΣCSL (Σ≦29) (%)
–	15	3	48.5	2.5	0.5	54.9
5	5	1.4	60.1	9.3	0.5	71.3
10		2.3	54.2	7.2	0.8	64.5
20		2.5	48.8	3.7	0.5	55.5
35		4.2	40.3	1.2	1.2	46.9
50		4	39.8	1.5	1.4	46.7

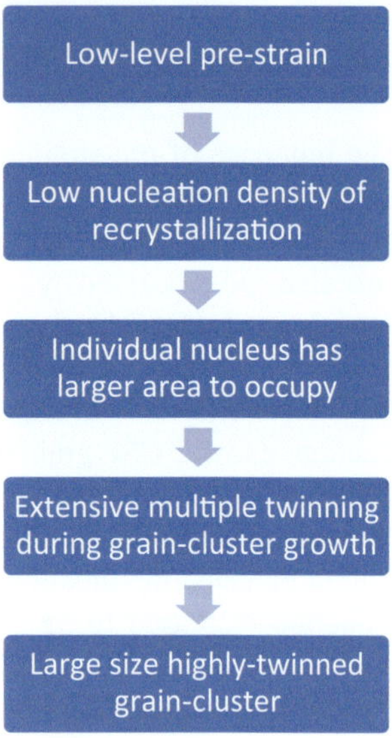

Fig. 6 Elucidation of the influence of small pre-strain on the development of large grain clusters

in a specimen with low pre-strain than in one with high pre-strain. Therefore, high proportion of $\Sigma 3^n$ GBs is formed in the specimen with low pre-strain. Low nucleation density of recrystallization is essential for achieving a GBE microstructure.

3.3 Effects of Annealing Temperature and Time

Recovery, recrystallization, and subsequent grain growth are likely to occur during annealing after deformation. The previous sections emphasized that multiple twinning events during recrystallization, originating from a single nucleus over a relatively large area, can result in the development of extensive, highly twinned grain clusters. Austenitic stainless steels and nickel-based alloys, which are suitable for the application of twinning induced GBE due to their low stacking fault energies, are required to be solution annealed as the final state of products to be delivered. The solution annealing temperatures for these materials are generally around 1020 °C or above. If lower temperatures are used, it is inevitable to cause incomplete dissolution or induce new precipitation of secondary phase particles, such as carbides, which would result in Cr depletion in the vicinity of GBs and then would cause intergranular sensitization. Therefore, annealing at solution annealing temperature is recommended to be used in the TMP of GBE.

The changes in Grain Boundary Character Distribution (GBCD) in nickel-based alloy 825 tubes, which were cold drawn by 3–10% and subsequently annealed at 1050–1125 °C for 10 min, are illustrated in Fig. 7 [31]. The fraction of low-Σ CSL boundaries was considerably enhanced after grain boundary engineering (GBE) processing. The fractions of low-Σ CSL boundaries decreased as pre-strain increased, which aligns with the trend discussed in earlier sections. The fraction of low-Σ CSL boundaries experienced only a modest decrease with increasing annealing temperature between 1050 and 1125 °C, especially in the case of the 5% pre-strained sample. In the meanwhile, the grain sizes increase. The minimal reduction in the proportion of low-ΣCSL boundaries suggests that twinning-induced GBE has a relatively broad range of annealing temperatures, making it advantageous for industrial applications.

However, during annealing after cold working, both recrystallization and subsequent grain growth are likely to occur [29], with their respective influences on GBCD development depicted in Figs. 8 and 9. The samples with 5% cold-drawing deformation underwent annealing at 1050 °C for 60 s ~ 420 s, respectively, and their GB, grain orientation, grain average misorientation and the recrystallization fraction maps are shown in Fig. 8. Figure 9 illustrates how annealing time influences the average grain size, recrystallization fraction, and proportion of low-Σ CSL grain boundaries in the specimen that underwent 5% strain before annealing at 1050 °C.

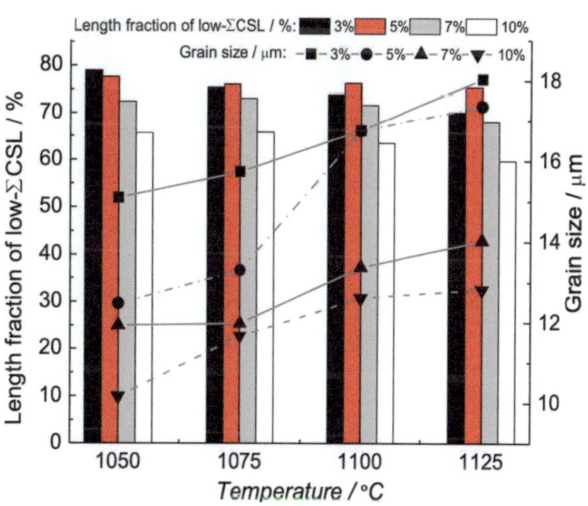

Fig. 7 The Influence of deformation and annealing temperature on low-Σ CSL boundary fractions and mean grain size in Alloy 825. The colored bars represent the length fraction of low-Σ CSL grain boundaries, while the points/line curves represent the grain size [31]

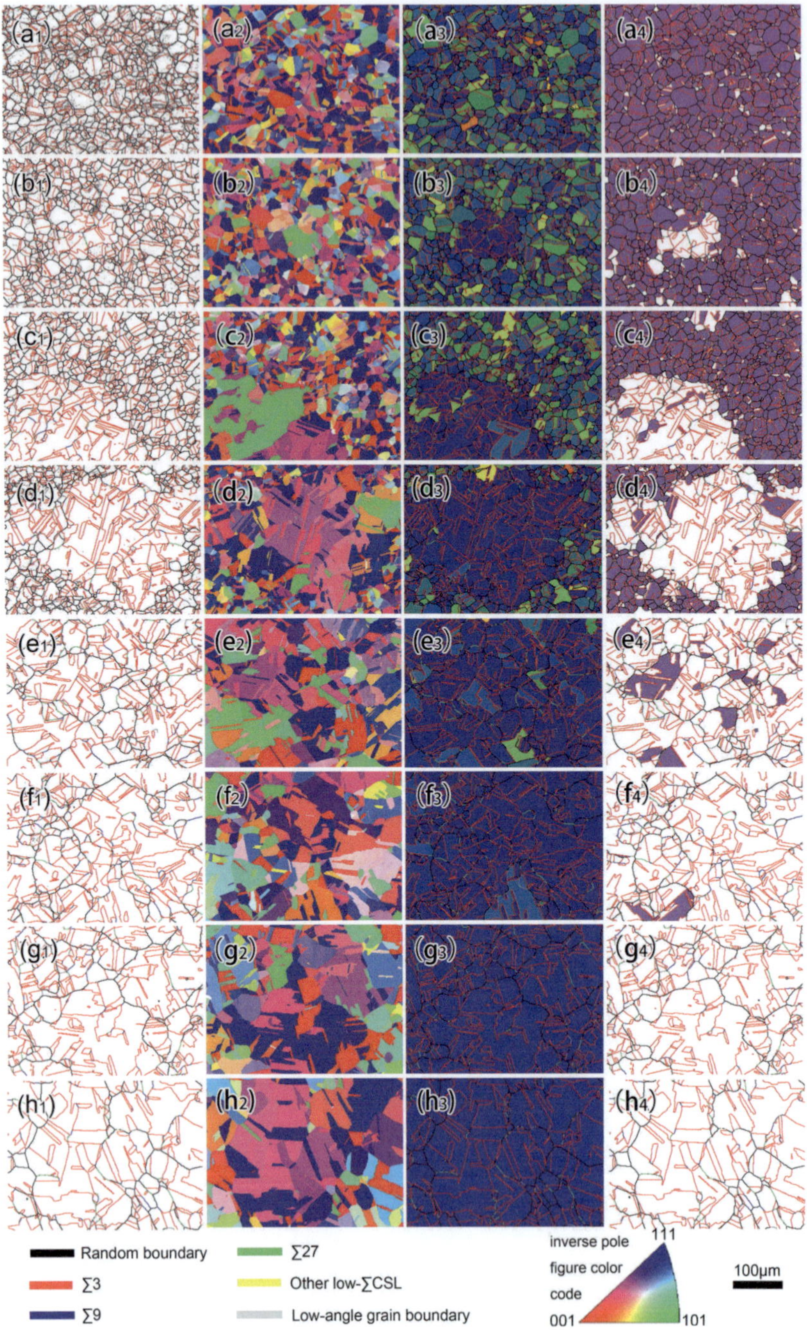
(a1)
(a2)
(a3)
(a4)
(b1)
(b2)
(b3)
(b4)
(c1)
(c2)
(c3)
(c4)
(d1)
(d2)
(d3)
(d4)
(e1)
(e2)
(e3)
(e4)
(f1)
(f2)
(f3)
(f4)
(g1)
(g2)
(g3)
(g4)
(h1)
(h2)
(h3)
(h4)
Random boundary
∑3
∑9
∑27
Other low-∑CSL
Low-angle grain boundary
inverse pole
figure color
code
111
001
101
100μm

◀**Fig. 8** EBSD examination of microstructural evolution during annealing at 1050 °C of alloy 825 tube samples subjected to 5% cold-drawing deformation. The annealing time are 60 s **a**, 90 s **b**, 120 s **c**, 150 s **d**, 180 s **e**, 210 s **f**, 240 s **g**, and 420 s **h**, respectively. Maps of various boundary types (a1–h1), grain orientation maps differentiated by inverse pole figure color (a2–h2), and maps colored by the mean grain misorientation (a3–h3) (The blue color indicates lower grain mean misorientation, signifying recrystallized grains), along with the component map showing the recrystallized fraction (a4–h4) (purple color represents non-recrystallized grains with a grain mean misorientation larger than 0.85°) [31]

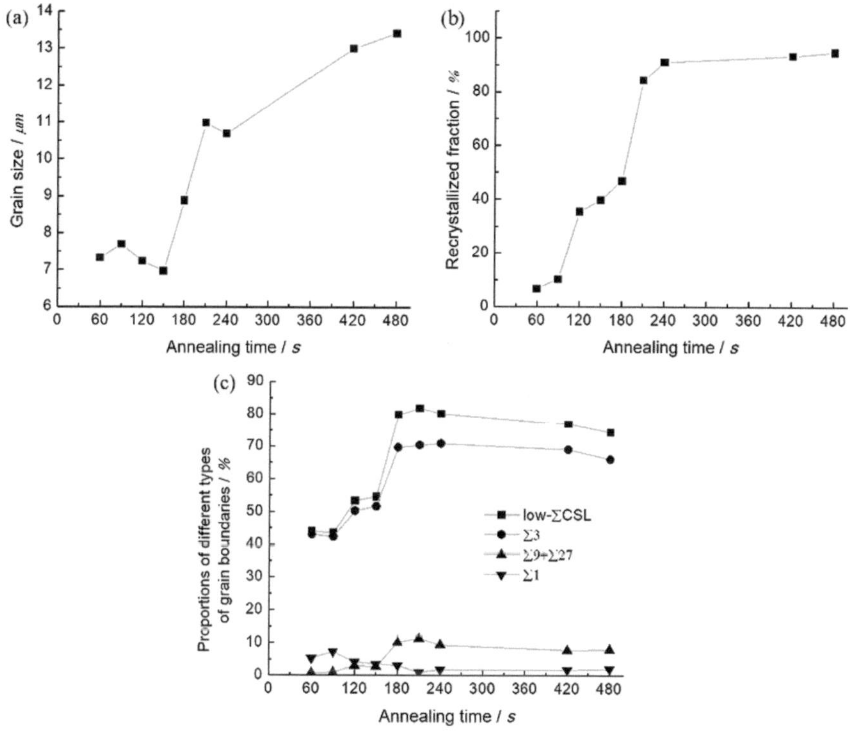

Fig. 9 Variation of average grain size **a**, recrystallization fraction **b**, and fraction of low Σ CSL GBs **c** in response to annealing time for specimens subjected to 5% strain followed by annealing at 1050 °C [31]

As depicted in the figures, increasing the annealing time from 60 to 420 s led to a clear progression in recrystallization and subsequent grain growth. In Fig. 8a1–a4, no recrystallization nuclei were present in the deformed microstructure. After extending the annealing time to 90 s, a small number of recrystallized grains appeared, with some Σ3 and Σ9 boundaries forming within the newly developed grain clusters. Recrystallized and non-recrystallized areas are separated by random GBs. An increase in annealing time leads to the growth of recrystallized grains, which progressively consume the deformed matrix. After 210 s, the new grains had fully replaced the microstructure of the specimen, signifying that the sample was nearly completely recrystallized. Meanwhile, as shown in Fig. 9, within the annealing time between 60 and 210 s, the fraction of Σ1 boundaries steadily decreased. However, the fraction of Σ3 grain boundaries increased significantly with longer annealing times, while the overall behavior of low-Σ CSL grain boundaries followed the same trend. The fraction of low-Σ CSL grain boundaries peaked at 210 s. After 210 s of annealing, the sample underwent a transition into the grain growth phase (post-recrystallization). As annealing time increased further, the proportions of Σ3 and low-Σ CSL grain boundaries steadily decreased.

Based on the analysis above, it can be further concluded that during recrystallization, $\Sigma 3^n$ grain boundaries are formed over multiple twinning linked to the expansion of recrystallization nuclei. After complete recrystallization, the fraction of low-Σ CSL grain boundaries reached its maximum. With extended annealing, the proportion of low-Σ CSL grain boundaries steadily decreased during the grain growth phase following recrystallization. In the course of grain growth process, the present low-Σ CSL grain boundaries are eliminated by the movement of general high-angle grain boundaries [30], resulting in a decrease in the proportion of low-Σ CSL grain boundaries as annealing time increases.

3.4 *Effects of Initial Microstructures*

The initial microstructures, including grain size, distribution of second-phase particles, and texture, significantly affect GBN development for the period of the TMP of GBE. This is also one of the reasons why different TMP approaches are applied to different materials. Hence, the TMP of GBE should be tailored to accommodate variations in initial microstructures.

3.4.1 Effects of Initial Grain Size

The size of the initial grains impacts the formation of grain clusters as well as the final grain (or twin) size after GBE [18]. It also has a combined influence on the proportion of low Σ CSL GBs, as depicted in Fig. 10. In the GBE microstructure, random GBs originate from the impingement of different grain clusters, while a grain cluster forms through multiple twinning initiated from a single recrystallization nucleus, as evident in the partially recrystallized microstructure discussed in previous sections. The nucleation density increases as the original grain size decreases, as GBs serve as preferred nucleation sites. Smaller initial grains lead to a higher density of nucleation sites. Consequently, the final GBN will exhibit smaller grain clusters and a higher density of random boundaries, which hinders the formation of a high proportion of low Σ CSL GBs. However, a smaller initial grain size also enhances the frequency of twinning. Through twinning, new orientations are introduced relative to the original recrystallized grain, facilitating the formation of high-mobility boundaries at the recrystallization front. As a result, in microstructures with smaller initial grain sizes,

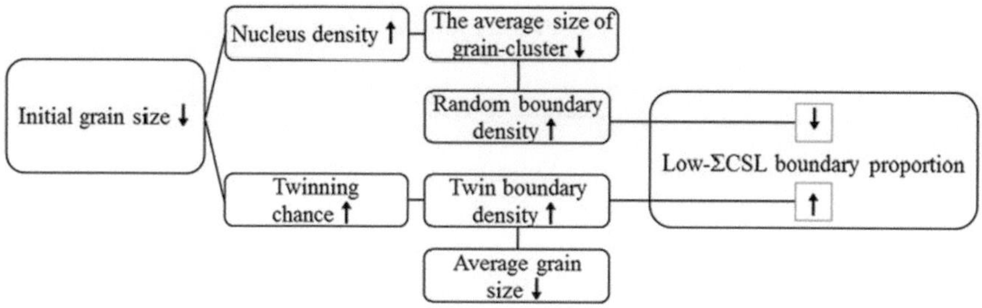

Fig. 10 Impact of initial grain sizes on GBE microstructures: grain-cluster size, individual grain size, and proportion of low ΣCSL GBs [18]

the misorientation of the advancing recrystallization front boundary changes more frequently during recrystallization [24], leading to a higher twin density. This factor favors the development of a high fraction of low Σ CSL GBs. Due to this collective influence, neither grain-cluster size nor twin boundary density alone is sufficient to determine the proportion of low Σ CSL GBs. It is evident that a GBN with larger grain clusters and smaller grains (smaller twins) is more likely to exhibit a higher proportion of low Σ CSL GBs. The ratio of grain-cluster size to grain size ultimately dictates the proportion of low Σ CSL GBs.

However, for materials intended for critical applications in demanding industrial environments, both the proportion of low Σ CSL GBs and the grain size must be taken into account, as excessively large grains can compromise mechanical properties. Hence, when applying the GBE technique, careful optimization is necessary to balance the fraction of low ΣCSL GBs with other microstructural parameters, such as grain size.

3.4.2 Effects of Second Phase Particles

(a) Secondary Carbides Precipitated at GBs

GBs serve as the preferred sites for the precipitation of second-phase particles, such as carbides in stainless steels and nickel-based alloys. GBE of these materials typically involves a small amount of deformation followed by annealing at a temperature similar to that used for solution annealing. When a material with pre-existing carbides at GBs undergoes GBE, the balance between carbide dissolution and GB movement determines the fraction of low Σ CSL GBs.

Aging-treated Ni-based Alloy 690 samples underwent GBE-type processing to examine the impact of pre-existing carbides at GBs on GBN development during GBE [32]. The results indicate that the recrystallization of the low-strain deformation microstructure is hindered or delayed by the carbides at the GBs in the course of annealing [32, 33]. In contrast to solution-annealed 690 samples, aging-treated samples containing pre-existing carbides at GBs need extended annealing times or elevated temperatures following low-strain deformation to increase the fraction of low-Σ CSL GBs. This is because the GBs are inhibited from moving until the carbides dissolve.

(b) Primary Carbides or Inclusions

Primary carbides are formed during solidification and have a much larger size than secondary carbides that precipitated at GBs. Furthermore, the primary carbides are very difficult to dissolve into the matrix upon heating. Hastelloy N, which is a high Mo containing nickel-based alloy, was subjected to GBE [34]. The microstructure of this material has many strings consisting of massive primary M_6C type carbide particles along the rolling direction. Concentrated smaller grains as a result of Particle Stimulated Nucleation (PSN) were found around the strings of primary carbide particles after the TMP of GBE, resulting in the increase of random GBs and the drop of the fraction of low Σ CSL GBs. The primary carbides not only hinder the migrations of GBs but also induce PSN, and hence have a detrimental effect on GBE.

4 Random Grain Boundary Connectivity

Random GBs are regarded as weak positions that are susceptible to intergranular degradations. Low Σ CSL GBs were regarded as "special GBs" as compared with random GBs. Many recent researches pointed out that only coherent Σ3 boundary (the coherent parts of annealing twin boundaries) is special due to its low index GB plane, i.e. on {111} planes [35]. Some metaphysical points of view believed that intergranular degradation can be terminated by segments of special boundaries which interrupt the connectivity of long paths of random GBs. However, many articles also show that highly-twinned grain-clusters or TRDs constitute the GBE microstructure [36–39]. The 3D characterization work of Lind and et al. revealed that the TRDs are developed independently from each other [40]. The evaluation of earlier sections also indicates that the fraction of low Σ CSL GBs increases due to the development of large, highly twinned grain clusters. These findings indicate that random boundaries originate from the impingement of different grain clusters. All twins and their $\Sigma3^n$ variant boundaries are inherently confined within the grain clusters rather than forming part of the external random GB network. As a result, they do not serve as interruptions to disrupt the connectivity of random GBs. The continuity of the random boundary network is highly dependent on the chosen CSL tolerance criterion [27]. When the Palumbo–Aust criterion is applied, the network remains largely intact, whereas it becomes significantly disrupted under the Brandon criterion. These two commonly used criterions for defining CSL misorientations are as follow:

Brandon criterion [2]: $\Delta\theta_{\max(\mathrm{B})} = 15^\circ\Sigma^{-1/2}$

Palumbo–Aust criterion [3]: $\Delta\theta_{\max(\mathrm{P\text{-}A})} = 15^\circ\Sigma^{-5/6}$

The difference between an experimentally determined misorientation and an exact CSL misorientation is represented as $\Delta\theta$, with $\Delta\theta max$ indicating the maximum allowable deviation. Therefore, the Brandon criterion allows for greater tolerances compared to the Palumbo–Aust criterion. The degree of deviation from the exact CSL can be expressed as a relative deviation, $\Delta\theta/\Delta\theta_{max}$ [1]. The relative deviations are denoted as Δθ/Δθmax(B) and Δθ/Δθmax(P–A) for the Brandon and Palumbo–Aust criteria, respectively. A misorientation is considered a CSL boundary under a given criterion only if its corresponding value is less than 1.

Figure 11a presents the GB map of a GBE-treated Alloy 690 sample, where CSL boundaries are identified using the Brandon criterion. It is evident that the continuity of random boundaries, located between grain clusters, is significantly disrupted by segments of low Σ CSL GBs upon applying the Brandon criterion. For instance, the random boundaries surrounding the two grain clusters I and J in Fig. 11a are clearly disrupted by several low Σ CSL GBs numbered 1–10. The deviations and relative deviations of these 10 boundaries are shown in Table 3. However, all the values of $\Delta\theta/\Delta\theta_{max(P\text{–}A)}$ of these 10 GBs are larger than 1, indicating these low Σ CSL GBs turn to random ones according to the Palumbo–Aust criterion. In this case, the connectivity of random GBs between grain-clusters is not broken up. For the purpose of statistical analysis, 60 low Σ CSL boundaries (identified using the Brandon criterion) that disrupt the continuity of the random boundary network are shown in Fig. 11b, along with the limits of both the Brandon and Palumbo–Aust criteria. It is clear that the deviations of most low Σ CSL boundaries (determined by the Brandon criterion) that disrupt the continuity of the random boundary network exceed the Palumbo–Aust criterion limit. This suggests that they are considered random GBs based on the Palumbo–Aust criterion, and thus, the connectivity of random GBs remains intact in this case.

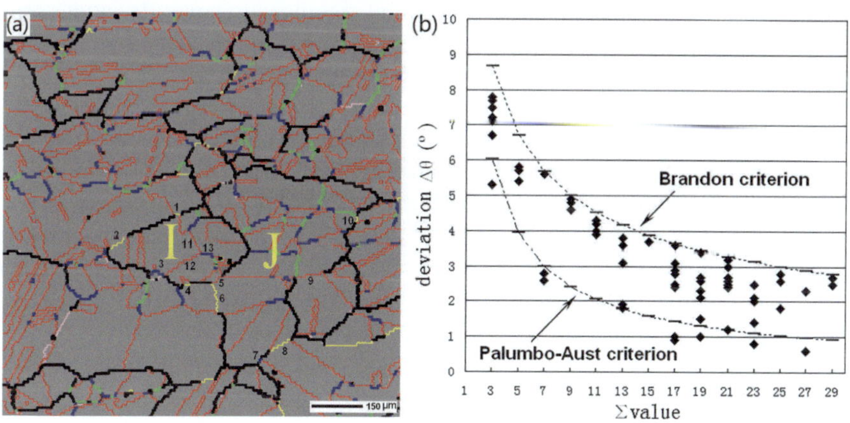

Fig. 11 The analysis of the connectivity of random grain boundaries (GBs) in a GBE-treated Alloy 690 sample: **a** A GB map with the CSL boundaries defined by the Brandon Criterion, where the connectivity of random GBs is significantly disrupted. The interrupters of the random boundary network surrounding two grain clusters, I and J, are labeled from 1 to 10. For comparison, some boundaries within cluster I are also labeled (11, 12, and 13). The deviations of all these GBs are provided in Table 3. **b** Deviations of several 60 boundaries (close to low Σ CSL misorientations) that appeared within the random boundary network [27]

Table 3 Misorientations of the grain boundaries numbered in Fig. 11a [27]

No	Misorientation	Closest CSL	Deviation ($\Delta\theta$)	$\Delta\theta/\Delta\theta_{max(B)}$	$\Delta\theta/\Delta\theta_{max(P\text{-}A)}$
1	46.9°[19 9 21]	Σ29b	2.5°	0.90	2.72
2	46.0°[−6 −7 −13]	Σ21b	2.5°	0.76	2.08
3	41.6°[−19 1 23]	Σ9	4.6°	0.92	1.90
4	47.6°[−20 −21 −1]	Σ11	3.6°	0.80	1.76
5	51.6°[−12 −13 12]	Σ3	8.5°	0.98	1.41
6	47.4°[7 −1 −16]	Σ15	3.6°	0.93	2.28
7	34.8°[−17 −2 −18]	Σ9	4.9°	0.98	2.02
8	38.4°[−1 −1 3]	Σ23	2.1°	0.67	1.89
9	55.1°[8 −9 10]	Σ3	7.1°	0.82	1.18
10	43.9°[0 1 19]	Σ29a	2.3°	0.82	2.50
Average value				0.86	1.97
11	59.8°[1 1 1]	Σ3	0.2°	0.02	0.03
12	59.2°[1 1 1]	Σ3	0.8°	0.09	0.13
13	39.5°[0 1 −1]	Σ9	0.4°	0.08	0.17
Average value				0.07	0.11

For comparison, three grain boundaries (GBs) numbered 11–13 within cluster I are also evaluated in Table 3, the relative deviations of which for both criteria are relatively small. This suggests that the low Σ CSL boundaries, which disrupt the connectivity of the random boundaries, exhibit significantly larger deviations than those found within the grain cluster.

As examined in earlier sections, random grain boundaries (GBs) are generated by the encountering of different grain clusters during recrystallization. Consequently, while the misorientations of some boundaries between grain clusters are near low Σ CSL, they merely fall within the acceptable range of the Brandon criterion. In contrast, the $\Sigma 3^n$ boundaries inside the clusters are geometrically necessary due to multiple twinning and hence have very low deviations. Therefore, the discontinuity of random boundaries observed in accordance with the Brandon criterion is not inherently required by geometry.

5 Morphologies of Twin Boundaries in 3D

Annealing twin boundaries in conventional (non-GBE) fcc materials have several typical morphologies as usually being observed on 2-dimensional sections [41]: corner twins, lamellar twins either grain spanning or terminated inside grain, and in some cases small steps appearing on twin boundary (TB). The morphology of the inter-connected annealing twins in GBE microstructure, as shown in Fig. 1a and b and Fig. 2a, differs very much from that of simple morphology of annealing twins in Non-GBE microstructures. However, 3D characterizations are needed to reveal their actual morphologies. References [42–44] documented the typical three-dimensional morphologies of annealing twins in 316L stainless steel specimens, characterized through serial sectioning and electron backscatter diffraction (EBSD) mapping. The 3D characterization of twin boundaries (TBs) reveals four distinct morphological types: single-face twin boundaries, tunnel twin boundaries, occluded twin boundaries, and parallel multi-planar twin boundaries [42]. Figure 12 shows an example of multi-planar twin boundaries in 3D, where lamella annealing twins are alternatively stacking together. The parallel multi-planar morphology of annealing twins is viewed as several paralleling straight lines in 2D characterizations. The 3D microstructure of conventional 316L stainless steel contains numerous parallel multi-planar twin boundaries, whereas their occurrence is significantly reduced after GBE treatment.

Following GBE treatment, some twin boundaries with intricate morphologies can be witnessed [42]. Figure 13 presents an example of a twin boundary exhibiting a highly complex structure. Its exceptionally intricate morphology sets it apart, particularly when compared to the simpler classifications derived from 2D observations. Notably, the entire interface between the adjacent grains is considered a single

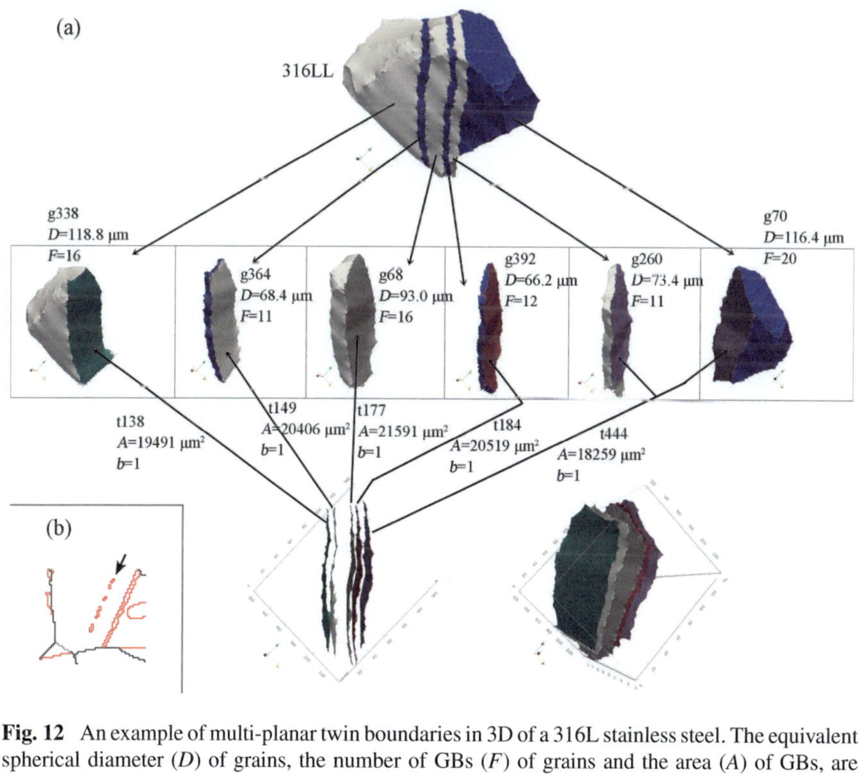

Fig. 12 An example of multi-planar twin boundaries in 3D of a 316L stainless steel. The equivalent spherical diameter (*D*) of grains, the number of GBs (*F*) of grains and the area (*A*) of GBs, are directly shown in the figure [42]

boundary. The morphologically intricate twin boundary in Fig. 13 represents the interface between two grains, indicated by gray and bright brown colors, respectively. The twin grains themselves exhibit intricate morphologies with various bar-shaped, lamellar-shaped, and torus-shaped branches. The twin boundary can be divided into several simpler segments, including parallel planar sections and tunnel-like segments. As shown in previous sections, large size highly-twinned grain-clusters may have many grains which belong to several orientations. Although the grains with a same orientation are apart from each other on the 2D section (for example, the grain-cluster in Fig. 2), they are highly probable a same grain in 3D. Twin boundary proportions measured as both area fraction and number fraction are greatly enhanced after GBE in most 2D studies. While, the 3D characterization shows that GBE specimen has more large-area twin boundaries (morphologically complex twin boundaries). As a result, the GBE-treated specimen exhibits a higher area fraction but a comparable number fraction of twin boundaries when compared to the conventional specimen [42]. These findings can be attributed to the exceptionally large and morphologically intricate twin boundaries, which are identified as a single boundary in 3D characterization but appear as distinct, separate boundaries in 2D analyses.

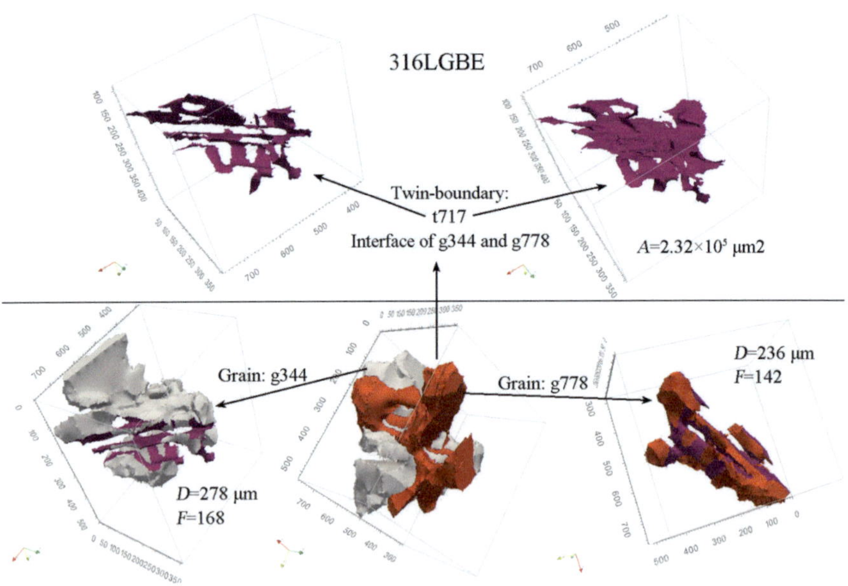

Fig. 13 An example of morphologically complex twin boundary (denoted by purple color) of a GBE-treated 316L stainless steel, representing the interface between grain A (indicated by the gray color) and grain B (denoted by bright brown color). Both of the two grains have very complex morphologies [42]

6 Effects of GBCD on GB Related Material Properties

6.1 The Effects of GB Character on GB Energy

Low ΣCSL GBs with different Σ values showed different susceptibility to element segregation, precipitation, corrosion and cracking. Not all of the low ΣCSL GBs has better properties than random GBs. West and Was observed that the Σ3 boundary exhibited a considerably lower cracking frequency compared to other low-Σ CSL boundaries (such as Σ9 and Σ27) and random GBs in their investigation of intergranular stress corrosion cracking susceptibility in 316L stainless steel and nickel-based Alloy 690 exposed to supercritical water [45]. In recent years, it has been more broadly acknowledged that only the coherent twin boundary (Σ3c) is unique due to its perfectly ordered atomic structure at the GB plane.

Thermal grooving has been regarded as a useful method for measuring relative GB energies in many metallic materials and ceramic materials [46, 47]. A developed groove can reflect the equilibrium between GB and free surface. By measuring the groove profile and the groove dihedral angle, the ratio of GB energy over the free surface energy can be calculated, with the assumption of isotropic free surface energy. Figure 14 shows developed thermal grooves of different character GBs in an Alloy 690 specimen, which was heat treated at 1100 °C in a high vacuum environment after carful polishing the surface [48]. It can be obvious seen that the widths and depths of the GBs are different from one to another. By correlating the GBs' groove morphologies with their misorientations obtained by EBSD, it can be roughly distinguished that the random GBs and Σ27 GBs have much wider and deeper grooves than that of Σ3c boundaries, and that of the Σ1, Σ3i (incoherent twin boundary) and the Σ9 GBs are in-between them.

It should be noted that, except Σ3c boundaries which have straight shape and lie on {111} planes, all other type boundaries have curved shape, for example the Σ9 GB in (e), Σ27 GB in (f) and random GB in (g) of Fig. 14. The crystallographic plane of a curved GB always changes along the GB rather than has an identical GB plane. Step segments can sometimes be observed on GBs. For example, in the Fig. 14e, the observed Σ27 GB have steps, and the groove of steps have different depth and width than that of other parts of this GB. This indicated that each step has different GB plane than other parts of that GB, and hence have different energies. Many articles pointed out the significant influences of GB plane orientation on the GB energy and properties. However, a curved GB or a GB with steps should be divided into small segments with each segment having a straight (or relatively straight) GB plane so that each part can be fully defined by the five parameters. When correlating GB properties with GB structures, it's better to analyze the GB segment by segment rather than just a whole GB. However, this kind of work is labor intensive and technically almost unfeasible. Therefore, only misorientation between adjoining grains is used to describe the structure of the GBs by ignoring the GB plane orientations in most cases.

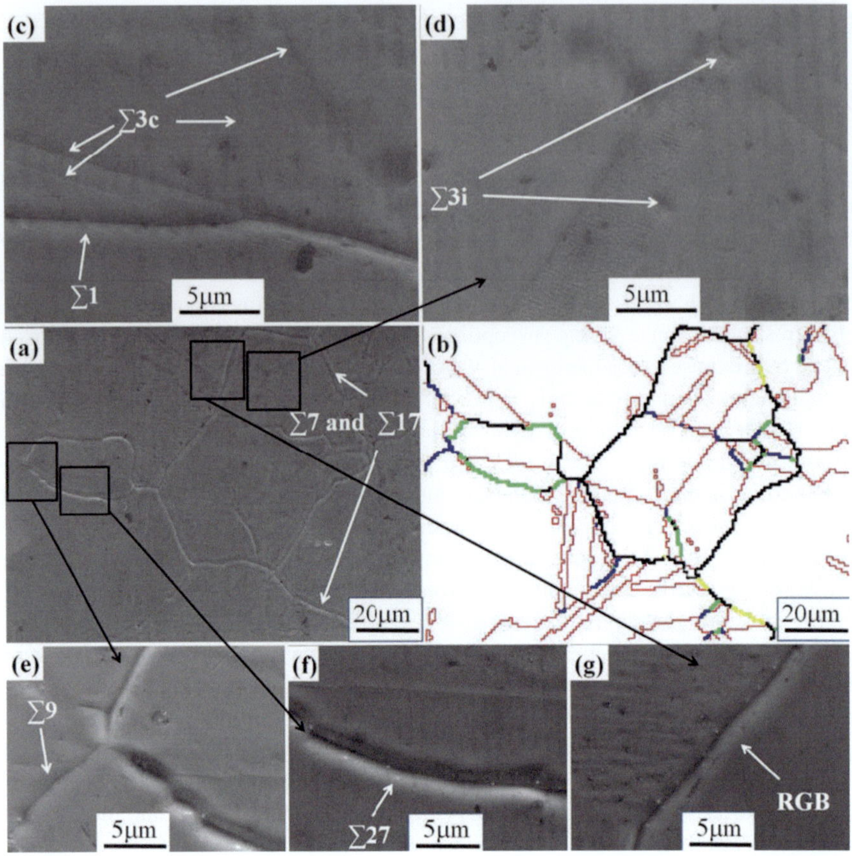

Fig. 14 The morphology of thermal grooves of different character GBs in an Alloy 690 specimen, which was heat treated at 1100 °C in high vacuum. SEM image **a** and its corresponding EBSD map **b** showing GB characters that were analyzed, and various colored lines representing different types of grain boundaries: red for Σ3, blue for Σ9, green for Σ27, yellow for other low Σ CSL, and black for random GB. **c–g** higher magnification SEM images showing the thermal grooves of different character GB [48]

The morphology profiles and dihedral angles of GB thermal grooves were measured with atomic force microscopy (AFM) and then the ratio of GB energy (γ_b) to surface energy (γ_s) are calculated following Mullins method for comparing the energy of different character GBs. At least 10 GBs for each character were chosen to be measured to get the average value, and the results are shown in Fig. 15. The GB energy of Σ3c, Σ3i and Σ9 GBs were almost one tenth, one half, and two thirds of that of the random GBs, respectively. The Σ27 GBs have a similar average value of γ_b/γ_s with that of random GBs. Σ1 has a GB energy between that of Σ3i and Σ9 GBs. The relative energy of GBs increased as the Σ value increase, except for random GBs and Σ1 GBs. Thus, it could be concluded that the energy of Σ3c GBs

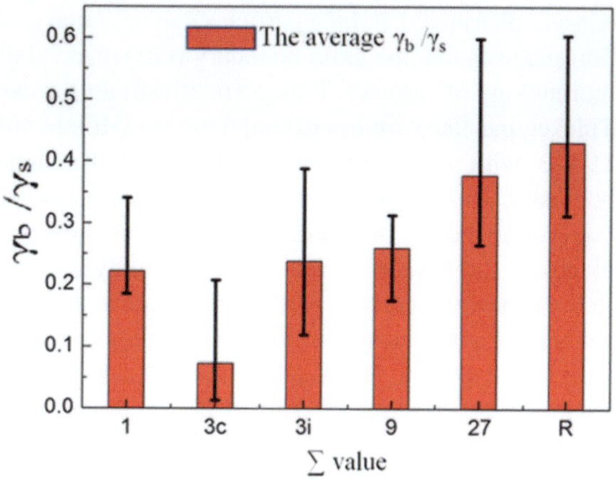

Fig. 15 The dependance of relative GB energies on the GB characters in an Alloy 690 specimen [48]

were much lower than that of others. The error bars show the ranges of the data that included in the calculation of the average value, revealing relatively large scattering of the data. The different inclination of the GB planes with respect to the surface and the non-identical surface energy of different surface orientations are the two most important factors that induce the scattering of the data.

6.2 *The Effects of GB Character on Carbide Precipitation*

Chromium depletion in the vicinity of grain boundaries plays a critical role in the intergranular corrosion (IGC) and intergranular stress corrosion cracking (IGSCC) in stainless steels and nickel-based alloys, especially when these materials are exposed to aggressive environments [49]. Subjecting the material to temperatures between 500 and 800 °C results in the precipitation of chromium-rich carbides at grain boundaries (GBs), which subsequently leads to the development of chromium depletion zones in the vicinity of these boundaries. The development and extent of chromium depletion zones are largely dictated by the morphology and distribution of carbides at grain boundaries, thereby impacting resistance to IGC and IGSCC.

The character of grain boundaries (GB) can have a profound impact on both carbide precipitation and the development of chromium depletion zones [50]. Figure 16 illustrates the carbide morphologies at different character grain boundaries of an Alloy 690 specimen that underwent thermal treatment at 715 °C for 15 h following solution annealing at 1100 °C [51]. The carbides present in the Alloy 690 specimen were characterized as $M_{23}C_6$, which display a cube-on-cube orientation relationship with one side of the matrix, as confirmed by TEM. A few small bar-like carbides can be found on the Σ3c boundary, while many similar bar-like carbides having a same growth direction precipitated on both sides of the Σ3i boundary which is at the nearby position of the Σ3c boundary, as shown in Fig. 16b. Whereas, tiny densely distributed carbide particles can be found on that Σ3i boundary. This variation suggests that the grain boundary (GB) plane plays a crucial role in shaping the morphology of carbides. Figure 16c illustrates the carbide morphology at a Σ9 GB. Thicker, bar-like carbides extend from the GB into only one side of the matrix at the Σ9 GB, with some continuous carbide particles, larger than those found at the Σ3i boundary, also present along the Σ9 GB. Carbide bars appear on both sides of the Σ3i boundary, whereas the Σ9 grain boundary is impacted on only one side. The {111} planes of the $M_{23}C_6$ carbide align optimally with the {111} planes of the matrix [52]. As a result, carbide particles nucleate at the grain boundaries (GBs) and align coherently with one side of the matrix, adopting the $\{111\}_{carbide}//\{111\}_{matrix}$ orientation. Due to the < 111 > /60° misorientation of the Σ3 boundary, one of the {111} planes of the grain on one side of the Σ3 boundary is parallel to one of the {111} planes on the grain of the opposite side of the boundary. As a result, the carbide bars can extend into the matrix on both sides of the Σ3i boundary while maintaining a coherent orientation relationship with the grains on either side. In contrast, the Σ9 GB has a misorientation of < 110 > /38.9°, meaning that none of the {111} planes of the grain on one side of the boundary are generally parallel to any {111} planes of the grain on the other side. Consequently, the carbide bars grow into the matrix only on one side of the Σ9 GB. The morphology of carbides precipitated at Σ27 GBs and random GBs is shown in Fig. 16d and e, respectively. No distinct carbide shapes are observed. The carbides formed at the Σ27 GB and random GB are significantly larger than those precipitated at the Σ3 and Σ9 GBs. Notably, no bar-like carbides are present near the Σ27 GBs and random GBs, likely due to their larger GB free volume, which provides sufficient space for carbide nucleation and allows adequate Cr diffusion to the carbide phase.

6.3 *The Effect of GBCD on IGC and IGSCC*

IGC and IGSCC are significant issues for stainless steels and nickel-based alloys, potentially causing unforeseen failures and resulting in substantial losses. GB character affects IGC and IGSCC in two manners: (1) through affecting intergranular carbide formation and Cr depletion; (2) directly affects IGC and IGSCC.

The IGC behavior of grain boundaries with different characteristics was examined in a grain boundary engineered (GBE) 304 stainless steel, which was aged at 650 °C for 12 h following solution annealing at 1100 °C. A solution of 1.4 mol/L HNO_3 and 0.8 mol/L HF at room temperature [53] was used to conduct IGC tests and examine the susceptibility of GBs to corrosion. Figure 17a presents an SEM image capturing the morphology of different GBs after 4 h of exposure, while Fig. 17b illustrates the corresponding GB map, providing insight into their distribution and characteristics. The results indicate that, despite being subjected to the same corrosive environment, the extent of intergranular attack varies significantly among different grain boundaries.

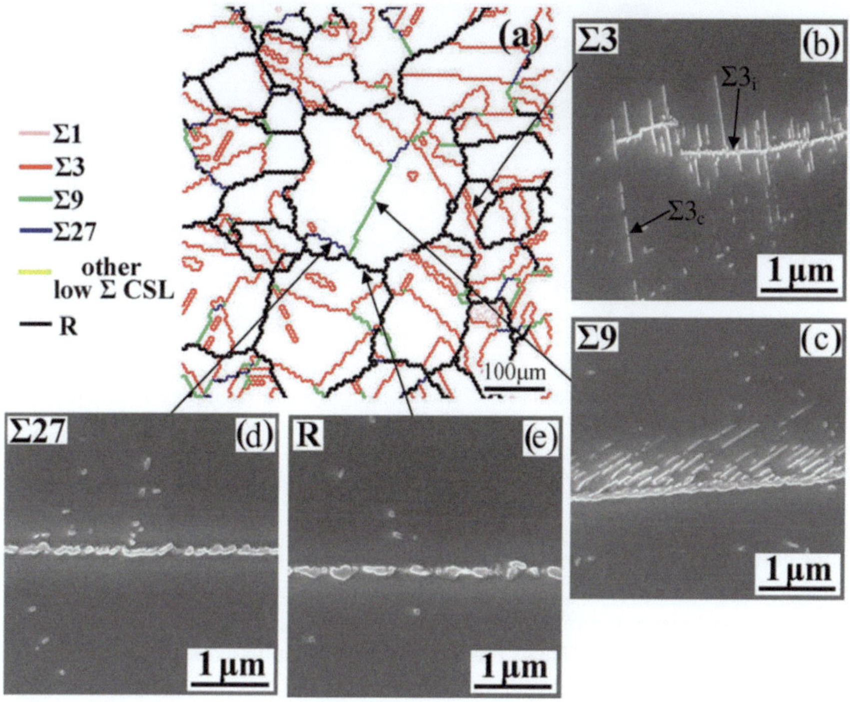

Fig. 16 The relationship between carbide morphologies and grain boundary (GB) characteristics in an Alloy 690 specimen aged at 715 °C for 15 h following solution annealing at 1100 °C. The GB map of the observed specimen is shown in **a**, with carbide morphologies precipitated at Σ3 GB **b**, Σ9 GB **c**, Σ27 GB **d**, and random GB **e**. Σ3i denotes the incoherent twin boundary, Σ3c refers to the coherent twin boundary, and R represents random GBs [51]

In Fig. 17a, all Σ3c boundaries displayed only minimal signs of corrosion, highlighting their exceptional resistance to intergranular corrosion (IGC). In contrast, Σ3i, Σ9, and Σ27 boundaries showed extensive corrosion, behaving similarly to random grain boundaries, which are known to be highly vulnerable to IGC. The results clearly indicate that only Σ3c boundaries consistently resist IGC, which justifies their classification as "special" grain boundaries. Σ3c boundaries uniquely exhibit low interfacial free energy, which inhibits precipitation and thereby prevents sensitization.

After sensitization treatment, carbide precipitation and the related chromium depletion near the boundaries were effectively suppressed at Σ3c boundaries, whereas these phenomena were more pronounced near other low-Σ CSL grain boundaries and random boundaries. Consequently, Σ3c boundaries exhibited significantly lower corrosion compared to other grain boundaries.

If the corrosion time is prolonged, IGC will propagate along the grain boundaries. Figure 18 depicts the IGC pathways on the cross-sectional surface of a GBE Alloy 690 sample, exposed for 24 days to a solution of 65% HNO3 + 0.4% HF at room temperature [9]. The specimen underwent thermal treatment at 715 °C for 2 h prior

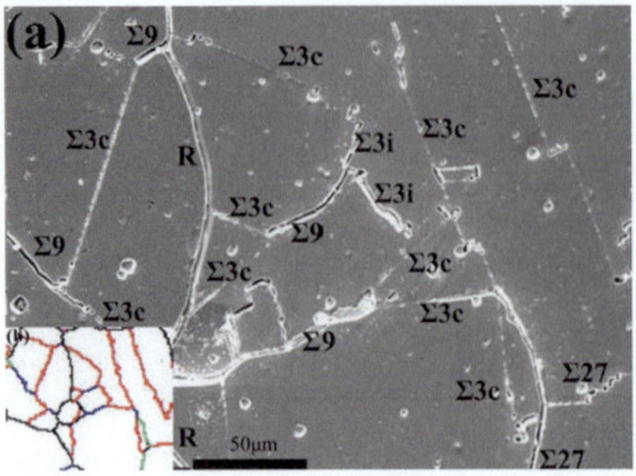

Fig. 17 Corrosion morphology of a sensitized 340 stainless steel GBE sample, aged at 650 °C for 12 h after solution annealing at 1100 °C. The intergranular corrosion test was performed in a solution of 1.4 mol/L HNO3 + 0.8 mol/L HF at room temperature for 4 h. **a** SEM image showing various corroded grain boundaries (GBs), **b** Corresponding GB map of **a**. R: random GB; Σ3c: coherent twin boundary; Σ3i: incoherent twin boundary [53]

to the corrosion test. In Fig. 18a–d, the SEM micrograph, optical micrograph, grain orientation distribution, and GB type map are shown in sequence. A large, highly-twinned grain cluster is present in this area, where all the boundaries within the cluster are of the $\Sigma3^n$ type. These boundaries establish a network of Σ3–Σ3–Σ9 and Σ3–Σ9–Σ27 triple junctions, as demonstrated in Fig. 18d. Six paths along which IGC penetrated into this area are shown in Fig. 18b, d and e. A close inspection of the figures shows that Σ9, Σ27, Σ81, and Σ3i boundaries do not display greater corrosion resistance than random boundaries. Only coherent twin boundaries (Σ3c) remained unaffected by corrosion and can thus be classified as special. All the IGC penetrations ended at Σ3–Σ3–Σ9 type triple junctions.

At a Σ3–Σ3–Σ9 triple junction, if the Σ9 boundary undergoes corrosion, the two adjoining coherent Σ3 boundaries will prevent further IGC penetration due to their resistance to sensitization and corrosion. Thus, the Σ3–Σ3–Σ9 triple junction acts as a fundamental barrier to intergranular corrosion penetration, as schematically depicted in Fig. 19a. Figure 19b illustrates the function of this basic unit within the GBN. If corrosion propagates from a Σ27 boundary to a Σ9 boundary, as illustrated in Fig. 19b, the Σ3–Σ3–Σ9 triple junction will obstruct its progression, ensuring the protection of downstream boundaries. Therefore, as shown in Fig. 19c, IGC propagating along Σ9, Σ27, or any higher-order $\Sigma3^n$ ($n > 3$) boundaries can be observed. Since all of these boundaries are part of the same grain-cluster, the corrosion will ultimately reach a Σ3–Σ3–Σ9 triple junction, stopping its further progression. Therefore, the grain-cluster with extensive twinning, generated by GBE, can obstruct the advancement of intergranular corrosion. The significantly improved resistance to intergranular corrosion in the GBE-treated sample can be attributed to this optimized microstructural feature, which is absent in the Non-GBE sample.

IGC also spreads along random boundaries between grain-clusters (or grains in conventional materials). IGC-induced grain detachment eventually occurred. Figure 20 shows the surface morphologies of the GBE and Non-GBE Alloy 690

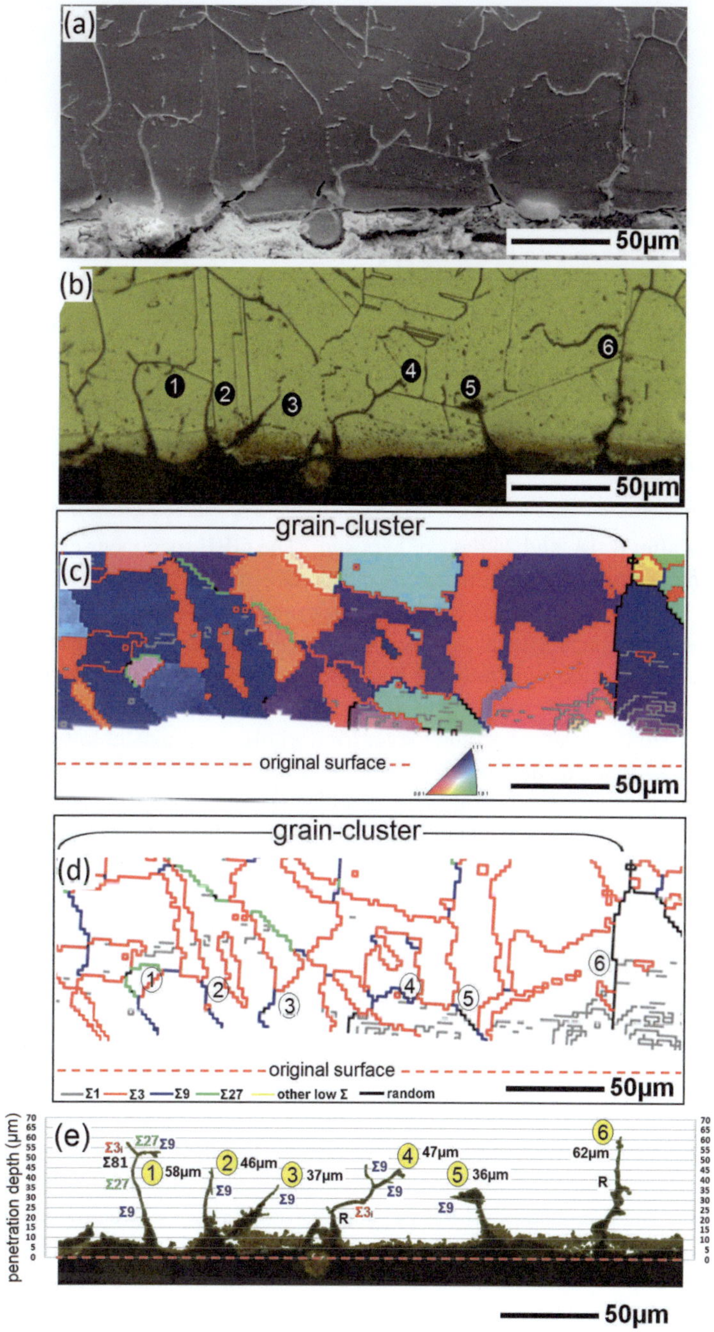

Fig. 18 EBSD analysis of a sensitized alloy 690 GBE specimen after 24 days of intergranular corrosion: **a** SEM image, **b** optical micrograph, **c** orientation distribution map with inverse pole figure colors, **d** map of grain boundary types, and **e** IGC path penetration depths. R: random grain boundary; Σ3c: coherent twin boundary; Σ3i: incoherent twin boundary [9]

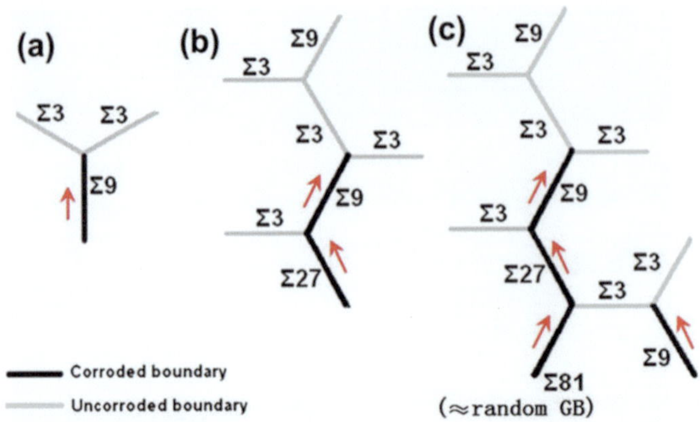

Fig. 19 Schematic representation of the role of a Σ3–Σ3–Σ9 triple junction in inhibiting IGC propagation [9]

specimens after 8, 14, and 24 days of exposure. The GBE specimens exhibited only intergranular etching with minimal grain detachment during the entire testing period, while the Non-GBE specimens experienced significant grain detachment.

The size and shape of grain-clusters play a significant role in the resistance to intergranular corrosion penetration as IGC spreads along random boundaries between them. The size effect is comparable to that of grain size: as the grains increase in size, the penetration paths become longer for a whole grain-cluster to separate from the matrix. Moreover, the irregular shape of larger grain-clusters contributes to a slower rate of grain detachment. The irregular shape causes the penetration path to be convoluted, which traps the grain-clusters and prevents them from detaching. Consequently, the underlying microstructures remain protected. It is only when the outermost grain-clusters detach that the underlying microstructure is exposed to the aggressive solution, just like the external ones. Thus, grains or grain-clusters in a microstructure with an irregular shape are less likely to detach during intergranular corrosion than those in a regular equiaxed grain microstructure. Therefore, the large, highly-twinned grain-clusters not only hinder IGC penetration but also shield the underlying microstructure, offering better IGC resistance compared to conventional microstructures.

For IGSCC initiation and propagation, GB characters also have decisive influences [54, 55]. Twin boundaries not only display significant resistance to IGSCC on their own but also contribute to the inhibition of cracking in neighboring grain boundaries. The propagation of IGSCC can be significantly slowed by the presence of triple or quadruple junctions in regions with twin boundaries. Triple junctions, where three grain boundaries meet, and quadruple junctions, where four grain boundaries intersect, play a significant role in enhancing the IGSCC resistance of materials. In particular, triple junctions with two twin boundaries can stop the propagation of

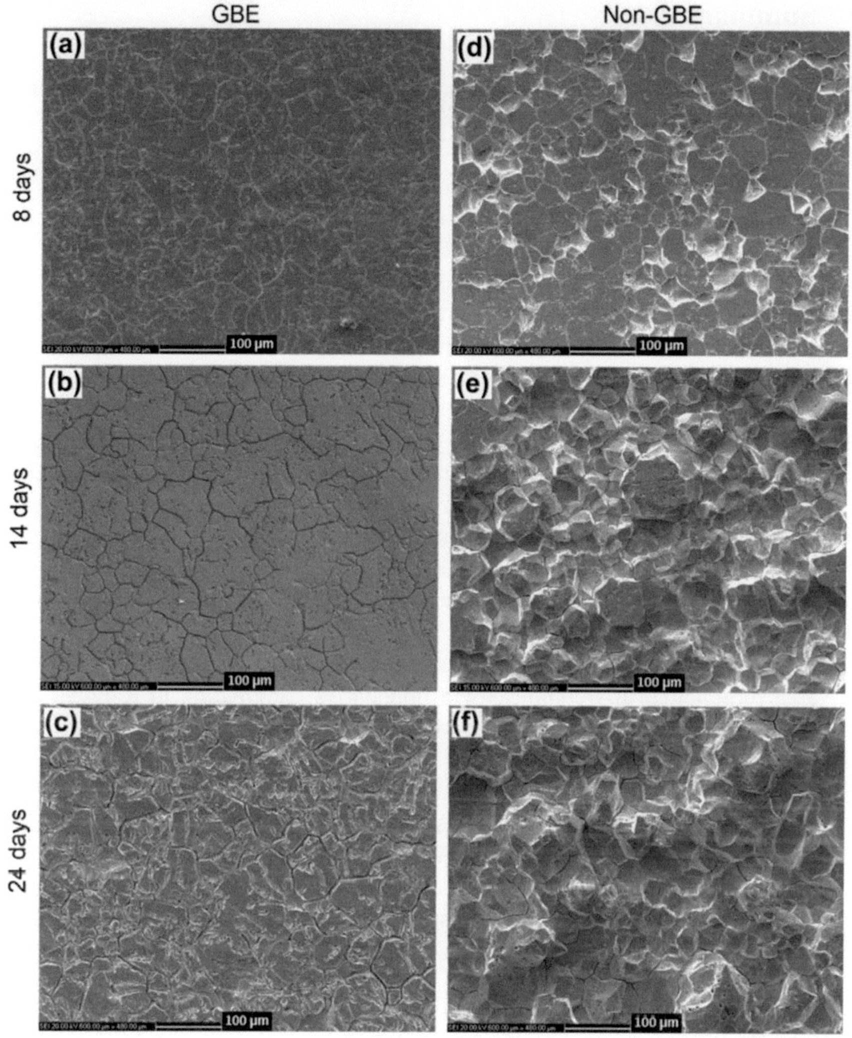

Fig. 20 SEM images of the surfaces for both sensitized GBE samples **a–c** and non-GBE samples **d–f** subjected to intergranular corrosion for 8, 14, and 24 days, respectively [9]

IGSCC, while quadruple junctions containing three twin boundaries can also effectively prevent the growth of IGSCC. Even quadruple junctions with only two twin boundaries show significant resistance to the propagation of IGSCC.

7 Summary

1. Twin-induced GBE greatly enhances the fraction of $\Sigma 3^n$ (n = 1, 2, 3…) type low ΣCSL ($\Sigma \leq 29$) GBs by promoting the formation of large-size highly-twinned grain-clusters (also termed as twin related domain). Grain-clusters arise from multiple twinning events starting from a single recrystallization nucleus in the course of the recrystallization process. Hence, low nucleation density as a result of slight deformation on the order of 5 ~ 10% before annealing yields large size of grain-clusters with each grain-cluster has a large area available to occupy.
2. All the twin grains inside a same grain-cluster can be joined into a same twin-chain, in which all the grains regardless of adjacent or not have $\Sigma 3^n$ mutual misorientations. Through twin-chain analysis, the random grain boundaries within grain-clusters are recognized as higher-order twin boundaries.
3. The grain boundaries between grain-clusters are crystallographically random, as they arise from the encountering of different grain-clusters. Some low Σ CSL GBs may interrupt the connectivity of the random GBs outside grain-clusters, however they have comparatively larger deviations, which happen to fall in the tolerances of Brandon criterion. If a more stringent criterion, such as the Palumbo–Aust criterion, is applied, the connectivity of random grain boundaries remains intact without fragmentation.
4. 3D-EBSD characterization shows that the GBE microstructure has more large-area morphologically-complex twin boundaries contributing to a higher area fraction but similar number fraction of twin boundaries as compared with conventional specimen.
5. Coherent twin boundaries have significantly lower GB energy than other types of GBs. The precipitation of carbides and Cr depletion are greatly suppressed at coherent twin boundaries, resulting in a much lower susceptibility to intergranular degradation. Intergranular attack penetration can be arrested at $\Sigma 3$–$\Sigma 3$–$\Sigma 9$ triple junctions once it enters a highly-twinned grain-cluster through higher order twin boundaries, resulting in a better IGC resistance of a GBE microstructure than a conventional one.

References

1. Randle V (1996) The role of the coincidence site lattice in grain boundary engineering. Cambridge University Press, Cambridge, England
2. Brandon D (1966) The structure of high-angle grain boundaries. Acta Metall 14:1479–1484
3. Palumbo G, Aust KT (1990) Structure-dependence of intergranular corrosion in high purity nickel. Acta Metall 38:2343–2352
4. Randle V (2004) Twinning-related grain boundary engineering. Acta Mater 52:4067–4081
5. Watanabe T (1984) Approach to grain boundary design for strong and ductile polycrystals. Res Mech 11:47
6. Lin P, Palumbo G, Erb U, Aust KT (1995) Influence of grain boundary character distribution on sensitization and intergranular corrosion of Alloy 600. Scr Metall Mater 33:1387–1392

7. Reed BW, Kumar M (2006) Mathematical methods for analyzing highly-twinned grain boundary networks. Scr Mater 54:1029–1033
8. Xia S, Zhou BX, Chen WJ (2009) Grain cluster microstructure and grain boundary character distribution in Alloy 690. Metall Mater Trans 40A:3016–3030
9. Xia S, Li H, Liu TG, Zhou BX (2011) Appling grain boundary engineering to alloy 690 tube for enhancing intergranular corrosion resistance. J Nucl Mater 416:303–310
10. Xia S, Zhou BX, Chen WJ, Wang WG (2006) Effects of strain and annealing processes on the distribution of Σ3 boundaries in a Ni-based superalloy. Scr Mater 54:2019–2022
11. Gottstein G (1984) Annealing texture development by multiple twinning in f.c.c. crystals. Acta Metall 32:1117–1138
12. Randle V (2006) The effect of twinning interactions up to the seventh generation on the evolution of microstructure. J Mater Sci 41:653–660
13. http://www.edax.com/
14. https://www.ebsd.com
15. Wang WG, Zhou BX, Gregory SR, Guo H, Cai ZX (2010) Textures and grain boundary character distributions in a cold rolled and annealed Pb–Ca based alloy. Mater Sci Eng A 527:3695–3706
16. Liu TG, Xia S, Li H, Zhou BX, Bai Q (2014) The highly twinned grain boundary network formation during grain boundary engineering. Mater Lett 133:97–100
17. Liu TG, Xia S, Wang BS, Bai Q, Zhou BX, Su C (2016) Grain orientation statistics of grain-clusters and the propensity of multiple-twinning during grain boundary engineering. Mater Des 112:442–448
18. Liu TG, Xia S, Li H, Zhou BX, Bai Q, Su C, Cai, ZG (2013) Effect of initial grain size on the grain boundary network during grain boundary engineering in Alloy 690. J Mater Res 28:1165–1176
19. Prithiv TS, Bhuyan P, Pradhan SK, Subramanya Sarma V, Mandal S (2018) A critical evaluation on efficacy of recrystallization vs. strain induced boundary migration in achieving grain boundary engineered microstructure in a Ni-base superalloy. Acta Mater 146:187–201
20. Kumar M, Schwartz AJ, King WE (2002) Microstructural evolution during grain boundary of low to medium stacking fault energy fcc materials. Acta Mater 50:2599
21. Shimada M, Kokawa H, Wang ZJ, Sato YS, Karibe I (2002) Optimization of grain boundary character distribution for intergranular corrosion resistant 304 stainless steel by twin-induced grain boundary engineering. Acta Mater 50:2331–2341
22. Barr CM, Leff AC, Demott RW, Doherty RD, Taheri ML (2018) Unraveling the origin of twin related domains and grain boundary evolution during grain boundary engineering. Acta Mater 144:281–290
23. Engelberg DL, Humphreys FJ, Marrow TJ (2008) The influence of low-strain thermo-mechanical processing on grain boundary network characteristics in type 304 austenitic stainless steel. J Microsc 230:435–444
24. Haasen P (1999) How are new orientations generated during primary recrystallization. Metall Trans B 24B:225–239
25. Kopezky CV, Andreeva AV, Sukhomlin GD (1991) Multiple twinning and specific properties of $\Sigma = =\Sigma 3^n$ boundaries in f.c.c. crystals. Acta Metall Mater 39, 1603
26. Berger A, Wilbrandt PJ, Ernst F, Klement U, Haasen P (1988) On the generation of new orientations during recrystallization: recent results on the recrystallization of tensile-deformed fcc single crystals. Prog Mater Sci 32:1–95
27. Xia S, Zhou BX, Chen WJ (2008) Effect of single-step strain and annealing on grain boundary character distribution and intergranular corrosion in Alloy 690. J Mater Sci 43:2990–3000
28. Fang X, Liu Z, Tikhonova M, Belyakov A, Wang W (2012) Evolution of texture and development of Σ3n grain clusters in 316 austenitic stainless steel during thermal mechanical processing. J Mater Sci 48:997–1004
29. Humhreys FJ, Hartherly M (2004) Recrystallization and related annealing phenomena. Elsevier, Oxford
30. Jin Y, Lin B, Brnacki M, Rohrer GS, Rollett AD, Bozzolo N (2014) Formation of annealing twins during recrystallization and grain growth in 304 L austenitic stainless steel. Mater Sci Eng A 597:295–303

31. Bai Q, Zhao Q, Xia S, Wang BS, Zhou BX, Su C (2017) Evolution of grain boundary character distributions in alloy 825 tubes during high temperature annealing: Is grain boundary engineering achieved through recrystallization or grain growth? Mater Charac 123:178–188
32. Liu TG, Xia S, Li H, Zhou BX, Bai Q (2014) Effect of the pre-existing carbides on the grain boundary network during grain boundary engineering in a nickel based alloy. Mater Charac 91:89–100
33. Katnagallu SS, Mandal S, Nagaraja AC, Boer BD, Vadlamani SS (2015) Role of carbide precipitates and process parameters on achieving grain boundary engineered microstructure in a Ni-based superalloy. Metall Mater Trans A 46A:4740–4754
34. Cao W, Xia S, Bai Q, Zhang WZ, Zhou BX, Li ZJ, Jiang L (2017) Effects of initial microstructure on the grain boundary network during grain boundary engineering in Hastelloy N alloy. J Alloy Compd 704:724–733
35. Gertsman VY, Bruemmer SM (2001) Study of grain boundary character along intergranular stress corrosion crack paths in austenitic alloys. Acta Mater 49:1589–1598
36. Field DP, Bradford LT, Nowell MM, Lillo TM (2007) The role of annealing twins during recrystallization of Cu. Acta Mater 55:4233–4241
37. Gertsman VY, Henager CH (2003) Grain boundary junctions in microstructure generated by multiple twinning. Interface Sci 11:403–415
38. Cayron C (2011) Quantification of multiple twinning in face centred cubic materials. Acta Mater 59:252–262
39. Bober DB, Lind J, Mulay RP, Rupert TJ, Kumar M (2017) The formation and characterization of large twin related domains. Acta Mater 129:500–509
40. Lind J, Li SF, Kumar M (2016) Twin related domains in 3D microstructures of conventionally processed and grain boundary engineered materials. Acta Mater 114:43–53
41. Mahajan S, Pande CS, Imam MA, Rath BB (1997) Formation of annealing twins in fcc crystals. Acta Mater 45:2633–2638
42. Liu TG, Xia S, Zhou BX, Bai Q, Rohrer GS (2018) Three-dimensional study of twin boundaries in conventional and grain boundary-engineered 316L stainless steels. J Mater Res 33:1742–1754
43. Liu TG, Xia S, Zhou BX, Bai Q, Rohrer GS (2017) Three-dimensional characteristics of the grain boundary networks of conventional and grain boundary engineered 316L stainless steel. Mater Charac 133:60–69
44. Liu TG, Xia S, Zhou BX, Bai Q, Rohrer GS (2018) Three-dimensional geometrical and topological characteristics of grains in conventional and grain boundary engineered 316L stainless steel. Micro 109:58–70
45. West EA, Was GS (2009) IGSCC of grain boundary engineered 316L and 690 in supercritical water. J Nucl Mater 392:264–271
46. Mullins WW (1957) Theory of thermal grooving. J Appl Phys 28:333–339
47. Saylor DM, Rohrer GS (1999) Measuring the influence of grain-boundary misorientation on thermal groove geometry in ceramic polycrystals. J Am Ceram Soc 82:1529–1536
48. Xia S, Yang HH, Zhang ZL, Zhao Q, Zhou BX, Bai Q (2015) The effect of grain boundary misorientation on the grain boundary energy in a highly twinned alloy 690. In: 17th International conference on environmental degradation of materials in nuclear power systems—water reactors August 9–13, 2015. Ottawa, Ontario, Canada paper No.:165, vol.1, ISBN: 978-1-5108-1395-3
49. Kai JJ, Yu GP, Tsai CH, Liu MN, Yao SC (1989) The effects of heat treatment on the chromium depletion, precipitate evolution, and corrosion resistance of INCONEL Alloy 690. Metall Trans 20A:2057–2067
50. Bi HY, Kokawa H, Wang ZJ, Shimada M, Sato YS (2003) Suppression of chromium depletion by grain boundary structural change during twin-induced grain boundary engineering of 304 stainless steel. Scr Mater 49:219–223
51. Li H, Xia S, Zhou BX, Chen WJ, Hu CL (2010) The dependence of carbide morphology on grain boundary character in the highly twinned alloy 690. J Nucl Mater 399:108–113

52. Lewis MH, Hattersley B (1965) Precipitation of M23C6 in austenitic steels. Acta Metall 13:1159–1168
53. Hu CL, Xia S, Li H, Liu TG, Zhou BX, Chen WJ, Wang N (2011) Improving the intergranular corrosion resistance of 304 stainless steel by grain boundary network control. Corros Sci 53:1880–1886
54. Zhang ZL, Xia S, Bai Q, Liu TG, Li H, Zhou BX, Wang LJ, Ma WJ (2019) Effects of 3-D grain boundary geometrical angles and the net normal stress on intergranular stress corrosion cracking initiation in a 316 stainless steel. Mater Sci Eng A 765:138277
55. Liu TG, Xia S, Shoji T, Bai Q, Zhou Zhou BX, Lu YH (2017) The topology of three-dimensional grain boundary network and its influence on stress corrosion crack propagation characteristics in austenitic stainless steel in a simulated BWR environment. Corros Sci 129:161–168

Chapter 8
Automated Reconstruction of Parent Phase Microtexture in a Displacive Phase Transformation

Sushil Kumar Giri, B. Sravan Kumar, and I. Samajdar

1 Introduction

Many metallic materials (example: steel, titanium (Ti) and zirconium (Zr)) undergo displacive phase transformation [1–3].The latter, often generalized as invariant plane strain or IPS [2, 4], involves no rotation of habit or interface plane plus uniform shear. In particular, the high temperature parent phase (example: austenite in steel) microstructure often controls the ambient temperature IPS product phase (example: martensite or bainite) structure and property. Traditional metallographic technique(s) of revealing prior austenite grain boundary only provides morphological information [5]. Further, such a method may not work for different grades of steel, especially low phosphorous carbon steel. Knowledge about the crystallography of the parent phase may also provide mechanistic details on IPS or phenomenological theory on martensite crystallography (PTMC), including selection of the crystallographic variants or the so-called variant selection [6–10].

Direct methods of orientation determination of parent phase are often difficult, as it requires corresponding high temperature diffraction experiments [11] or retaining the high temperature phase to room temperature [12, 13]. In all such cases, reconstruction of the parent phase from the microtexture measurements of the IPS product can be effectively exploited, see Fig. 1. Reconstruction of parent microstructures were successfully carried out in Ti and Zr alloys, undergoing bcc to hcp displacive transformation during cooling [14–17]. The same technique was then extended to austenite reconstruction in steel [18–31]. The process of automated reconstruction involves

S. K. Giri
R&D, Tata Steel Limited, Jamshedpur, India
e-mail: sushil.giri@tatasteel.com

B. S. Kumar · I. Samajdar (✉)
Department of Metallurgical Engineering and Materials Science, IIT Bombay, Mumbai, India
e-mail: indra@iitb.ac.in

S. Suwas and D. P. Field (eds.), *Advances in Texture, Microtexture, and Allied Techniques*, Materials Horizons: From Nature to Nanomaterials,
https://doi.org/10.1007/978-981-96-5346-1_8

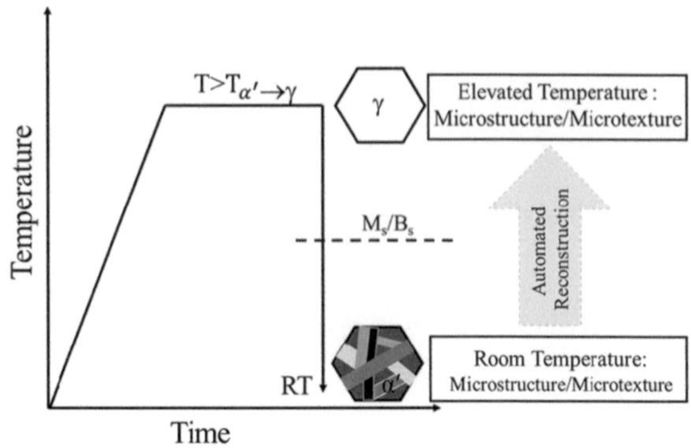

Fig. 1 Schematic describing automated reconstruction of parent phase microstructure from ambient temperature microtexture measurements in IPS (invariant plane strain) product phase

replacing orientations of the IPS product, in an electron backscattered diffraction (EBSD) map or data-set, with appropriate orientation(s) of the parent phase. This naturally involves an algorithm to calculate the parent orientations. But before getting into the actual algorithm, it is important to introduce the reader to the aspects of orientation and orientation relationship [32–35].

2 Orientation of a Crystal

Orientation of a crystal is normally expressed by a rotation, g, which transforms the sample's external coordinate system into crystal coordinate system (Fig. 2a). The g can be specified in different ways [32]. The most frequently used is the one expressed in terms of miller indices (h k l) [u v w], where (h k l) is the crystal lattice plane parallel to rolling plane and [u v w] is the crystallographic direction parallel to the rolling direction. It can also be expressed in terms of Euler angles in Bunge notation $\boldsymbol{\varphi}_1, \boldsymbol{\varphi}, \boldsymbol{\varphi}_2,$ which are shown in the Fig. 2b. First rotation, $\boldsymbol{\varphi}_1,$ is about ND to bring the RD direction in coincidence with a line of intersection between RD-TD plane and the [1 0 0]-[0 1 0] plane. This is followed by rotation, about RD such that ND direction becomes parallel to [0 0 1] crystallographic direction. Then, a rotation, $\boldsymbol{\varphi}_2,$ is made to bring the RD and TD directions in coincidence with [1 0 0] and [0 1 0] crystallographic directions respectively.

The orientation can also be expressed in terms of orientation matrix or tensor, which is mathematically the most convenient to use.

$$g = \left[g_{ij}\right] \tag{1}$$

where $g_{ij} = \cos\left(\boldsymbol{x}_i, \boldsymbol{x}_j^{'}\right)$. It consists of the direction cosines of the axes, $\boldsymbol{x}_i$ of the crystal coordinate system**,** with respect to the axes, $\boldsymbol{x}_j^{'}$ of the sample coordinate system. The nine components of the matrix are not independent, and they are corelated with the three Euler angles ($\boldsymbol{\varphi}_1, \boldsymbol{\varphi}, \boldsymbol{\varphi}_2$) as follows.

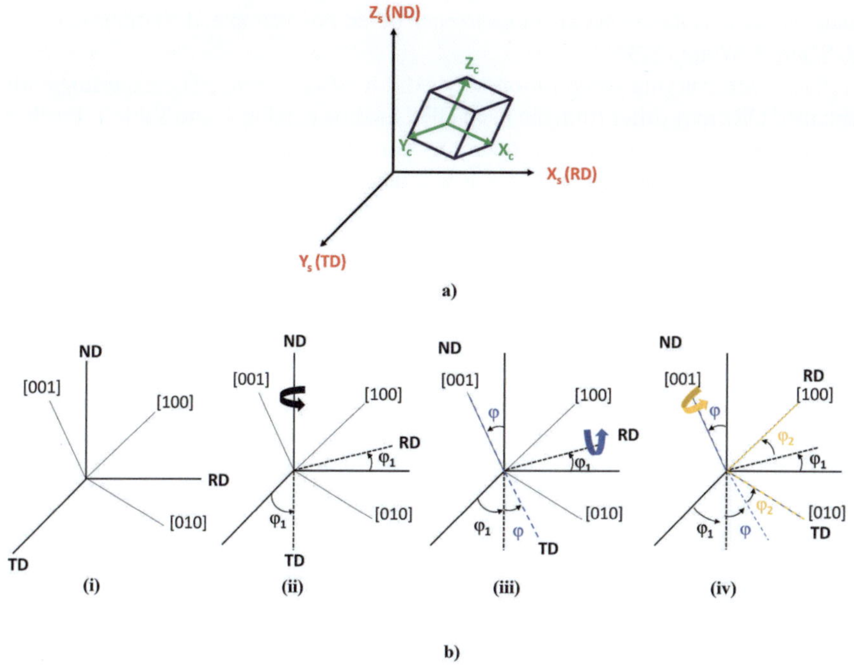

Fig. 2 **a** Orientation g of crystal – representing rotation of sample coordinate system (RD-TD-ND), to crystal coordinate system (<1 0 0>–<0 1 0>–<0 0 1>). **b** Definition of Euler angles $\varphi_1, \varphi, \varphi_2$ in the Bunge convention

$$g(\varphi_1, \varphi, \varphi_2) = \begin{bmatrix} \cos\varphi_1\cos\varphi_2 - \sin\varphi_1\sin\varphi_2\cos\varphi & \sin\varphi_1\cos\varphi_2 + \cos\varphi_1\sin\varphi_2\cos\varphi & \sin\varphi_2\sin\varphi \\ -\cos\varphi_1\sin\varphi_2 - \sin\varphi_1\cos\varphi_2\cos\varphi & -\sin\varphi_1\sin\varphi_2 - \cos\varphi_1\cos\varphi_2\cos\varphi & \cos\varphi_2\sin\varphi \\ \sin\varphi_1\sin\varphi & -\cos\varphi_1\sin\varphi & \cos\varphi \end{bmatrix} \quad (2)$$

3 Orientation Relationship

Orientation relation (OR) between a parent orientation and product orientation is conventionally expressed in terms of family of planes and directions that remain parallel or almost parallel in both phases. They can also be expressed as a rotation in terms of Euler angles. This rotation corresponds to rotation from coordinate system of parent phase to coordinate system of product phase. Table 1 shows commonly reported orientation relationships [6, 33] for steels. As shown in Fig. 3 and also listed in Table 1, these ORs are close to each other. For example, deviations between K-S and N-W are ~5.3°.

Steels with varying composition may exhibit different ORs. The experimentally obtained ORs may differ from the ideal ORs, as shown in Fig. 3 and Table 1. Further, a significant spread may exist around any particular OR [22]. In displacive phase transformation, orientation of all pixels within α' (martensite or bainite for example) grains are not same, and some in-grain orientation spread often exists [22]. Using any particular OR for automated reconstruction brings ambiguity. It is better to use an average OR for reliable reconstruction [20, 22, 36]. Numerical fitting using orientations of variants formed inside a large austenite grain [34, 35] has been used to calculate average OR very successfully. Such an approach does not require the orientation of the parent austenite (γ) phase.

In brief, there are two aspects of algorithms for reconstruction of parent phase microstructure (Fig. 4); algorithm (i) to determine parent orientation calculation [20, 23, 37–40] and (ii) for reconstruction [18–31]. These are being discussed in the next sections.

Table 1 Orientation Relationships between austenite (γ) and product phase(s). While K-S, N-W and Bain are used for displacive (martensite or bainite) phase transformation, Pitsch has been related to tempered microstructures and carbide formation

Orientation relationship	Planes and directions	Euler angles ($\alpha \rightarrow \gamma$)
Kurdjumov–Sachs (KS)	$\{1\overline{1}1\}\gamma // \{1\overline{1}0\}\alpha$, $\{\overline{1}\overline{1}0\}\gamma // \{\overline{111}\}\alpha$	(5.77°, 48.19°, 5.77°)
Nishiyama-Wassermann (NW)	$\{1\overline{1}1\}\gamma // \{1\overline{1}0\}\alpha$, $< 011 > \gamma // < 001 > \alpha$	(0°, 45°, 9.74°)
Pitsch	$\{001\}\gamma // \{0\overline{1}1\}\alpha$, $< 110 > \gamma // < 111 > \alpha$	(9.74°, 45°, 0°)
Bain	$\{001\}\gamma // \{0\overline{1}1\}\alpha$, $< 100 > \gamma // < 100 > \alpha$	(0°, 45°, 0°)

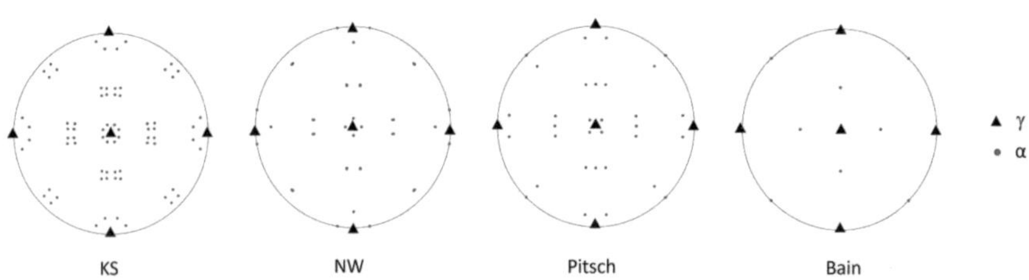

Fig. 3 $\{001\}$ poles of bcc variants projected onto the $(001)_{\gamma}$ plane corresponding to Bain, K-S, N-W, Pitsch orientation relationship

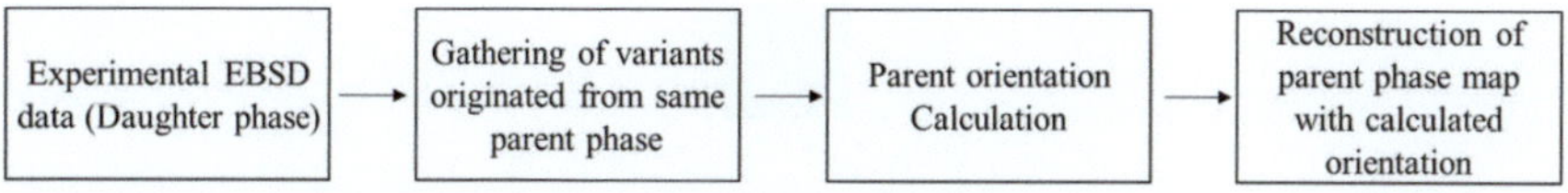

Fig. 4 Steps for automated reconstruction of parent microstructure

4 Algorithm to Determine Parent Phase Orientation

During displacive phase transformation [10, 37, 41], a parent grain generates several grains or orientations of the product. These are called variants. It is to be noted that an OR exists between the orientations of the variants and the orientation of the parent. All possible orientations of the product phase need to be calculated (see Fig. 5).

$$g_\alpha^i = \Delta g.S_i.g_\gamma \quad (3)$$

However, the parent orientation calculation (Fig. 6) is more of a reverse calculation of that mentioned above (see Fig. 5). This takes into consideration of orientations of the variants formed inside an austenite grain. From orientation of one product phase variant, a set of all possible parent orientations are calculated as,

$$g_\gamma^i = \Delta g^{-1}.S_i.g_\alpha \quad (4)$$

Δg^{-1} transforms the coordinate system of the product phase to the coordinate system of the parent phase. In particular, each product variant or orientation generates a set of parent crystal lattice orientations. The trick is to find a common parent orientation from a large number of possibilities. In order to calculate the common parent orientation (Fig. 7), all parent orientations corresponding to each product variant (e.g.: g_α^5) are kept as references. A comparison between different sets of

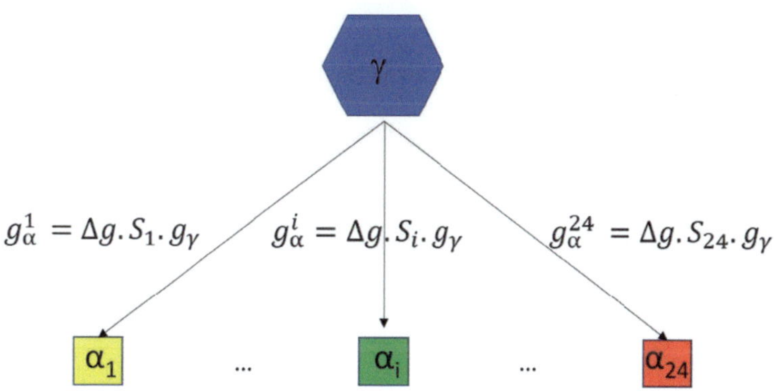

Fig. 5 Schematic showing calculation of orientations of product (martensite or bainite) variants from austenite orientations

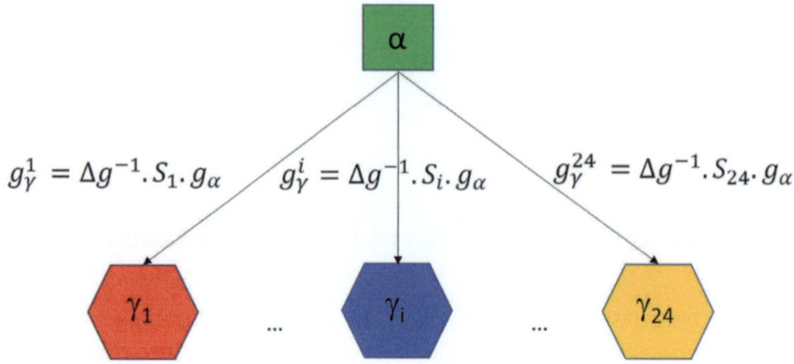

Fig. 6 Schematic showing calculation of possible orientations of parent variants from orientations of the product phase

neighboring parent orientations yields misorientations $(\Delta\theta_\gamma)$ between them, number of variants (n_V) and $p = (\frac{n_v\, for\, solution\, 1}{n_v\, for\, solution\, 2})$. Based on a cut-off for these three parameters, a common parent orientation is then derived. In a transformed parent grain, all variants may not form. A minimal number of variants (n_V) are hence required for retrieving the parent orientation. This number vary with alloy systems and orientation relationship [22]. In steel, the minimum number of variants required to unambiguously calculate the parent orientation was reported to be 7 for KS OR and 4 for NW OR [22] (see Table 2). Further, it is also claimed that the minimum number of variants can be lowered down to only 2 with an average OR [20].

Though procedures for all reconstruction methods are nearly similar, calculations adopted by Miyamoto et al. [20] and Pham et al. [36] are slightly different. In this, austenite orientation is guessed first, and subsequently more accurate orientation is calculated by rigorous fitting with all variants. 'Guessed' austenite orientations are obtained using orientation of one pixel in a selected area of variants [36]. Each of these austenite orientations is then checked to see if they are having OR with all pixels in

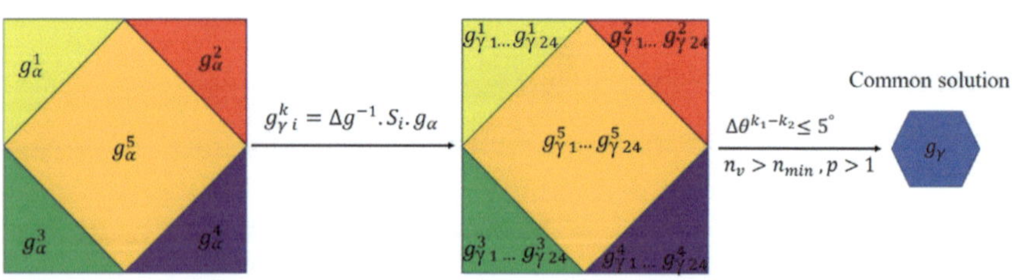

Fig. 7 Schematic showing α (martensite) grains in a transformed (austenite (γ) -to-martensite (α) transformation) microstructure. For each α grain (example: g_α^{5}), and its immediate neighbors (g_α^{1}-g_α^{4}), potential γ orientations were calculated with an average orientation relationship (OR). The common solution or the pre-transformed γ orientation ($g_{\gamma j}^1$) was considered successful when: $(\Delta\theta_\gamma < 5°)$, $n_V \geq 4$ and $p > 1$. $\Delta\theta_\gamma$ = misorientation between the potential near-neighbor solutions for γ, n_V = number of α variants and $p = (\frac{n_v\, for\, solution\, 1}{n_v\, for\, solution\, 2})$

Table 2 Probability (%) of finding a common orientation unambiguously as a function of available variants N_v and OR [22]

OR	NV						
	1	2	3	4	5	6	> 6
Kurdjumov–Sachs	0	69.6	93.5	99.2	99.9	99.67	100
Nishiyama-Wassermann	0	74.2	98.2	100	100	100	100

the selected area. Once, the orientation guessed is finalized, the accurate orientation is calculated by fitting with all pixel's orientations using Newton–Raphson method [36].

5 Algorithm for Reconstruction

There are two steps in the reconstruction of parent microstructure: (i) gathering of variants belonging to a prior austenite grain and (ii) calculation of parent orientation from the orientation of these variants. If the boundaries of parent phase are clearly visible in the ESBD map of the product phase, then gathering of the variants belonging to a parent phase/grain becomes obvious. The orientation of parent phase can then be easily determined using the procedure mentioned in the earlier literature [20, 23, 37–40]. However, often the parent phase grain boundaries are not clearly visible due to complexities of the displacive microstructures. In such cases, one may use appropriate pole figures to figure out the inherited variants [23]. Poles corresponding to all three Bain zone variants (24 variants) in prior austenite grains are arranged in a specific way. The three rings of poles are 90° apart in connecting great circles. This information of such an arrangement of poles, for example in a < 001 > pole figure, is sought out using 'interactive' mode of an EBSD analysis software. This 'manual' user-defined interactive analysis may enable identification of different austenite grains. Once the variants belonging to an austenite grain is gathered or identified, austenite orientation can then be calculated using an appropriate OR. Such a manual mode of reconstruction becomes strenuous, especially when large area reconstruction is needed. In such cases, automated reconstruction approach needs to be adopted (Fig. 8). However, a critical challenge in any such automated reconstruction is to identify prior austenite grains in complex steel microstructure with highly symmetric parent and product cubic phases.

Several strategies have emerged for gathering the variants, see Fig. 9a–c. The first method of variant gathering (Fig. 9a) has originated from the reconstruction work by Cayron et al. [18]. This method uses groupoid concept. First, it searches for triplets of adjacent grains which satisfy the coherence conditions of pre-calculated groupoid data. Coherent conditions are related to inter-variant misorientation which are fixed for a specific OR. Region occupied by each of such triplets are considered part of same parent grain and is called nucleus. Then, this nucleus is allowed

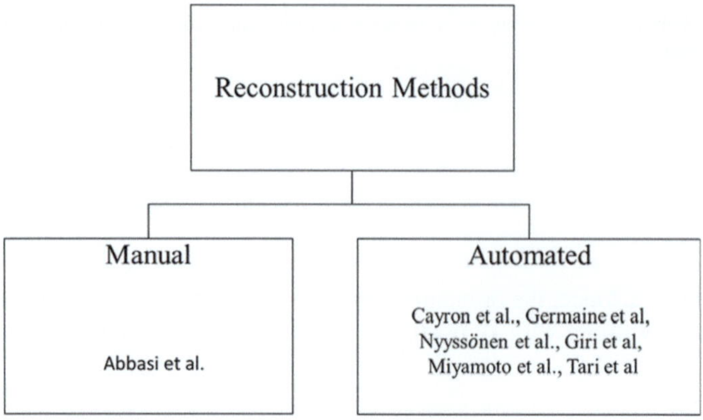

Fig. 8 Reconstruction methods reported in literature

to grow by merging with the neighbors which meets the conditions of coherence. This nucleation or growth process is continued for all nucleus till no more eligible grains or regions are left for merging with the parent nucleus. This process results in the automated reconstruction of the parent microstructure. More recently, clusters of grains belonging to same prior austenite grain are created through the use of Markov cluster algorithm [26, 28, 29, 31]. Once the cluster of grains originating from same prior austenite grain are identified, orientation of parent grain can easily be calculated. Method developed by Nyssonnen [28] is claimed to be very fast and has been implemented in MTEX shareware [43, 44]. Downside of this method is that it still cannot eliminate ambiguous prior austenite orientation calculation.

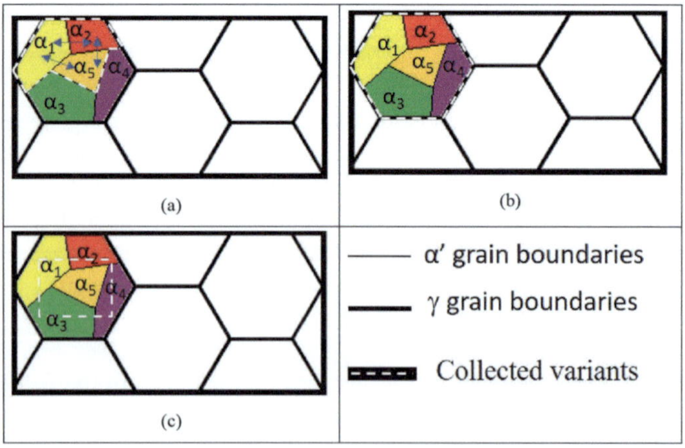

Fig. 9 Gathering of variants for calculating orientation of austenite. **a** Collecting neighboring grains which have specific misorientation compatibility with respect to each other [17, 18, 28, 29] **b** Collecting grains and all its neighbors [22, 30]. **c** Pixels in a domain independent of the shape of grains [20, 25]

Second method (Fig. 9b) of variant gathering consists of selection of sets of neighboring grains or variants [22, 30]. The algorithm subsequently tries to retain those grains in the set, which are in OR with a unique austenite grain. The set should contain minimum number of variants, specific to the OR used (example: four variants for KS OR), in order to obtain a unique austenite orientation for the set. Following the calculation for all sets, some variants remain unassigned with any parent austenite orientation. If the unassigned variant is found to be in orientation relationship with the prior austenite orientation corresponding to a neighboring grain or variant within a higher tolerance angle, then these variants are clubbed with neighboring variant to which prior austenite orientation is already assigned. However, there are some ambiguous domains which can originate from multiple austenite orientations. At location of such variants, austenite orientation is manually assigned taking additional clue from corresponding image quality (IQ) map [22, 30]. Example of a neighbor-to-neighbor reconstruction is given in Fig. 10.

The third approach (Fig. 9c) of variant gathering comes from Miyamoto et al. [20], Bernier et al. [25]. Orientations of all pixels belonging to a small square shaped domain are gathered. Instead of using an average orientation of α' variants, the calculation considers orientations of all α' pixels in the domain. It is to be noted that the size of the selected domain must contain at least two variants and needs to be much smaller than the size of the prior austenite grain. The local austenite orientation for each of this domain is then recovered. The process makes it possible to know the orientation spread within each prior austenite grain. In this method, ideal ORs (like KS, NW) are not used. Instead, an average OR is calculated by numerical fitting of the orientation data of all pixels belonging to a large prior austenite grain [34, 35]. The accuracy of the average OR is claimed to be very high. Further, with the new average OR, the minimum number of variants required to retrieve unique austenite orientation is claimed to be as low as two [20]. This method of Miyamoto et al. [20] or Pixel-by-Pixel reconstruction (see Fig. 11 for example) is, however, computationally far more expensive than the earlier methods. It is often advisable to use the earlier methods (Figs. 9a,b and 10) for statistics and Pixel-by-Pixel (Figs. 9c and 11) for in-depth analysis.

6 Challenges Associated with Automated Reconstruction

There are several challenges associated with automated reconstruction. These are generalized in the following categories.

- Ambiguous Solutions: This may range from certain portions of parent microstructure not providing a definite solution to orientation ambiguities. The former may arise from insufficient number of variants - in the set of variants gathered. Possible parent orientation then may exceed one. Orientation ambiguities, on the other hand, arises from the fact that some α' variants can be related to multiple austenite orientations. Twins in austenite, for example, can have 6 variants (α') in

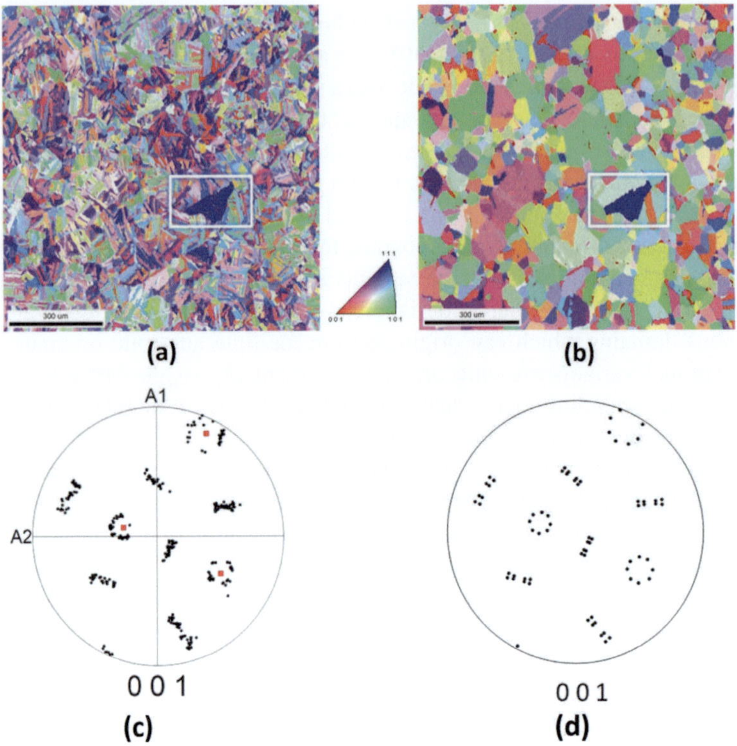

Fig. 10 Neighbor to neighbor reconstruction: EBSD IPF (inverse pole figure) maps of **a** martensite (α) and **b** reconstructed austenite (γ) using neighbor to neighbor reconstruction strategy. (001) pole figure showing **c** measured orientations of martensite variants transformed from a single prior austenite grain and **d** orientations of martensite variants calculated based on orientation relationship and orientation of reconstructed austenite. Martensite variants of this prior austenite grain and corresponding reconstructed austenite grain are highlighted in **a** and **b**. Red colored squares superimposed in Fig. 10**c** represent (001) poles of calculated orientation of this austenite grain.

common. Reconstruction of any such regions/orientations thus becomes ad-hoc, ranging from dilation or spreading of neighbor austenite orientation and manual intervention at later stage.

- Variation in OR: OR may vary spatially within the same austenite grain. Reconstruction with a theoretical OR thus may turn out to be inadequate. Potential solution is to use an average OR.
- Reconstruction in Deformed Austenite: It is best to use reconstruction in domain of reduced scale or size (example: Pixel-Pixel reconstruction) with respect to prior austenite grain, repeating the same for each domain in the entire EBSD map and to use average OR.

Finally, there may not be a universal protocol for all challenges associated with automated reconstruction. These three points highlights potential difficulties and

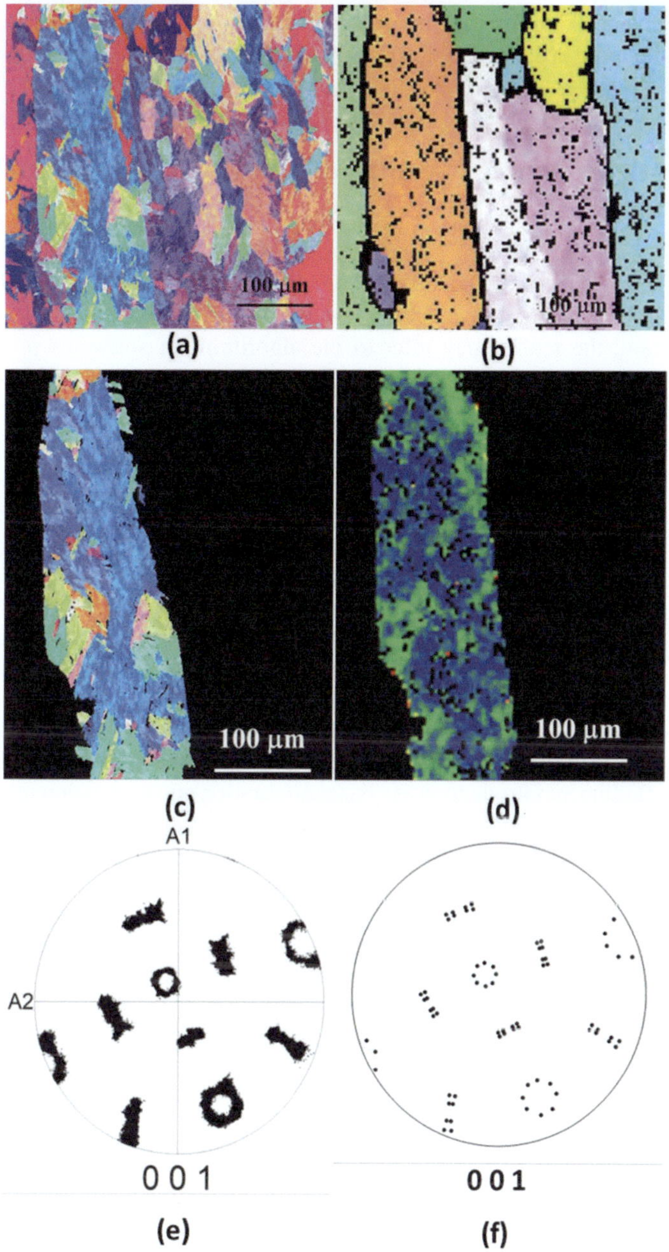

Fig. 11 Pixel by pixel reconstruction. EBSD IPF (inverse pole figure) maps of **a** martensite (α) and **b** reconstructed austenite (γ). Also shown are **c** martensite variants and **d** grain reference orientation deviation (GROD) in a single austenite grain. These also include the respective (001) pole figures of measured and calculated orientations of martensite corresponding to this austenite grain (figure **e** & **f**)

possible remedies (and some of them can be ad-hoc), though the challenge faced by a user can be more specific and needs to be addressed on a case by case basis.

7 Potential Resources: Available Reconstruction Software/ Shareware

There are several software(s) and shareware(s) available. These include ARPGE™, based on groupoid method developed by Cayron et al. [42]. For neighbor-to-neighbor reconstruction, the reader may refer to the algorithms or softwares proposed by Germain et al. [22], Tari et al. [40] or Giri et al. [30]. Similarly, for Pixel-by-Pixel reconstruction one may adopt the algorithm of Miyamoto et al. [20]. Some of these may not be available as a shareware (being restricted for use by respective research groups), but the algorithms can easily be adopted in writing user specific computer programs. Finally, Mtex™ shareware now includes a reconstruction algorithm developed by Nyyssonnen et al. [43, 44].

8 Summary

This book chapter highlights the procedure for automated reconstruction of parent phase microstructures in a displacive phase transformation. It starts with a brief introduction, including an introduction to crystal orientation and orientation relationship, and then summarizes the algorithms for calculating parent phase orientation and of reconstruction. Finally, a summary of potential challenges and resources are made.

References

1. Krauss G (2005) Steels: processing, structure and performance, 1st ed., ASM International, Materials Park, OH 44073–0002
2. Bhadheshia HKDH, Honeycombe RWK (2017) Steels, 4th edn. Elsevier, Butterworth-Heinemann, Oxford, UK
3. Banerjee S, Mukhopadhyay P (2007) Phase transformations–Examples from Titanium and Zirconium Alloys of, Elsevier, volume 12
4. Bhadheshia HKDH (2001) Bainite in Steels, 2nd edn. IOM Communications Ltd., UK
5. Brechet S, Beaujard L (1955) New reagent for the micrographical demonstration of the austenite grain of hardened or hardened-tempered steels. Rev Met 52:830–836
6. He Y (2005) Grain-scale characterization of FCC/BCC correspondence relations and variant selection, PhD Thesis, McGill University
7. Woo KD (2011) Variant selection of Allotriomorphic Ferrite in Steels, PhD diss. Pohang Univ Sci Technol 2011:10–86
8. Miyamoto G, Iwata N, Takayama N, Furuhara T (2012) Quantitative analysis of variant selection in ausformed lath martensite. Acta Mater 60:1139–1148

9. Abbasi M, Nelson TW, Sorensen CD (2013) Analysis of variant selection in friction-stir processed high-strength low-alloy steels. J Appl Cryst 46:716–725
10. Kundu S, Verma AK, Sharma V (2012) Quantitative Analysis of Variant Selection for displacive transformations under stress. Metall Mater Trans A 43A:2552–2565
11. Humbert M, Blaineau P, Germain L, Gey N (2012) Determination of the orientation relation from variants inherited by phase transformation. Mat Sci For 702:862–865
12. He Y, Godet S, Jacques PJ, Jonas JJ (2006) Crystallographic relations between face–and body-centred cubic crystals formed under near-equilibrium conditions: Observations from the Gibeon meteorite. Acta Mater 54:1323–1334
13. Gardiola B, Esling C, Humbert M, Hensger KE (2003) EBSD study of the γ to α phase transformation in an CSP-HSLA steel. Adv Eng Mat 5(8):583–587
14. Germaine L, Gey N, Humbert M (2007) Reliability of reconstructed β orientation maps in titanium alloys. Ultramic 107:1129–1135
15. Glavicic MG, Kobryn PA, Bieler TR, Semiatin SL (2003) A method to determine the orientation of the high-temperature beta phase from measured EBSD data for the low-temperature alpha phase in Ti-6Al-4V. Mat Sci Eng A 346:50–59
16. Glavicic MG, Kobryn PA, Semiatin SL (2004) Validation of an automated EBSD method to deduce the β-phase texture in Ti–6Al–4V with a colony-α microstructure. Mat Sci Eng A 385:372–376
17. Mani Krishna KV, Tripathi P, Hiwarkar VD, Pant P, Samajdar I, Srivastava D, Dey GK (2010) Automated reconstruction of pre-transformation microstructures in zirconium. Scr Mater 62:391–394
18. Cayron C, Artaud B, Briottett L (2006) Reconstruction of parent grains from EBSD data. Mater Char 57:386–401
19. Blaineau P, Germain L, Humbert M, Gey N (2010) A new approach to calculate the γ orientation maps in steels. Sol Sta Phen 160:203–210
20. Miyamoto G, Iwata N, Takayama N, Furuhara T (2010) Mapping the parent austenite orientation reconstructed from the orientation of martensite by EBSD and its application to ausformed martensite. Acta Mater 58:6393–6403
21. Miyamoto G, Iwata N, Takayama N, Furuhara T (2011) Reconstruction of parent austenite grain structure based on crystal orientation map of bainite with and without ausforming. ISIJ Inter 51:1174–1178
22. Germain L, Gey N, Mercier R, Blaineau P, Humbert M (2012) An advanced approach to reconstructing parent orientation maps in the case of approximate orientation relations: application to steel. Acta Mater 60:4551–4562
23. Abbasi M, Nelson TW, Sorensen CD, Wei L (2012) An approach to prior austenite reconstruction. Mater Charact 66:1–8
24. Abbasi M, Kim DY, Nelson TW, Abbasi M (2014) EBSD and Reconstruction of pre–transformation microstructures, examples and complexities in steels. Mater Charact 95:219–221
25. Bernier N, Bracke L, Malet L, Godet S (2014) An alternative to the crystallographic reconstruction of austenite in steels. Mat Char 89:23–32
26. Gomes E, Kestens LAI (2015) Fully automated orientation relationship calculation and prior austenite reconstruction by random walk clustering. In IOP Conf Ser: Mat Sci Eng 82:012059
27. Karthikeyan T, Dash MK, Saroja S, Vijayalakshmi M (2015) Estimation of martensite feature size in a low-carbon alloy steel by microtexture analysis of boundaries. Mic 68:77–90
28. Nyyssönen T, Isakov M, Peura P, Kuokkala VT (2016) Iterative determination of the orientation relationship between austenite and martensite from a large amount of grain pair misorientations. Metal Mat Trans A 47:2587–2590
29. Nyyssönen T, Peura P, Kuokkala VT (2018) Crystallography, morphology, and martensite transformation of prior austenite in intercritically annealed high-aluminum steel. Metal Mat Trans A 49:6426–6441
30. Giri SK, Durgaprasad A, Manikrishna KV, Anoop CR, Kundu S, Samajdar I (2019) Exploring the origin of variant selection through martensite-austenite reconstruction. Phil Mag 19:699–717

31. Araujo EGD, Pirgazi H, Sanjari M, Mohammadi M, Kestens LA (2021) Automated reconstruction of parent austenite phase based on the optimum orientation relationship. J App Crysts 54:569–579
32. Bunge HJ (1987) Three-dimensional texture analysis. Int Mater Rev 32:265–291
33. Bunge HJ, Weiss W, Klein H, Wcislak L, Garbe U, Schneider JR (2003) Orientation relationship of Widmannstätten plates in an iron meteorite measured with high-energy synchrotron radiation. J App Cryst 36:137–140
34. Miyamoto G, Takayama N, Furuhara T (2009) Accurate measurement of the orientation relationship of lath martensite and bainite by electron backscatter diffraction analysis. Scr Mater 60:1113–1116
35. Humbert H, Germaine L, Gey N, Boucard E (2015) Evaluation of the orientation relations from misorientation between inherited variants: application to ausformed martensite. Acta Mater 82:137–144
36. Pham AH, Ohba T, Morito S, Hayashi T (2017) Application of Newton's method for precise calculation of parent orientation and orientation relationship from orientations of daughter phase. Mater Char 132:108–118
37. Humbert M, Moustahfid H, Wagner F, Phillippe MJ (1994) Evaluation of the high temperature texture of the beta phase of a TA6V sample from the individual orientations of grains of the low temperature alpha phase. Scr Mater 30:377–382
38. Humbert M, Wagner F, Moustahfid H, Esling C (1995) Determination of the orientation of a parent β grain from the orientations of the inherited α plates in the phase transformation from body-centred cubic to hexagonal close packed. J Appl Cryst 28:571–576
39. Humbert M, Gey N (2002) The calculation of a parent grain orientation from inherited variants for approximate (b.c.c.-h.c.p.) orientation relations. J Appl Cryst 35:401–405
40. Tari V, Rollett AD, Beladi H (2013) Back calculation of parent austenite orientation using a clustering approach. J App Cryst 46:210–215
41. Humbert M, Wagner F, Esling C (1992) Numbering the crystallographic variants in phase transformation. J App Cryst 25:724–730
42. Cayron C (2007) ARPGE: a computer program to automatically reconstruct the parent grains from electron backscatter diffraction data. J Appl Cryst 40:1183–1188
43. Hielscher R, Nyyssönen T, Niessen F, Gazder AA (2022) The variant graph approach to improved parent grain reconstruction. Materi 22:101399
44. Niessen F, Nyyssönen T, Gazder AA, Hielscher R (2022) Parent grain reconstruction from partially or fully transformed microstructures in MTEX. J App Cryst 55:180–194

Chapter 9
Microtextured Regions (MTRs) in (α + β) Titanium Alloys

Arjun Mahato, Saumya Gupta, Souvik Sahoo, and Shibayan Roy

1 Introduction

Titanium (Ti) and its alloys are widely used as structural materials in automotive, aerospace, chemical and biomedical industries owing to their excellent high-temperature specific strength, corrosion resistance and biocompatibility [1–4]. Pure titanium undergoes an allotropic phase transformation from α phase having hexagonal close packed (hcp, $P6_3/mmc$) at room temperature to β phase with body-cantered cubic (bcc) crystal structure (Im3m) at ~ 882 ± 2 °C (Fig. 1a) [1–3]. The temperature at which α → β transformation occurs on heating or β → α on cooling is referred to as the β-transus (T_β) [1–3, 5]. The α and β phases crystallographically stay related by Burgers orientation relationship (BOR) as $(110)_\beta||(0001)_\alpha$ and $< 111 >_\beta \; || < 11\bar{2}0 >_\alpha$ [2–4]. The nucleation of product α phase within the parent β phase proceeds through the formation of ledges at the α/β interfaces wherein α and adjacent β phases connect via the ledge plane or terrace planes, side plane or *a*-type ledge and *c*-type ledge [6–8]. The α and β phases also stays related by lattice invariant line $< 533 >_\beta \; || < 513\overline{18}0 >_\alpha$ and macroscopic broad faces, $\{111113\}_\beta||\left\{11\overline{15}40\right\}_\alpha$ [9, 10]. The tip of the α phase remains incoherent with the adjacent parent β phase, while the broad face stays semi-coherent [11, 12].

Based on the concentration of substitutional alloying elements (e.g. α stabilizers Al, Sn and β stabilizers V, Mo, Fe, Cr etc.) as well as interstitial alloying elements (e.g. O, C, H, N), titanium alloys are further classified into α, (α+β) and β alloys according

Arjun Mahato, Saumya Gupta and Souvik Sahoo are equal contributions.

A. Mahato
School of Nano Science and Technology, Indian Institute of Technology, Kharagpur, India

S. Gupta · S. Sahoo · S. Roy (✉)
Materials Science Centre, Indian Institute of Technology, Kharagpur, India
e-mail: shibayan@matsc.iitkgp.ac.in

S. Suwas and D. P. Field (eds.), *Advances in Texture, Microtexture, and Allied Techniques*, Materials Horizons: From Nature to Nanomaterials,
https://doi.org/10.1007/978-981-96-5346-1_9

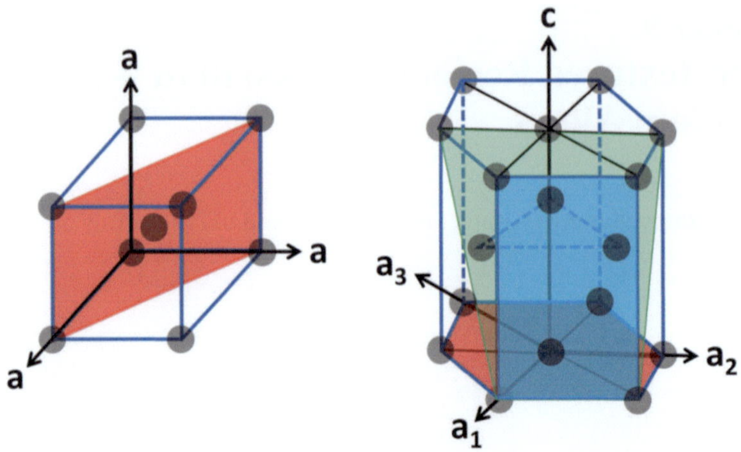

(a)

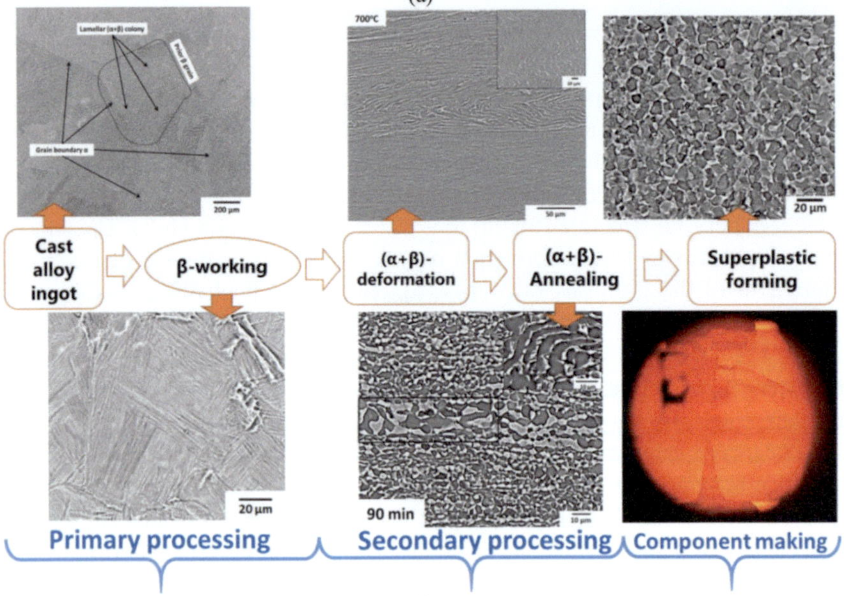

(a)

Fig. 1 **a** Phases of titanium showing unit cell of hcp α (left) and bcc β (right); **b** typical processing schedule for Ti–6Al–4V alloy emphasizing the microstructural transformation through various processing stages (reproduced from PhD thesis of SR)

to their phase stability [1, 2, 3]. Two-phase (α + β) titanium alloys, e.g. Ti–6Al–4V are distinctive due to their ability to tailor various microstructures, enabling a large combination of strength, toughness and high-temperature properties for specific property requirements [2, 4]. They are especially attractive for aerospace applications that demand superior properties at temperatures ~ 600 °C. By varying TMP conditions, lamellar, basketweave, equiaxed, bimodal and even trimodal microstructures can be produced for various ($\alpha + \beta$) titanium alloys (Fig. 1b) [4]. The microstructure evolution and in turn the processing window for ($\alpha + \beta$) titanium alloys is primarily decided by the beta transus T_β as well as other factors like deformation temperature, time, strain and strain rate, deformation mode and deformation ratio. In as-cast condition or after primary β processing (deformation above T_β), a slow-cooled microstructure is characterized by lamellar ($\alpha + \beta$) colonies at room temperature having alternating α phase in plate like morphology and separated by thin β lamellae [13–16]. Basketweave microstructure on the other hand contains acicular α and β lamellae and forms under extreme cooling following casting and/or β deformation. The purpose of primary β processing is to refine the coarse microstructure in castings by initiating dynamic recrystallization followed by comparatively fast cooling from β-phase field [17]. Within a single ($\alpha + \beta$)-colony, constituent α-lamellae are characterized by identical orientation following one of the 12 possible variants in $\beta \rightarrow \alpha$ phase transformation [18].

The transformed lamellar or basketweave ($\alpha + \beta$) microstructures is subsequently converted to a bimodal or fully equiaxed one through secondary TMP that consists of prior deformation, usually by low temperature ($\alpha + \beta$) warm working which is effective to avoid twinning but not sufficient to initiate dynamic spheroidization, followed by static annealing near the β transus [19]. In the bimodal microstructure, initial α phase retained after ($\alpha + \beta$) deformation or annealing and appearing as equiaxed grains is the primary α_p-phase [20]. On the other hand, lamellar α phase derived after secondary TMP from reverse $\beta \rightarrow \alpha$ transformation on cooling are recognized as secondary α_s phase. Equilibrium β phase is retained as thin layer between α_s lamella and together form basketweave or lamellar ($\alpha_s + \beta$) colonies within the transformed β regions in room-temperature bimodal microstructure. The two types of α phases (α_p and α_s) are crystallographically similar but only varies in respective origin, morphology and composition [20]. At times, coarsening of equiaxed α_p grains is promoted by long-term annealing in lower ($\alpha + \beta$) phase field that completely removes ($\alpha_s + \beta$) colonies; this process is termed as recrystallization annealing [21–23]. A complete equiaxed microstructure containing only α_p grains with prior β phase confined as isolated islands is essential for component making, for example, by superplastic forming [24].

During casting, primary β working or secondary TMP, micro-textured regions (MTR) develop for two-phase ($\alpha + \beta$) titanium alloys within the room-temperature microstructure. They manifest as α phase clusters oriented along a single texture component, often spanning over several millimetres and typically larger than equiaxed α_p grains (Fig. 2) [25]. The formation of MTR has garnered significant attention due to their influence on fatigue crack initiation in components made out of ($\alpha + \beta$) titanium alloys, especially by secondary TMP route [26–29]. Numerous field failure reports within the aviation industry have linked MTRs to cold dwell fatigue in titanium alloy engine components [30–32]. A notable example is the uncontained engine failure of AIRBUS A380-861 equipped with Engine Alliance GP7270 engines and operated by Air France on September 30, 2017, while cruising over Greenland (Denmark) [33].

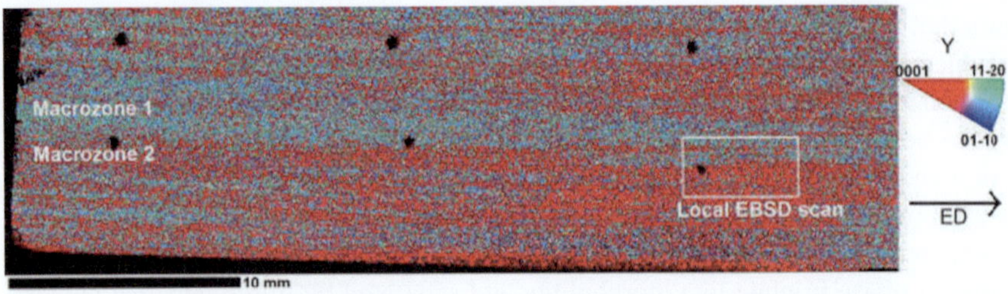

Fig. 2 Inverse pole figure map from electron backscatter diffraction (EBSD) scan showing MTRs (designated as macrozones) in IMI 834 alloy billet after secondary TMP [25]

2 MTR Formation in β → α Phase Transformation

One of the primary reasons for the existance of MTRs at the end of secondary TMP is their formation in β → α phase transformation during casting or primary β processing and even in additive manufacturing. MTRs in these latter cases consist of multiple α variants from adjacent prior β grains sharing a common orientation or a common $(0001)_{\alpha}$ plane across a large microstructural region [10, 34, 35]. As an example, Fig. 3a illustrates two adjacent β grains, β1 and β2; β → α phase transformation leads to several grain boundary (GB) α variants below T_{β}, e.g. a relatively longer GB α1 at the boundary between adjacent β1 and β2 grains (Fig. 3b). As the temperature continues to decrease, lamellar α phase begin to originate from GB α phase and grow towards the transforming β grain interior (Fig. 3c). Lamellar α2 and α3 variants develop from GB α1 variant within β2 and β1 grains, respectively, and share an equivalent orientation between them and with GB α1 variant. Consequently, a large MTR region forms within the transformed microstructure having a common orientation between lamellar α variants from adjacent β grains and GB α variant between them. The size and morphology of MTRs is influenced by the sequence of GB α variant nucleation during β → α transformation [10, 34]. For an early nucleated GB α variant, corresponding lamellar α variants also nucleate early in the adjacent β grains following Burgers OR. This growth advantage then allows the lamellar α variants to expand and cover a larger microstructural area within the transforming β grain, ultimately resulting in MTRs within the transformed microstructure [10, 34–36].

3 Influence of β Grain Boundary Misorientation

High angle β grain boundaries (misorientation > 15°) typically encourage early nucleation of GB α variants in contrast to low angle boundaries (misorientation < 15°) [35, 37, 38]. Under certain extreme conditions, GB α phase may even be absent along low angle β grain boundaries due to their delayed formation in β

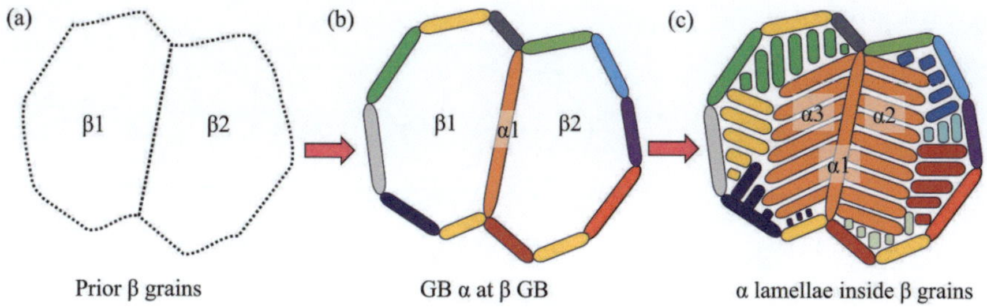

Fig. 3 Schematic showing the process of microstructure development during β → α phase transformation in titanium alloys

→ α transformation [38]. MTR formation at β grain boundary is further facilitated when adjacent β grains are misoriented by specific angle/axis pairs, e.g. $10.5°/<110>_{\beta}$ (**Type A**), $49.5°/<110>_{\beta}$ (**Type B**), $60°/<110>_{\beta}$ (**Type C**) and $60°/<111>_{\beta}$ (**Type D**) [39–42]. They promote a common orientation between GB α and lamellar α variants on either sides when both maintain Burgers OR with the parent β grains (Fig. 4a–d) [42]. For instance, when the common $(110)_{\beta}$ planes from adjacent β grains are misoriented by $49.5°/<110>_{\beta}$ (illustrated by the green and purple rectangles for β1 and β2 grains, respectively, in Fig. 4b), $[\overline{1}1\overline{1}]_{\beta1/\beta2}$ or $[\overline{1}11]_{\beta1/\beta2}$ directions also remain misoriented by 49.5° with respect to the common $(110)_{\beta1/\beta2}$ plane. The $(0001)_{GB\ \alpha}$ plane (shown as an orange hexagon in Fig. 4b) follows Burgers OR as $(110)_{\beta1/\beta2}||(0001)_{GB\ \alpha}$, $[\overline{1}1\overline{1}]_{\beta1}||[\overline{2}110]_{GB\ \alpha}$ and $[\overline{1}11]_{\beta2}||[\overline{1}2\overline{1}0]_{GB\ \alpha}$ with the adjacent β grains. Similarly, lamellar α variants evolve following Burgers OR as $(110)_{\beta1}||(0001)_{lamellar\ \alpha}$ and $[\overline{1}1\overline{1}]_{\beta1}||[\overline{2}110]_{lamellar\ \alpha}$ with β1 grain, and $(110)_{\beta2}||(0001)_{lamellar\ \alpha}$ and $[\overline{1}11]_{\beta2}||[\overline{1}2\overline{1}0]_{lamellar\ \alpha}$ with β2 grain. A common $(0001)_{\alpha}$ plane therefore emerges between the GB α and lamellar α variants from adjacent β1 and β2 grains, ultimately producing MTR within the transformed microstructure.

Out of the high and low angle grain boundaries between adjacent prior β grains, GB α phase nucleates early at the specific β grain boundaries, i.e. **Types A-D** [35]. The special β grain boundaries feature either a common close-packed $\{110\}_{\beta}$ plane and/or $<111>_{\beta}$ direction between adjacent β grains which facilitates GB α nucleation more effectively than random β grain boundaries. Additionally, the nucleation of GB α phase is supported by a favourable balance between high β grain boundary energy and low GB α/β interfacial energy at the special β boundaries. Notably, the energy of a semi-coherent α/β interface (0.05 J/m^2) is significantly lower than that for an incoherent α/β interface (0.4 J/m^2) [13]. GB α phase nucleation at special β grain boundaries promotes an exact or near Burgers OR between GB α variant and adjacent β grains, hence the semi-coherent interface facilitates an early nucleation [35]. Furthermore, GB α phase nucleates earlier at high angle **Types B–D** boundaries in β → GB α transformation than the low angle **Type A** β boundary. A single GB α variant tends to form more frequently along **Types B & C** boundaries than along

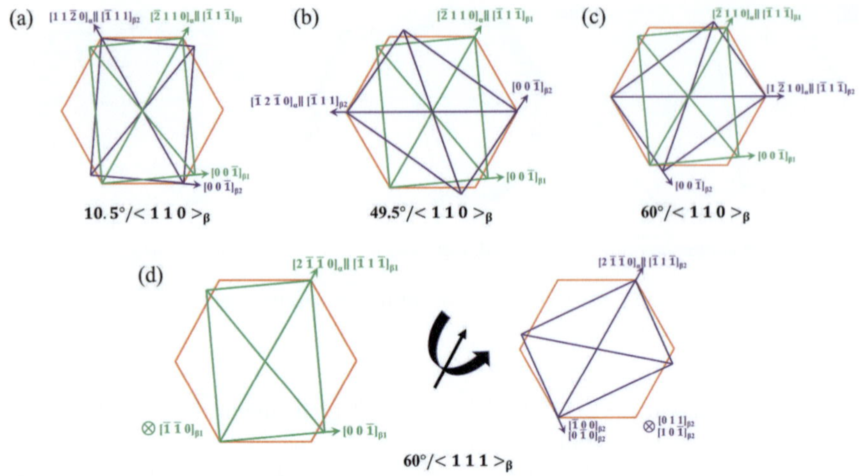

Fig. 4 Schematic showing the orientation relationship between α and β phases in the formation of special prior β grain boundaries [42]

Type D boundaries. Overall, the sequence of GB α phase nucleation is ranked as **Types B/C > Type D > Type A** [35, 43].

4 Effect of β Grain Orientation

The frequency of special β grain boundaries (**Types A–D**) associated with MTR formation is influenced by TMP conditions in β phase field. Liu et al. [35] demonstrated that the variations in TMP conditions lead to alterations in bulk texture components, which in turn affects the frequency of special β grain boundaries. Figure 5a–f represents $(001)_\beta$, $(011)_\beta$ and $(111)_\beta$ pole figures under different TMP conditions, along with corresponding frequencies of **Types A–D** special β grain boundaries. Apparently, the total frequency of **Types B–D** boundaries (mentioned as "sum" in Fig. 5) is critical because of their connection with early GB α phase nucleation and large MTR formation. In as-cast alloy, solidification microstructure contains columnar grains beside the peripheral chilled zones and equiaxed grains at the ingot interior. The columnar grains possess $< 001 >_\beta$ fibre texture due to preferential growth perpendicular to the ingot axis, whereas equiaxed grains are characterized by a near random texture [2]. The frequency of **Types B–D** β grain boundaries are considerable and accounting for 18% of the sum for equiaxed grains (Fig. 5a). In contrast, **Types B–D** special boundaries nearly remain absent (frequency of only 0.1%) for columnar grains (Fig. 5b). However, the frequency of **Type A** β grain boundaries is much higher for columnar grains than within the random textured equiaxed grains.

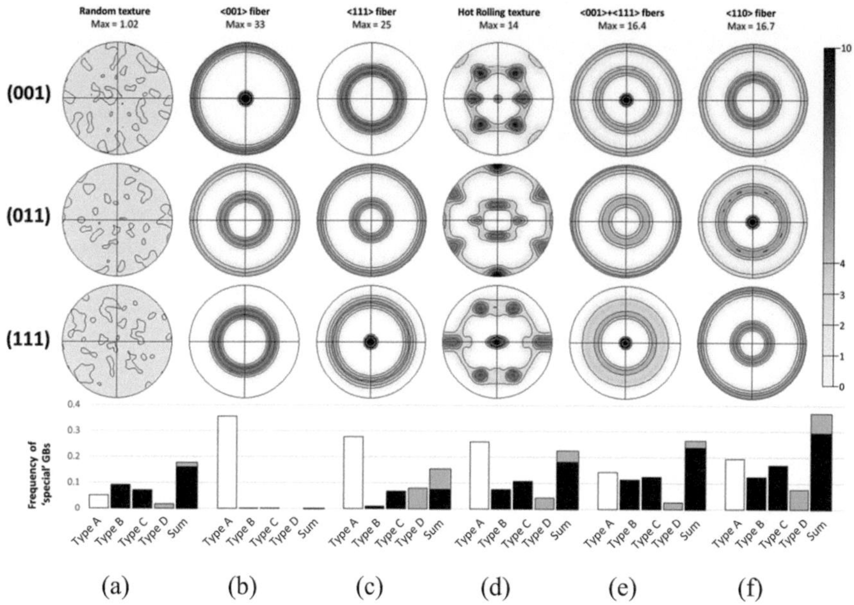

Fig. 5 The influence of different β textures on special grain boundary formation under various processing conditions [35]. The 'sum' represents the total frequency of **Types B–D** pairs at each condition

Similar in castings, prior β grains appears in columnar morphology in additive manufacturing and possess a strong $< 001 >_\beta$ fibre texture along the build direction [4, 44, 45]. The frequency of **Type A** β grain boundary is again significantly higher (>35%) compared to **Types B–D** having a frequency of only 0.1% (Fig. 5b). The propensity of large MTR formation within the transformed microstructure is henceforth much reduced in additive manufacturing. Uniaxial compression above β transus results in strong $< 001 >_\beta$, $< 111 >_\beta$, or a combination of $< 001 >_\beta + < 111 >_\beta$ fibre texture parallel to the compression direction (CD) [46]. Subsequent annealing produces $< 001 >_\beta$||CD fibre texture (Fig. 5b). The likelihood of special β grain boundary formation then becomes similar to the as-cast and additive manufactured alloys having a strong $< 001 >_\beta$ fibre texture. Uniaxial compression at a high strain rate on the other hand produces $< 111 >_\beta$||CD fibre texture with the fraction of **Types B–D** special β grain boundaries (15.5%,) being lower than **Type A** (>25%, Fig. 5c). Finally, $< 001 >_\beta + < 111 >_\beta$||CD fibre texture in β compression increases the frequency of **Types B–D** β grain boundaries up to 26.5% compared to ~ 15% for **Type A** (Fig. 5e). This suggests that the $< 001 >_\beta$ grains adjacent to $< 111 >_\beta$ grains within the deforming microstructure significantly enhances **Types B–D** special β grain formation.

Similarly, hot rolling above the β transus temperature increases the fraction of specific β grain boundaries (22.5% for **Types B–D** and 25% for **Type A**, Fig. 5d). With increase in rolling temperature, the frequency of special β grain boundaries also

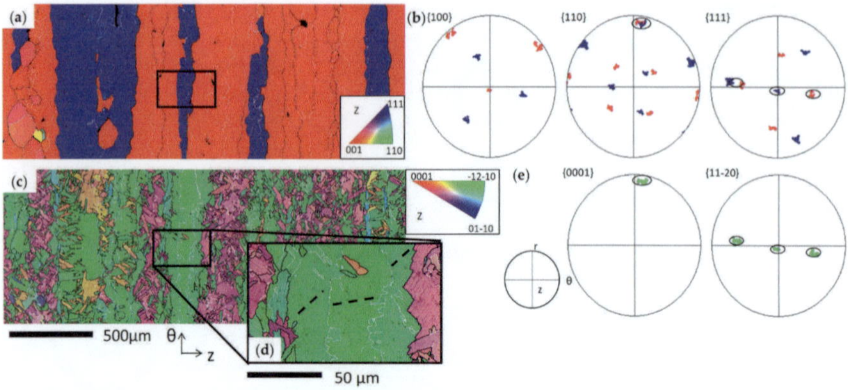

Fig. 6 Inverse pole figure (IPF) map of **a** reconstructed β phase and **b** $(001)_\beta$, $(011)_\beta$ and $(111)_\beta$ streograms; **c** IPF map of α phase from the black rectangular region of β IPF map and **d** $(0001)_\alpha$ and $(11\bar{2}0)_\alpha$ streograms [46]

increases [47]. Extrusion above the β-transus temperature results in $< 110 >_\beta$ fibre texture [48]. In this scenario, the fraction of critical grain boundaries (37% for **Types B–D**) is the highest among all TMP conditions, as shown in Fig. 5f. The fraction of **Type A** grain boundaries is also relatively high (~20%). This increase occurs because all the special β grain boundaries are misoriented with respect to $< 110 >_\beta$ axis, including those at $60°/ < 111 >_\beta$, which aligns with $70.5°/ < 110 >_\beta$ orientation.

Furthermore, the reconstructed inverse pole figure (IPF) map for β phase in case of a β-forged Ti–6Al–2Sn–4Zr–2Mo alloy displays $< 001 >_\beta$ grains (shown in red colour) adjacent to $< 111 >_\beta$ grains (shown in blue colour) within the black rectangle in Fig. 6a [46]. Discrete streograms from this region contain overlapping and closely spaced $(011)_\beta$ and $(111)_\beta$ poles corresponding to two β grains (marked by the black ellipse in Fig. 6b). A one to one correspondence between the black rectangular regions in α IPF map (Fig. 6c–d) and β IFP map (Fig. 6a) suggests a common α variant (in green colour, Fig. 6c) within adjacent β grains. This is further confirmed by the common $(0001)_\alpha$ and $(11\bar{2}0)_\alpha$ poles of α variants (in green colour) in corresponding streograms (Fig. 6e). The common $(0001)_\alpha$ and $(11\bar{2}0)_\alpha$ poles again correspond to poles shown by black ellipse in $(011)_\beta$ and $(111)_\beta$ streograms (Fig. 6b). Altogether, it represents a large MTR across three adjacent β grains wherein a $< 001 >_\beta$ grain is adjacent $< 111 >_\beta$ grain.

5 GB α and Lamellar α Phase Nucleation

More often than not, β → α transformation leads to α variant selection depending upon several factors, e.g. TMP conditions, β grain orientation, retained residual stress, and defects (dislocation, twins, etc.) [47]. At triple junction [34, 49] or other locations of

β grain boundary [4, 49, 50] and grow until impinge on each other. The GB α phase formation depends on adjacent β grain orientation and in particular, on the number of common $\{110\}_\beta$ planes. None or more than one common $\{110\}_\beta$ planes produce multiple GB α variants, whereas a single common $\{110\}_\beta$ plane results in a single GB α variant at β grain boundary [10, 50]. GB α variant formation also depends on the inclination angle between β grain boundary plane and habit plane between GB α and β phases. In case of misorientation angle between GB α and adjacent non-Burgers OR-related β grain (θ_m) > 15°, β grain boundary plane inclination decides the formation of GB α phase, while when θ_m < 15°, the number of common $\{110\}_\beta$ planes adjacent β grains plays crucial role for GB α phase evolution [42]. Such GB α variants follow Burgers OR with one, both, and none of the adjacent prior β grains orientation; in the last case, GB α variants are called as non-Burgers GB α phase [10, 51, 52].

The lamellar α phase nucleates beside GB α phase by either sympathetic nucleation or interfacial instability which decides the orientation being the same or different, respectively between the GB α and lamellar α variants [53]. Similar to GB α phase, lamellar α phase formation depends on the number of common $\{110\}_\beta$ planes between adjacent β grains. A single common $\{110\}_\beta$ plane results in lamellar α variants following $(0001)_\alpha||\{110\}_\beta$ orientation relationship with the common $\{110\}_\beta$ plane, while several or none of the common $\{110\}_\beta$ planes leads to multiple lamellar α variants within adjacent β grains following $(0001)_\alpha||\{110\}_\beta$ with either the common or any of the twelve $\{110\}_\beta$ planes, respectively (Fig. 7) [50].

As mentioned earlier, special β boundaries with $10.5°/[110]_\beta$, $49.5°/[110]_\beta$, $60°/[110]_\beta$ and $60°/[111]_\beta$ misorientation combinations ensure that GB α and lamellar α variants maintain Burgers OR with adjacent β grains on both sides [42]. In the process, specific β boundaries influence variant selection for both GB α and lamellar α phases during $\beta \rightarrow \alpha$ transformation [10]. For example, these misorientation combinations create a common orientation for GB α and adjacent lamellar α variants since $(0001)_\alpha$ planes for them stay parallel to the common $(1\bar{1}0)_\beta$ planes of two adjacent parent β grains [10]. The special β boundaries therefore promote lamellar α variant formation exclusively from the interfacial instability wherein it inherits the orientation of GB α variant and maintains Burgers OR with parent β grain. Therefore, a significant fraction of β boundaries with specific misorientations can substantially contribute to the variant selection for both GB α and lamellar α phases and in turn lead to MTR formation. Furthermore, special β boundaries also dictate the order of formation for GB α variants during $\beta \rightarrow \alpha$ transformation [34, 43]. The GB α phase tends to nucleate first at β grain boundaries having $49.5°/[110]_\beta$ and $60°/[110]_\beta$ misorientations, followed by $60°/[111]_\beta$ boundaries and finally at $10.5°/[110]_\beta$ boundaries. Following this order, lamellar α phase also forms at β boundaries having $49.5°/[110]_\beta$ and $60°/[110]_\beta$ misorientations at the earliest stages of $\beta \rightarrow \alpha$ transformation. These initial lamellar α variants thereafter grow and cover larger microstructural area and may lead to MTRs in the transformed microstructure [35, 43].

On and above the misorientation between adjacent β grains, the size and orientation of lamellar α variants depends on their nucleation from GB α variants. The

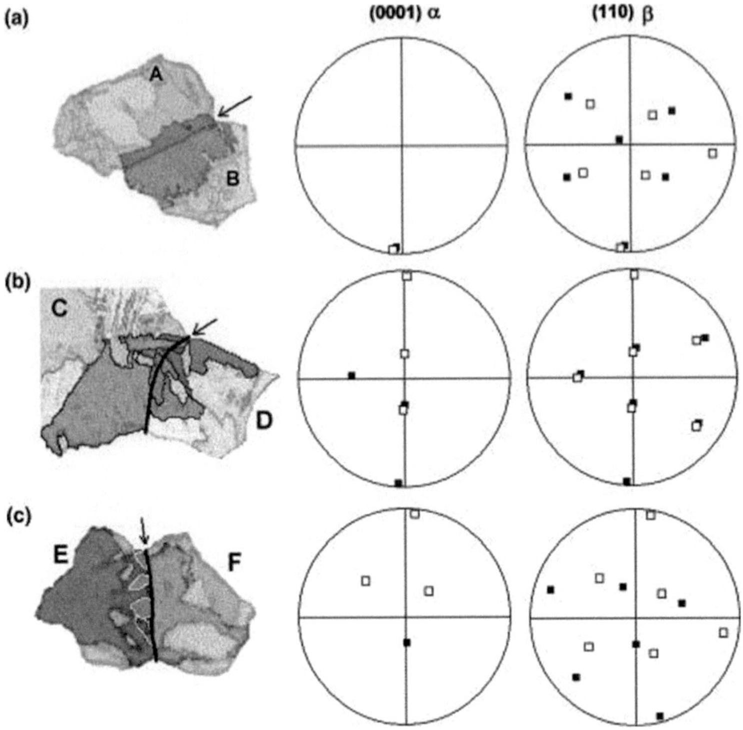

Fig. 7 EBSD band contrast map of adjacent prior β grains are shown on the left. The prior β grain boundaries have been arrowed, and the α variants across the boundary highlighted. The orientation of the prior β grains are shown by (110) stereograms on the right. The orientation of the highlighted lamellar α variants formed beside the boundary are shown in the (0001) stereograms in the centre. Three different scenarios are shown, **a** adjacent β grains share one common (110) pole, **b** adjacent β grains share a number of common (110) poles and **c** adjacent β grains have no close crystallographic relation [50]

lamellar α variants that nucleate earlier in β → α transformation grow unhindered within the transforming β grain and form in the largest sizes in the transformed microstructure [34]. Out of the two nucleation processes (interfacial instability vs. sympathetic nucleation), the former is more likely for lamellar α phase and proceeds earlier due to easier accommodation of transformation strain while maintaining Burgers OR with parent β grain. Sympathetically nucleated lamellar α phase creates additional interfaces between GB α, lamellar α and remnant β phases and is less likely from energy considerations [53–55]. The GB α and adjacent β interface is semi-coherent while lamellar α tip and adjacent β phase forms incoherent interface. The overall energy minimization required in sympathetic nucleation is therefore much higher (both interfacial plus elastic strain energy) than that in nucleation via interfacial instability. Absence of any additional interface makes the energy minimization easier in the latter case when direct growth of lamellar α phase occurs from GB α variants. Due to the growth advantage, lamellar α variants forming via interfacial instability are much larger and often lead to MTR formation within the transformed microstructure.

6 MTR Formation in Secondary TMP

Various secondary operations, e.g. hot rolling, forging, compression, and extrusion are employed in the commercial component manufacturing schedule of titanium alloys. They alter the properties by introducing variety of deformation modes that ultimately modify the microstructure [56]. Secondary TMP operations are classified into dynamic and static depending upon the conditions and number of steps involved. Under dynamic conditions, the alloy undergoes microstructural modification/s concurrent to hot deformation step, while static conditions involve warm working followed by annealing treatment to mobilize remaining effects from prequal dynamic TMP. The purpose of secondary TMP is to transform the initial lamellar or basketweave microstructures into fully equiaxed or bi-modal/duplex ones [57]. Equiaxed microstructure improves the strength and ductility, provides resistance to fatigue crack initiation and enhances low-cycle fatigue life. However, it exhibits limited fracture toughness and lack of fatigue crack growth resistance. Bimodal microstructure on the other hand is suitable especially for fracture critical applications, e.g. involving fatigue because of high resistance to crack growth and propagation [58]. It also increases the strength and ductility but falls short in fracture toughness.

7 Spheroidization of α-Phase

In secondary TMP, deformed α lamellae converts to equiaxed morphology by spheroidization or in simple industrial terms, globularization [17, 20, 59–62]. The micro-mechanisms of spheroidization are predominantly capillary driven and leads to completion through various instabilities (structural defects, e.g. dislocations and sub-boundaries and morphological features like kinks and striations) either already present in the initial lamellar structure or form during secondary TMP [58, 63–66]. These mechanisms include thermal grooving and boundary splitting, Rayleigh perturbation instability, direct cylinderization, edge spheroidization, termination migration and fault migration etc. [58, 63–66]. Important to note that the spheroidization mechanisms are not unique to two-phase titanium alloys but also operates in other two-phase materials like pearlitic steel and magnesium alloys having an initial lamellar morphology [67–71]. The main driving force for microstructural conversion of α_p phase from lamellar to equiaxed morphology is the reduction in α/β interfacial energy as most of the spheroidization mechanisms (thermal grooving and coarsening) are diffusion controlled [58, 64]. Since Burgers OR is followed during $\beta \rightarrow \alpha$ transformation, α/β interface remains semi-coherent in the β transformed microstructure with occasional ledges to minimize the lattice parameter mismatch between the two phases [72–74]. The imposed deformation on warm deformation elevates α/β interface incoherency by piling up dislocation arrays at the boundary plane [13]. This eventually increases α/β interfacial energy and promotes interfacial diffusion on static annealing which escalates spheroidization [63].

The spheroidization mechanisms act as one at a time or in a combination during the course of secondary TMP (deformation and subsequent annealing). Out of them, dynamic spheroidization (concurrent to hot deformation) proceeds via boundary formation and thermal grooving [62, 63, 75], while static annealing after warm working proceeds by both, boundary splitting and termination migration [17, 60]. In addition, continued coarsening occurs by direct cylinderization when aspect ratio of α_p lamellae sufficiently decreases below a critical limit due to termination migration [64, 66]. Boundary splitting proceeds by dislocation substructure formation as intra-lamellar boundaries or sub-boundaries upon recovery [16]. During hot deformation or static annealing, like signed dislocations form low angle intra-lamellar α/α boundary (misorientation angle $\theta < 15°$). The adjacent β phase then penetrates through α lamellae in order to set local equilibrium of surface tensions between the inter-lamellar α/β and intra-lamellar α/α boundaries at the triple point (Fig. 8a) [14, 18, 19]. Materials transport from curved interfaces at the groove area to flat interfaces of the same or adjacent lamella afterwards destroys local equilibrium, increases dihedral angle and drags the triple point deeper until local equilibrium is re-established. The thermal grooving process thus continues and sub-boundary is eventually eliminated. The cusp that forms on either side of the lamella thickness finally joins and a new α/β interface forms. This is to say, α lamellae divides at several places and create new terminations.

Termination migration is initiated by the difference in curvature of microstructural features resulting in atomic diffusion between the terminations and flat interfaces (Fig. 8b) [76]. The diffusion increases in presence of dislocations from hot deformation and is influenced by amount of applied strain. The mass transfer from termination edge to flat interface results in broken lamellae with curved edges. In direct cylinderization, instability initiates due to the difference in curvature leading to mass transfer between edge of the plate and adjacent flat surface (Fig. 8c) [77]. This results in a recession of edge and compensatory accumulation over the flat interface and continues with formation and growth of the ridges along the length of plate. The ridges merge and ultimately create cylindrical shapes. This process is particularly prevalent in plates with low aspect ratio wherein ongoing edge recession and resulting ridge formation are indicative of balanced mass transport producing cylindrical morphology. Edge spheroidization is similar to direct cylinderization but more favourable to structures with large aspect ratio (Fig. 8d) [64]. The mass transfer again occurs from the edge to the interface out of capillary forces leading to edge recession and formation of ridges. Additionally, edge spheroidization promotes growth of the ridges and introduce perturbation (Rayleigh instability) along the length of the interface that classifies it from direct cylinderization.

Further to the microstructural mechanism/s, a study by Roy and Suwas [78] sheds new insights about the spheroidization response of α lamellae from an orientation

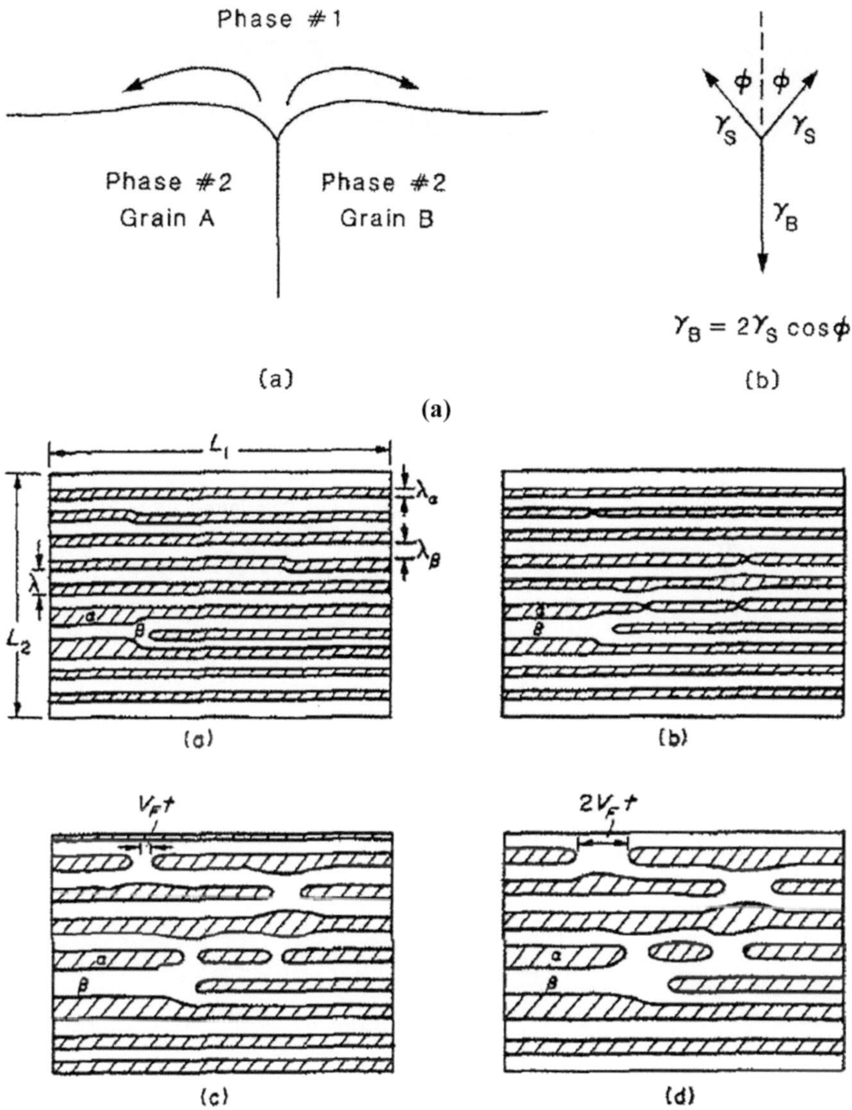

Fig. 8 Schematic showing the processes of **a** boundary splitting, **b** termination migration [64, 76], **c** direct cylinderization [77] and **d** edge spheroidization [64]

perspective (Fig. 9). They have correlated the orientation of lamellar α colonies with respect to the straining directions during prior (α + β) deformation (warm rolling) to their spheroidization response on subsequent static (α + β) annealing. The α-colonies from any deforming microstructure were classified into several orientation types. This dictates slip activation and ultimately varies dislocation generation in warm

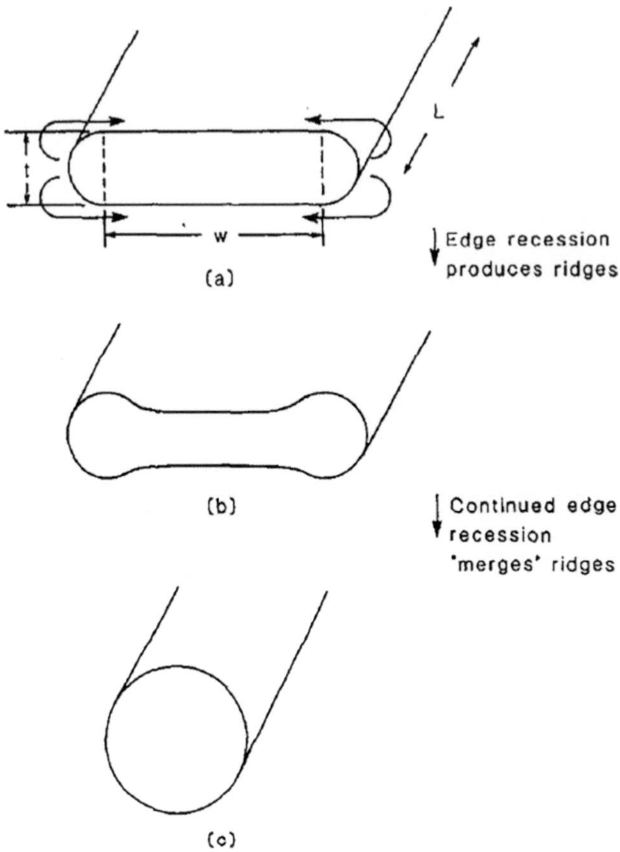

(c)

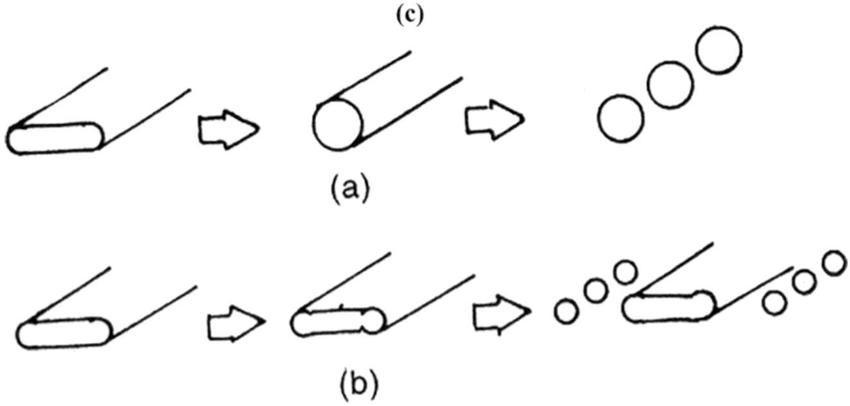

(d)

Fig. 8 (continued)

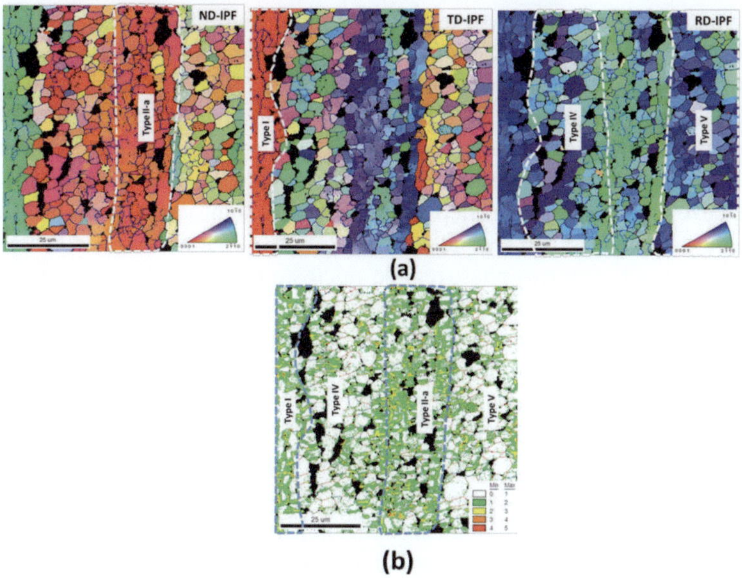

Fig. 9 **a** Inverse pole figure (IPF) maps with principal directions from prior rolling as normal direction, ND (left, ND-IPF), transverse direction, TD (centre, TD-IPF) and rolling direction, RD (right, RD-IPF) pointing normal of the plane of observation (low angle grain boundary, LAGB, $\theta = 5$–$15°$ in blue and high angle grain boundary, HAGB, $\theta > 15°$ in black) and **b** Kernel average misorientaion, KAM map (LAGB, $\theta = 5$–$15°$ in blue and HAGB, $\theta > 15°$ in red) after 60 min of static annealing [78]

rolling. Later, spheroidization response of different α colony orientations differ on subsequent static (α + β) annealing as per their initial orientations since slip activation and dislocation generation influences the mechanisms of morphological conversion (e.g. boundary splitting, thermal grooving, and termination migration). The same applies to other forms of deformation e.g. hot compression as well.

8 MTR Formation Mechanisms

A direct consequence of spheroidization during secondary TMP is the formation of MTRs in the room temperature bimodal microstructure of two-phase titanium alloys. It signifies for a sharp texture heterogeneity within the bimodal microstructure and is usually more significant than the equiaxed α_p grains [25]. MTRs are reported and investigated for various two-phase titanium alloys, e.g. IMI834, Ti-6Al-4V, Ti6242, TA15, and Ti6321 (Fig. 10) [79]. Figure 11 represents example of several MTRs with different crystallographic orientations within the bimodal microstructure of TIMETAL 834 billet at the end of hot compression and subsequent static annealing [80]. Understanding the mechanisms of MTR formation and their development in secondary TMP remains an active field of research. Interestingly, production of spheroidized α_p grains within bimodal microstructure was originally the aim of billet conversion without any connection with MTRs [81]; they were recognized only after the advent of EBSD [82].

Formation of MTRs within the bimodal microstructure is related in the first place to the orientation proximity between secondary α_s and primary α_p phases [80]. They start to develop in deformation step and continues in some cases during subsequent annealing treatment. When secondary TMP is carried out at temperatures near T_β although well within the two-phase regime, microstructure contains a large fraction of β phase from the forward $\alpha \rightarrow \beta$ transformation upon heating. The high-temperature β phase preferentially transforms from α colonies which contain remnant strain from prior ($\alpha + \beta$) deformation (warm working) [83]. For them, additional dislocations help to accommodate $\alpha \rightarrow \beta$ transformation strain better. The product α_s phase within transformed β regions then follows the orientation of adjacent α_p phase during reverse $\beta \rightarrow \alpha$ transformation on cooling in order to reduce the elastic (transformation) strain energy at α_p/prior β phase boundary. As a result, $\beta \rightarrow \alpha$ transformation leads to strong variant selection for α_s phase so much so that transformed β regions also acquire the orientation of adjacent α_p phase. Together, α_p and ($\alpha_s + \beta$) regions form MTRs with a single orientation component expanding over a large microstructural area [84].

In a more recent study, Roy and Suwas [78] pointed out that not the entire α_p phase, as previously understood [25, 80], rather a particular orientation of initial α colonies

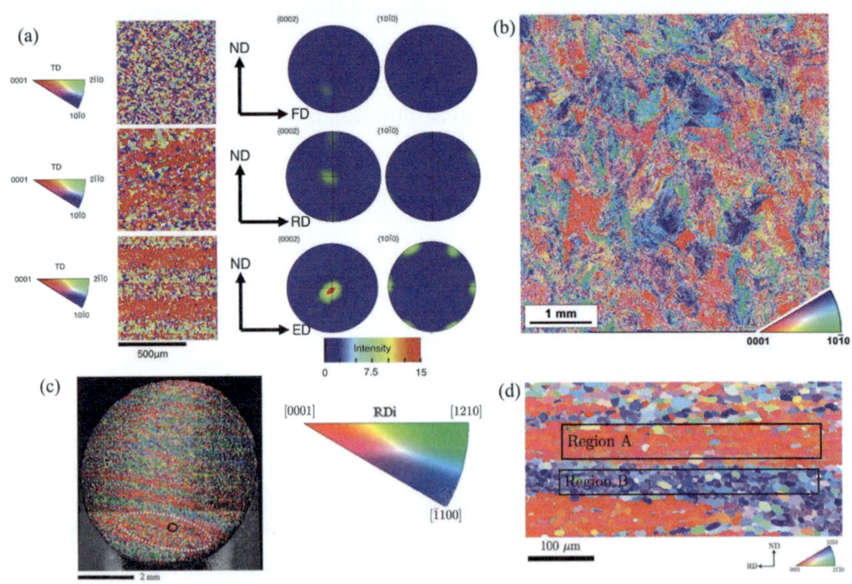

Fig. 10 Experimentally observed **a** macrozone free, intermediate and strong MTR microstructures from top to bottom within Ti–6Al–4V alloy, **b** large-scale MTRs within Ti–6Al–2Sn–4Zr–2Mo alloy, **c** crack nucleation site within a MTR in IMI 834 alloy and **d** neighbouring macrozones with different crystallographic orientations in Ti–6Al–4V alloy [79]

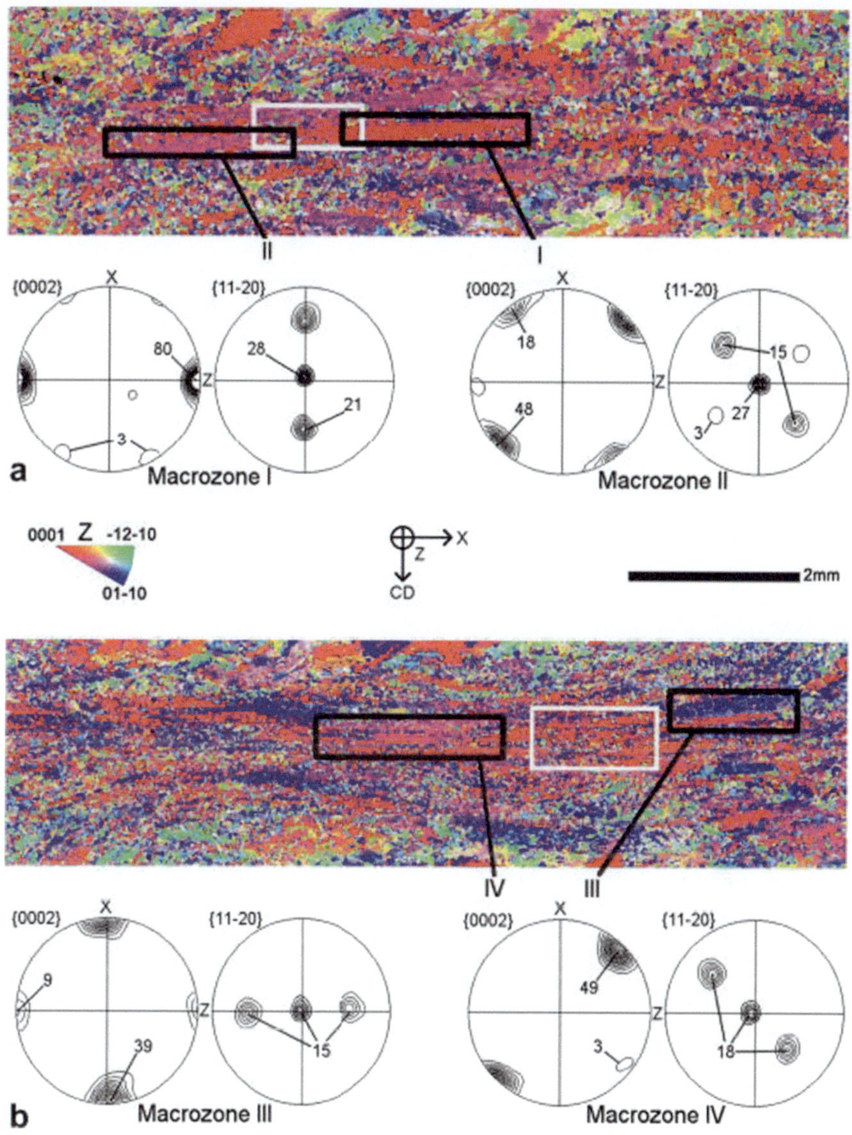

Fig. 11 MTRs formed after **a** hot compression and **b** subsequent annealing of TIMETAL 834 alloy billet [80]

is responsible for MTR formation after secondary TMP (Fig. 12). For this initial α phase, orientation (c-axis||transverse direction) is such that prior warm rolling fails to initiate spheroidization on subsequent annealing. Their stability against spheroidization and additional strain during prior deformation influences adjacent α_s orientation from $\beta \rightarrow \alpha$ transformation upon cooling. Basal-oriented α_p grains are generally

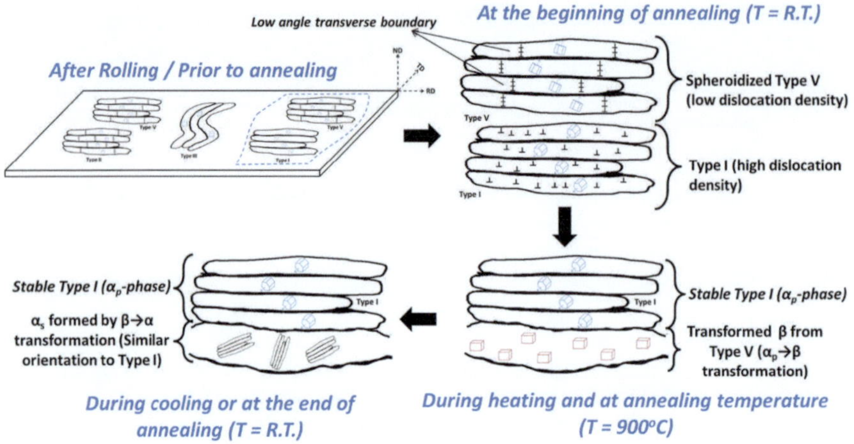

Fig. 12 Schematic showing the interrelation between orientation-dependent spheroidization and MTR formation during static annealing after prior warm rolling [78]

harder than prism-oriented grains, making them more favourable for MTR formation [85]. In case of uniaxial secondary hot deformation, lamellar α colony having their *c*-axis oriented from 45° to 60° with respect to principal loading direction breaks MTR by multiple basal $< a >$ plus prism $< a >$ slip activation [75]. $c - axis||RD$ texture on the other hand impedes spheroidization for hard α lamellae and allow MTR to retain within bimodal microstructure [86]. The geometric orientation of initial lamellar α colony relative to deformation direction also plays a decisive role in MTR formation after secondary TMP [87]. The α colonies having their *c*-axis and broad face initially aligned along transverse direction assist in MTR formation in hot (α + β) rolling [88].

9 Effect of MTR on the Fatigue Properties

The presence of MTRs as clusters of α phase with minimal misorientation within the bimodal microstructure of wrought titanium alloys is detrimental for fracture critical applications, especially under cold-dwell fatigue conditions [89]. Dwell fatigue involves prolonged period of high stresses at room temperature up to 200 °C. It is known to markedly decrease the fatigue life by an order of magnitude or decrease fatigue strength by more than 100 MPa in comparison to regular cyclic fatigue with typical loading–unloading cycles. Only as recently as 1997, an engine failure incident occurred due to dwell fatigue, underscoring its significance for critical safety-related parts, e.g. rotating parts like fan discs and fan blades in aircraft engines [90]. It is encountered by the cold part of aircraft engines during cruising, take-off or landing when significant amount of stress is maintained for a prolonged period. To ensure high reliability of titanium alloys in aerospace applications, extensive research has been carried out to explore the mechanisms by which dwell fatigue induces damage. These studies focused on microstructural factors such as grain size and texture, effect of test conditions for dwell fatigue and approaches for predicting material lifespan [81, 89, 91].

Several field failure reports have linked MTRs with dwell fatigue of titanium alloy engine components used in the aerospace industries [31, 82, 83]. The fatigue lifetime of aero-engine parts e.g. discs in the gas-turbine engines depends on the number density, size and orientation of MTRs [92]. While initial crack size dominantly controls fatigue performance for most titanium alloys, cold-dwell fatigue lifetime is dependent on occurrence of MTR zones within the microstructure [93]. In testing for cold-dwell fatigue, cracks almost always initiate from MTR sites primarily due to basal slip activation, more often so when the loading direction coincides with c-axis or remains within 15°–40° from it within MTRs [58, 94]. MTRs are classified as hard or soft depending on the loading direction being along c-axis ($< 0001 >_{\alpha}$) or perpendicular to it ($< 11\bar{2}0 >_{\alpha}$ or $< 10\bar{1}0 >_{\alpha}$), respectively [95]. The presence of adjacent favourable or soft and unfavourable or hard orientations allow more facet failure within hard MTR zones. A hard MTR can initiate crack easily that reduce fatigue life of the component, whereas slip transfer can happen easily in soft MTRs that enhance creep under high temperature loading conditions [42, 95].

Moreover, MTR having similar orientation of GB α and adjacent lamellar α variants can ease slip transfer between two adjacent β grains through the GB α phase [42]. Additionally, local creep processes within favourably oriented α_p grains causes stress redistribution on hard ones during dwelling [96]. High grain boundary stresses due to localized creep even accelerate crack nucleation [97]. MTRs with more accessible basal or prismatic slip generate numerous small cracks, often scaling with α_p grain sizes throughout the neighbourhood and contribute to crack growth. On the contrary, MTR with no favourable basal and prism slips does not produce any crack [98]. In addition to the morphology and size, geometry of MTRs also plays an important role in controlling stress redistribution between the soft and hard orientations via creep processes. MTRs with high aspect ratio are more detrimental than equiaxed ones due to redistribution of stresses, especially when their long axes are orientated near-normal to the principal loading direction, such that their basal planes are oriented within ~ 15° to the principal stress direction (Fig. 13) [79].

In case of hot (α + β) forged Ti–6Al–4V alloy with bimodal microstructure, both experiments and crystal plasticity finite elements modelling (CPFEM) reveal simultaneous transgranular fatigue crack initiations on basal and prismatic planes, which was cross-verified by employing Tanaka–Mura model [99]. Intergranular crack initiation on the other hand is observed in α grain pairs that share the same basal planes. Hence, plastic deformation in fatigue regime, especially in high cyclic fatigue exhibits significant scatter over the forged alloy part. The effect is more remarkable wherever the stresses are lower than macroscopic yield point making the plasticity relatively lower and localized. For bimodal Ti–5Al–4V alloy however, internal cracks initiate at basal twist grain boundaries although transgranular crack growth is preferred over them [100]. In case of a near-α TIMETAL 834 alloy, a quasi in-situ experiment was performed to track slip traces as well as transgranular vs. intergranular crack

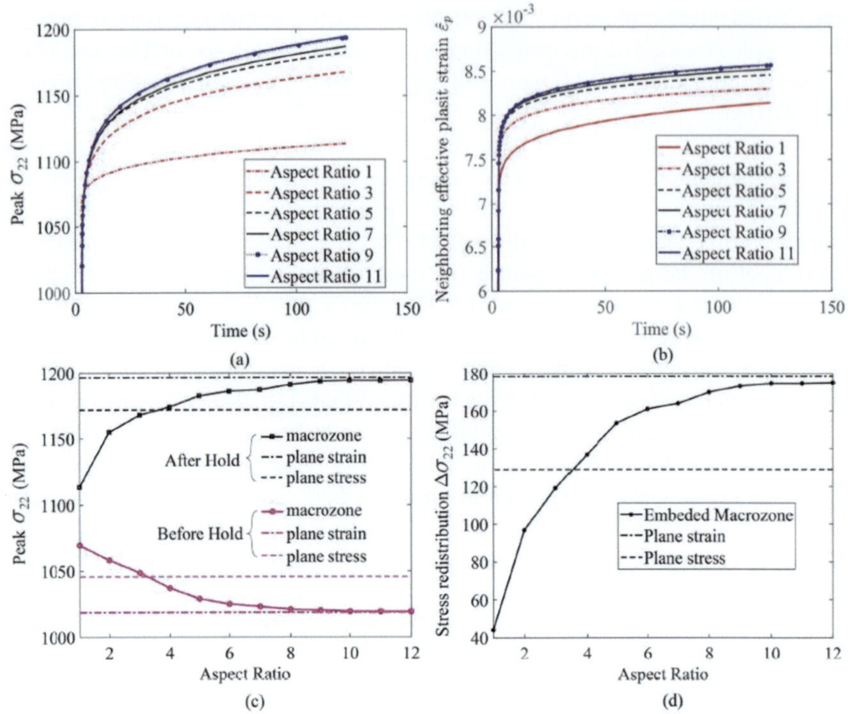

Fig. 13 The effect of MTR aspect ratio on **a** hard grain stress, and **b** adjacent soft grain strain evolution during stress dwell; **c** peak hard grain stress prior to and just after the applied remote stress hold against MTR aspect ratio and **d** stress redistribution on to the hard grain with increasing MTR aspect ratio. The results for quasi-2D plane strain and plane stress analyses are also shown in (**c**) and (**d**) [79]

formation (Fig. 14) [101]. The intergranular crack initiation preceded transgranular one while increasing fraction of α_p grains within the bimodal microstructure was effective in significantly decreasing the propensity of both kind of cracks. Instead of sharp MTRs, smooth regions show low dwell fatigue response and stress sensitivity due to the lack of rapid crack growth pathway.

Several studies focused on the crack growth behaviour in ($\alpha + \beta$) titanium alloy with equiaxed or bimodal microstructure after secondary TMP [101–104]. The mechanism of crack growth was found relatively less influenced by α_p grain size, MTR size and distribution, β phase fraction as well as elemental distribution post-deformation in case of Ti6242 and Ti6246 alloys processed in sub-β transus regime and subsequently heat treated to generate different microstructures [105]. The cracks formed at (0001) and $\{10\bar{1}1\}$ twist boundaries wherein a higher resolved shear stress on the slip plane leads to localized strain and crack formation. In case of TC11 alloy, dwell fatigue sensitivity of equiaxed microstructure was notably higher than the bimodal one on

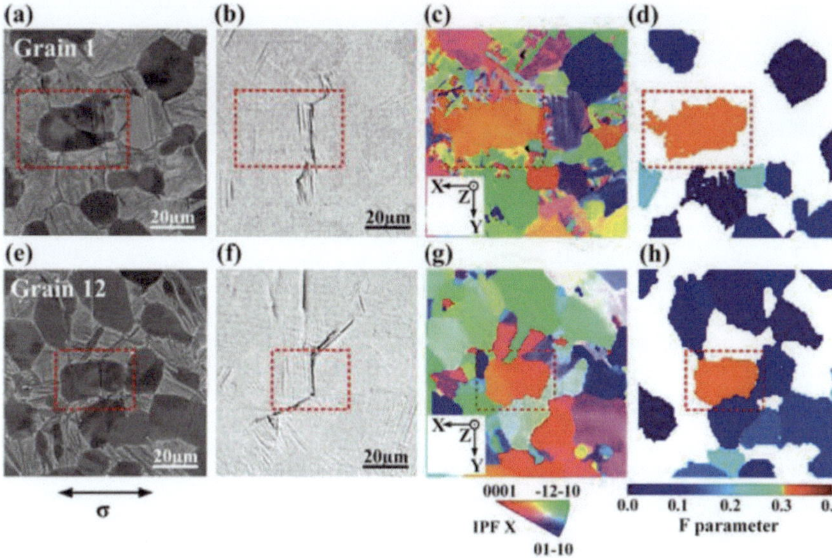

Fig. 14 Typical transgranular crack pathway and their systematic study in different grains for forged TIMETAL®834 ingot [101]

application of identical normalized stresses [106]. For a newly developed Ti–Al–Mo–Zr–Nb–V alloy, effect of MTRs on the static globularization kinetics was systematically investigated during static annealing post unidirectional and cross-rolling through an analysis of slip activation and transmission (Fig. 15) [107]. Boundary splitting is hindered within MTRs due to similar orientation of α phase while deformation modes undergo different kinetics. MTRs are characterized by [0001]||*TD* orientation and results in significantly anisotropic mechanical properties between rolling direction (RD) and transverse direction (TD) loading. In cross-rolling and static annealing, two different orientations evolve for MTRs wherein cracks nucleate at their interfaces and propagates along the length adjacent to each another.

Further, Wang et al. [108] compared the impact of MTR scale and its variable intensity on the low cycle vs. dwell fatigue and proposed two different mechanisms for them. The cracks nucleated in low cycle fatigue along either basal slip bands or through basal twist grain boundaries of the grains having the highest Schmidt factor (Fig. 16). On the contrary, crack nucleation in dwell fatigues is independent of Schmidt factor for basal slip that includes unfavourable or hard grains as well. It is further concluded that a large area of faceted fracture morphology and associated crack initiation and growth along (0001) $< 11\bar{2}0 >$ direction is closely related to basal slip initiation, whereas MTRs perpendicular to the principle deformation direction acts as a barrier towards faceted crack growth [27].

Additionally, a significant research progress is made towards developing quantitative models for MTRs to interpret their role in dictating the fatigue performance of (α + β) titanium alloys. For example, EBSD image-based crystal plasticity (CP)

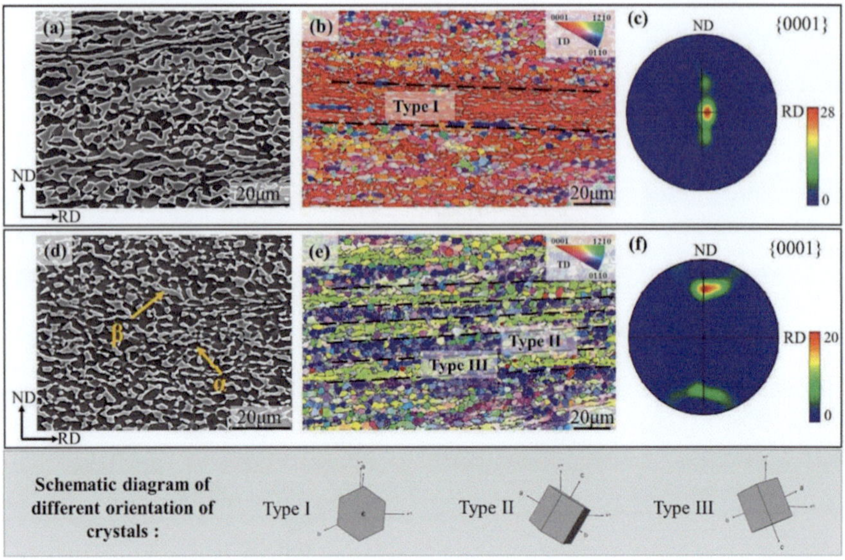

Fig. 15 Scanning electron microscope image in secondary electron mode, inverse pole figure maps and pole figures of α phase in the rolled sheet after annealing at 850 °C for 8 h; **a**, **b**, **c** unidirectional and **d**, **e**, **f** multistage cross-rolling [107]

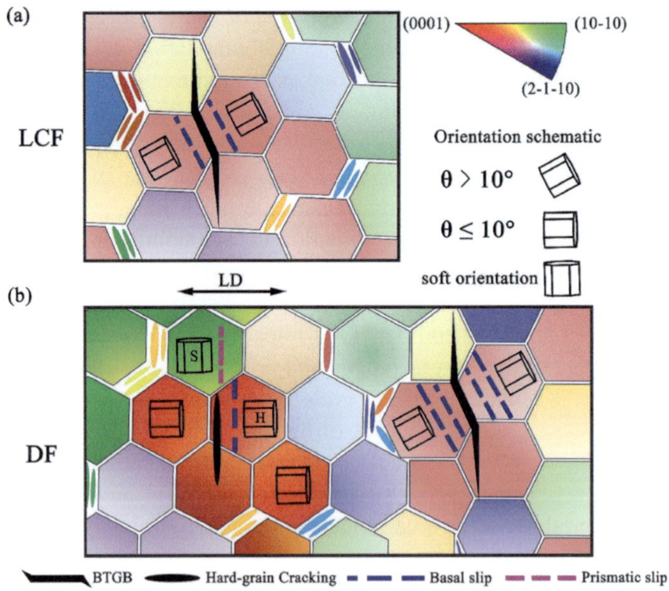

Fig. 16 Schematic showing the mechanism for **a** low cycle (LCF) and **b** dwell fatigue (DF) in a forged Ti60 alloy [108]

model is established for computing localized stresses and plastic strain accumulation in the proximity of crack initiation site [109]. Here, the presumption that the grain boundaries are perpendicular to the surface helps to generate realistic crystallographic orientations and grain morphologies and thereby facilitates to simulate the intra and inter-MTR deformation. The fatigue cracks are found to initiate at a cluster of hard grains surrounded by two soft MTRs because of stress redistribution and accumulation of high plastic shear strain. A study by Chen et al. [110] involving crystal plasticity elasto-viscoplastic full-field model with integrated FFT solver (CP-EVPFFT) reveals that fatigue response of MTRs is subject to change with its distribution wherein the intensity is crucial than morphology for a fixed volume fraction (Fig. 17). The simulated MTR morphologies possess a cylindrical and spherical shape with different axial inclinations. Further, a MATLAB-based computation tool was constructed by Fernández et al. [111] for automatic detection of MTRs through feeding c-axis tilting and different crystallographic rotations. Its effectiveness was compared with the previously reported data by Pilchak et al. [112] using open-access DREAM3D™ software package which showed relatively restrictive approach when provided a more well-defined pole figure. Structural-scale fatigue crack nucleation was studied by Shen et al. [113] for Ti–6Al–4V alloy using a parametric upscaled consecutive and crack nucleation modelling (PUCM/PUCNM). It simulates a general engine blade under dwell loading by incorporating input factors like MTR size, crystallographic orientation, principal loading direction, and average angular misorientation with respect to MTRs.

In case of a lamellar β forged Ti6242 alloy, cold-dwell fatigue crack initiation sites were analysed using EBSD and related to crystallographic features of faceted initiation sites along radial direction [46]. Facetted cracks seemingly initiate in large feather-like α regions consisting of Widmanstätten colonies on either side of prior β grain boundary and having close orientations with their c-axis at 18–22° to the loading

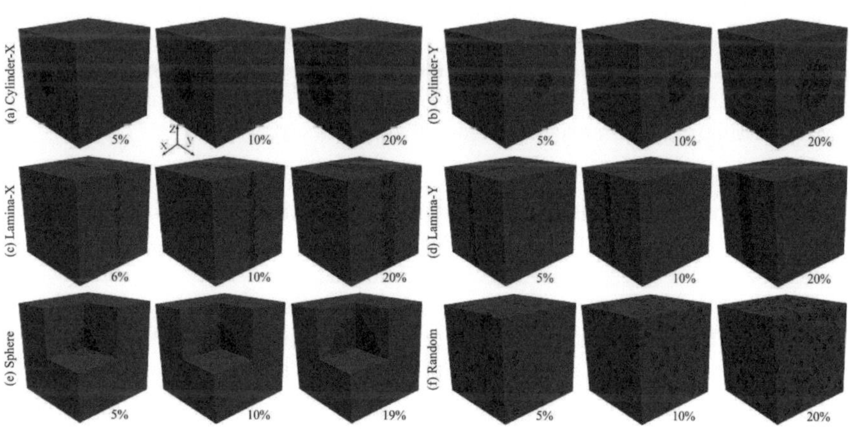

Fig. 17 Different MTR morphologies with varying volume fractions simulated by CP-EVPFFT. Red regions represent MTRs and blue as non-MTR zone [110]

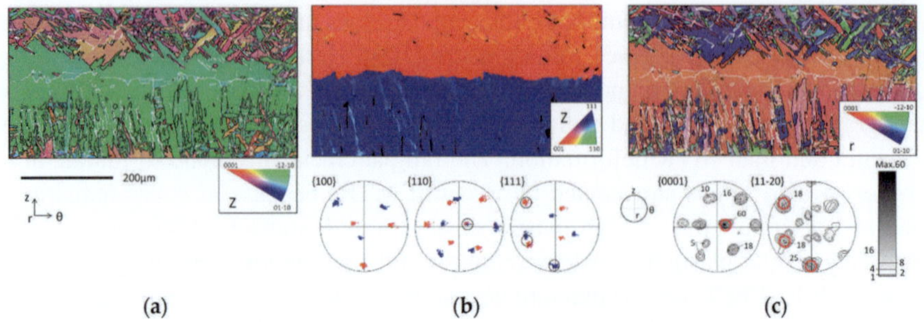

Fig. 18 EBSD map acquired below the facet from fatigue crack initiation and propagation, **a** measured α IPF||CD map (angular misorientations between 3 and 15°: in white); **b** reconstructed β IPF||CD map and stereograms of β orientations close to the boundary; **c** measured α-IPF||RD map (angular misorientations between 3–15°: in white) and corresponding α pole figures. The β/α poles implicated in GB variant selection according to Burgers' OR are encircled on corresponding pole figures [46]

axis (Fig. 11). Adjacent to the initiation site, facet formation is assisted by a stress redistribution from a 'weak' to 'strong' region due to elastic and plastic anisotropy of titanium alloys. The α texture in the most deformed part of the disk is characterized by $< 11\bar{2}0 > ||\mathit{compression\ direction}(CD)$ during β forging, thereby producing a strong α regions having c-axis along radial directions of the disk (Fig. 18). This texture is inherited from prior β forging texture fibres characterized by $< 100 > ||CD$ and $< 111 > ||CD$.

10 Summary

In two-phase (α + β) titanium alloys, MTR formation has attracted considerable attention due to their role in the initiation of fatigue cracks that ultimately leads to failure of critical components. Various mechanisms of MTR formation and their impact on fatigue performance is discussed as follows:

- Microstructure of two-phase (α + β) titanium alloys exhibits lamellar or basketweave (α + β) colonies at room temperature after casting or β processing which then transform to equiaxed, bimodal or duplex microstructure during secondary TMP via spheroidization of α lamellae.
- MTR formation in casting and β processing is related to microstructure development during β → α transformation. Special prior β grain boundaries promote early nucleation of GB α phase, while certain orientation similarity between adjacent β grains results in strong variant selection in β → GB α transformation. Lamellar α phase then nucleates from GB α phase; when the nucleation occurs via interfacial instability, a strong variant selection can take place for them. Altogether, certain orientation similarity between prior β grains and in turn, certain specific β grain

boundaries translate into large lamellar α variants on either sides having identical orientation between them and with GB α phase. This results in MTRs within the transformed microstructure.

- Spheroidization during secondary TMP is driven by various thermodynamic instability mechanisms, such as direct cylindrization, edge spheroidization, boundary splitting and termination migration, which can occur individually or in combination.
- MTR formation within the bimodal microstructure is related to orientation proximity between α_s and α_p phases. The microstructure exhibits significant β phase from $\alpha_p \rightarrow \beta$ transformation during heating which on cooling results in strong variant selection in $\beta \rightarrow \alpha_s$ transformation within the transformed β regions such that they follow the orientation of adjacent α_p phase. Together, MTRs form out of remaining α_p grains and α_s colonies having similar orientations.
- Cold-dwell fatigue occurs in high static applications stresses at moderate to low temperatures and recognized to significantly reduce fatigue life compared to conventional cyclic fatigue. Numerous previous field failure reports as well as experimental and modelling works linked MTRs as the crack initiation sites during dwell fatigue of titanium alloy engine components primarily from aerospace industries.

References

1. Leyens C, Peters M (2006) Titanium and titanium alloys: fundamentals and applications. Wiley Online Library
2. Banerjee D, Williams JC (2013) Perspectives on titanium science and technology. Acta Mater 61:844–879
3. Lutjering G, Williams JC (2007) Titanium. Springer, New York
4. Sahoo S, Roy S (2021) Chapter 8–Additive manufacturing of titanium alloys: microstructure and texture evolution, defect formation and mechanical response. Addit Manufac 151–182
5. Roy S, Tungala V, Suwas S (2011) Effect of hypoeutectic boron addition on the β Transus of Ti-6Al-4V alloy. Metall Mater Trans A 42:2535–2541
6. Zheng Y (2016) The role of the ω phase on the non-classical precipitation of the α phase in metastable β-titanium alloys. Scr Mater 111:81–84
7. Furuhara T, Howe JM, Aaronson HI (1991) Interphase boundary structures of intragranualr proeutectoid alpha plates in a hypoeutectoid Ti-Cr alloy. Acta Metall Mater 39:2873–2886
8. Li D, Zhu L, Shao S, Jiang Y (2016) First-principles based calculation of the macroscopic α/β interface in titanium. J App Phy 119:225302
9. Tong V, Joseph S, Ackerman AK, Dye D, Britton TB (2017) Using transmission Kikuchi diffraction to characterise α variants in an α+β titanium alloy. J Microsc 267:318–329
10. Shi R (2014) Variant selection during alpha precipitation in titanium alloys a simulation study, in materials science and engineering. The Ohio State University, 1–364
11. Laird C, Aaronson HI (1969) The growth of y plates in an Al-15% A8 alloy. Acta Metall 17:505–519
12. Li T, Ahmed M, Sha G, Shi R, Casillas G, Yen HW, Wang Y, Pereloma EV, Carney JM (2015) The influence of partitioning on the growth of intragranular α in near-β Ti alloys. J Alloy Comp 643:212–222

13. Zherebtsov S, Salishchev G, Semiatin SL (2010) Loss of coherency of the alpha/beta interface boundary in titanium alloys during deformation. Phil Mag Lett 90:903–914
14. Mironov S, Murzinova M, Zherebtsov S, Salishchev GA, Semiatin SL (2009) Microstructure evolution during warm working of Ti–6Al–4V with a colony-α microstructure. Acta Mater 57:2470–2481
15. Zherebtsov S, Murzinova M, Salishchev G, Semiatin SL (2011) Spheroidization of the lamellar microstructure in Ti–6Al–4V alloy during warm deformation and annealing. Acta Mater 59:4138–4150
16. Nath P, Marandi L, Sen I (2022) Processing-microstructure-property correlation in thermo-mechanically processed Ti-6Al-4V alloys: A comparative study between conventional and novel approaches. J Alloy Comp 927:167039
17. Kalinyuk AN, Trigub NP, Zamkov VN, Ivasishin OM, Markovsky PE, Teliovich RV, Semiatin SL (2003) Microstructure, texture, and mechanical properties of electron-beam melted Ti-6Al-4V. Mater Sci Eng A 346:178–188
18. Zhu ZS, Gu JL, Chen NP (1995) Variant selection and phase transformation texture in titanium. J Mater Sci Lett 14:1153–1154
19. Lütjering G (1998) Influence of processing on microstructure and mechanical properties of (α + β) titanium alloys. Mater Sci Eng A 243:32–45
20. Salem AA, Glavicic MG, Semiatin SL (2008) A coupled EBSD/EDS method to determine the primary–and secondary-alpha textures in titanium alloys with duplex microstructures. Mater Sci Eng A 494:350–359
21. Semiatin SL, Kirby BC, Salishchev GA (2004) Coarsening behavior of an alpha-beta titanium alloy. Metall Mater Trans A 35:2809–2819
22. Semiatin SL, Corbett MW, Fagin PN, Salishchev GA, Lee CS (2006) Dynamic-coarsening behavior of an α/β titanium alloy. Metall Mater Trans A 37:1125–1136
23. Warwick JLW, Jones NG, Bantounas I, Preuss M, Dye D (2013) In situ observation of texture and microstructure evolution during rolling and globularization of Ti–6Al–4V. Acta Mater 61:1603–1615
24. Roy S, Karanth S, Suwas S (2013) Microstructure and texture evolution during sub-transus thermo-mechanical processing of Ti-6Al-4V-0.1B alloy: Part II. Static annealing in (α + β) regime. Metall Mater Trans A 1–15
25. Germain L, Gey N, Humbert M, Bocher P, Jahazi M (2005) Analysis of sharp microtexture heterogeneities in a bimodal IMI 834 billet. Acta Mater 53:3535–3543
26. Sackett EE, Germain L, Bache MR (2007) Crystal plasticity, fatigue crack initiation and fatigue performance of advanced titanium alloys. Inter J Fat 29:2015–2021
27. Bantounas I, Dye D, Lindley TC (2010) The role of microtexture on the faceted fracture morphology in Ti–6Al–4V subjected to high-cycle fatigue. Acta Mater 58:3908–3918
28. Bantounas I, Dye D, Lindley TC (2009) The effect of grain orientation on fracture morphology during high-cycle fatigue of Ti–6Al–4V. Acta Mater 57:3584–3595
29. Bantounas I, Lindley TC, Rugg D, Dye D (2007) Effect of microtexture on fatigue cracking in Ti–6Al–4V. Acta Mater 55:5655–5665
30. Woodfield AP, Gorman MD, Sutliff JA, Corderman RR (1999) Effect of microstructure on dwell fatigue behavior of Ti-6242. In: Conference: TMS Fall Meeting '98, symposium on fatigue behavior of titanium alloys, Chicago, IL (US), United States. Minerals, Metals and Materials Society, Warrendale, PA (US)
31. Law CC, Blackburn MJ (1976) Influence of subsurface defects on fatigue properties of titanium alloys. Pratt and whitney aircraft groupeast hartford CT. Semiannual rept 3:77–112
32. NTSBNSRA-t. l (1998) Inspection recommendations, and discussion of areas with abnormal high-oxygen microstructure, in response to uncontained failure of GEAE CF6–50, -8OA, and -8OC2 HPC stage 3–9 spools. National Transportation Safety Board (NTSB), 6975
33. BdEedABI R (2020) Accident to the AIRBUS A380–861 equipped with engine alliance GP7270 engines registered F-HPJE operated by air France on 30 September 2017 in cruise over Greenland (Denmark)

34. Salib M, Teixeira J, Germain L, Lamielle E, Gey N, Gautier EA (2013) Influence of transformation temperature on microtexture formation associated with α precipitation at β grain boundaries in a β metastable titanium alloy. Acta Mater 61:3758–3768
35. Liu T, Germain L, Teixeira J, Gautier EA, Gey N (2020) Criteria for fast and selective α precipitation at β grain boundaries in Ti-alloys consequence for in-service microstructures. Matec Web Conf 321:12040
36. Balachandran S, Tripathi A, Banerjee A, Chinara M, Teja R, Suresha SJ, Choudhuri D, Banerjee R, Banerjee D (2020) Transformations, recrystallization, microtexture and plasticity in titanium alloys. Matec Web Conf 321:11020
37. Dixit V (2013) Grain-boundary parameters controlled allotriomorphic phase transformations in beta-processed titanium alloys, in materials science and engineering. The Ohio State University, AAT 3673350. ISBN: 9781321510003
38. Davis AE, Donoghue J, Kennedy JR, Byres N, Prangnell PB (2021) In-situ observation of single variant α colony formation in Ti-6Al-4V. Acta Mater 220:117315
39. Bhattacharyya D, Viswanathan GB, Fraser HL (2007) Crystallographic and morphological relationships between β phase and the Widmanstätten and allotriomorphic α phase at special β grain boundaries in an α/β titanium alloy. Acta Mater 55:6765–6778
40. Cayron C (2008) Importance of the α → β transformation in the variant selection mechanisms of thermomechanically processed titanium alloys. Scr Mater 59:570–573
41. Whittaker R, Fox K, Walker A (2010) (2010), Texture variations in titanium alloys for aeroengine applications. Mater Sci Tech 26:676–684
42. Shi R, Dixit V, Fraser HL, Wang Y (2014) Variant selection of grain boundary α by special prior β grain boundaries in titanium alloys. Acta Mater 75:156–166
43. Liu T, Germain L, Teixeira J, Gautier EA, Gey N (2017) Hierarchical criteria to promote fast and selective αGB precipitation at β grain boundaries in β-metastable Ti-alloys. Acta Mater 141:97–108
44. Sahoo S, Joshi A, Balla VK, Das M, Roy S (2021) Site-specific microstructure, porosity and mechanical properties of LENS™ processed Ti–6Al–4V alloy. Mater Sci Eng A 820:141494
45. Nagase T, Hori T, Todai M, Sun SH, Nakano T (2019) Additive manufacturing of dense components in beta-titanium alloys with crystallographic texture from a mixture of pure metallic element powders. Mater and Des 173:107771
46. Germain L, Samih Y, Delaleau P, Gilgert J, Gey N (2020) Analysis of cold dwell fatigue crack initiation site in a β-Forged Ti-6242 disk in relation with local texture. Metal 10:1–14
47. Obasi GC (2011) Variant selection and its effect on texture in Ti-6Al-4V. In school of materials University Manchester 1–181
48. Roy S, Suwas S, Tamirisakandala S, Srinivasan R, Miracle DB (2012) Microstructure and texture evolution during β extrusion of boron modified Ti–6Al–4V alloy. Mater Sci Eng A 540:152–163
49. Sun Z, Li X, Wu H, Yang H (2016) A unified growth model of the secondary grain boundary α phase in TA15 Ti-alloy. J Alloy Comp 689:693–701
50. Stanford N, Bate PS (2004) Crystallographic variant selection in Ti-6Al-4V. Acta Mater 52:5215–5224
51. Nag S (2008) Influence of beta instabilities on the early stages of nucleation and growth of alpha in beta titanium alloys, in materials science and engineering, The Ohio State University, AAI3300131. ISBN: 9780549448778
52. Zhao J (2015) The deformation and texture evolution of Ti-10V-2Fe-3Al. In materials engineering Monash University 1–199
53. Tang B, Kou H, Zhang X, Gao P, Li J (2016) Study on the formation mechanism of α lamellae in a near β titanium alloy. Prog Nat Sci Mater Inte 26:385–390
54. Sun Z, Guo S, Yang H (2013) Nucleation and growth mechanism of α-lamellae of Ti alloy TA15 cooling from an α + β phase field. Acta Mater 61:2057–2064
55. Appolaire B, Héricher L, Gautier EA (2005) Modelling of phase transformation kinetics in Ti alloys–Isothermal treatments. Acta Mater 53:3001–3011

56. Verlinden B, Driver J, Samajdar I, Doherty RD (2007) Thermo-mechanical processing of metallic materials. Elsevier. ISBN: 9780080974439
57. Lütjering G (1998) Influence of processing on microstructure and mechanical properties of (α+β) titanium alloys. Mater Sci Eng A 243:32–45
58. Bartholomeusz MF, Wert JA (1994) The effect of thermal exposure on microstructural stability and creep resistance of a two-Phase TiAl/Ti3Al lamellar alloy. Metall Mater Trans A 25:2371–2381
59. Stefansson N, Semiatin SL, Eylon D (2002) The kinetics of static globularization of Ti-6Al-4V. Metall Mater Trans A 33:3527–3534
60. Stefansson N, Semiatin NSL (2003) Mechanisms of globularization of Ti-6Al-4V during static heat treatment. Metall Mater Trans A. 34 A, 691–698
61. Weiss I, Froes FH, Eylon D, Welsch GE (1986) Modification of α morphology in Ti-6Al-4V by thermomechanical processing. Metall Mater Trans A 17:1935–1947
62. Mironov S, Murzinova M, Zherebtsov S, Salishchev GA, Semiatin SL (2009) Microstructure evolution during warm working of Ti-6Al-4V with a colony α microstructure. Acta Mater 57:2470–2481
63. Zherebtsov S, Murzinova M, Salishchev G, Semiatin SL (2011) Spheroidization of the lamellar microstructure in Ti-6Al-4V alloy during warm deformation and annealing. Acta Mater 59:4138–4150
64. Sharma G, Ramanujan RV, Tiwari GP (2000) Instability mechanisms in lamellar microstructures. Acta Mater 48:875–889
65. Tian Y, Kraft R (1987) Mechanisms of Pearlite Spheroidization. Metall Mater Trans A 18:1403–1414
66. Courtney TH, Kampe JCM (1989) Shape instabilities of plate-like structures–II. Analysis, Acta Metall 37:1747–1758
67. Brien JMO, Hosford WF (2002) Spheroidization cycles for medium carbon steels. Metall Mater Trans A 33:1255–1261
68. Maity J, Saha A, Mondal DK, Biswas K (2013) Mechanism of accelerated spheroidization of steel during cyclic heat treatment around the upper critical temperature. Phil Mag Lett 93:231–237
69. Meng Y, Fukushima S, Sugiyama S, Yanagimoto J (2015) Cold formability of AZ31 wrought magnesium alloy undergoing semisolid spheroidization treatment. Mater Sci Eng A 624:148–156
70. Yu HC, Wang HY, Chen L, Zha M, Wang C, Li C, Jiang QC (2017) Spheroidization of primary Mg2Si in Al-20Mg2Si-4.5Cu alloy modified with Ca and Sb during T6 heat treatment process. Mater Sci Eng A 685:31–38
71. Zhang J, Li H, Zhan M (2020) Review on globularization of titanium alloy with lamellar colony. Manufac Rev 7:1–14
72. Furuhara T, Aaronson HI (1991) Crystallography and interfacial structure of proeutectoid α grain boundary allotriomorphs in a hypoeutectoid Ti-Cr alloy, Acta Metall. Mater 39:2887–2899
73. Furuhara T, Lee HJ, Menon ESK, Aaronson HI (1990) Interphase boundary structures associated with diffusional phase transformations in Ti-base alloys. Metall Mater Trans A 21:1627–1643
74. Furuhara T, Ogawa T, Maki T (1995) Atomic structure of interphase boundary of an a precipitate plate in a β Ti-Cr alloy. Phil Mag Lett 72:175–183
75. Li Q, Cai J, Chen L, Li C, Yu Y, Yang X, Wang W, Wan K (2024) High-temperature deformation behavior and microstructural evolution of TC17 alloy. Mater Today Commu 40:109383
76. Lin LY, Courtney TH, Ralls KM (1977) Deformation induced microstructural instability in the Pb Sn eutectic. Acta Metall 25:99–106
77. Kampe JCM, Courtney TH, Leng Y (1989) Shape instabilities of plate-like structures—I. Experimental observations in heavily cold worked in situ composites. Acta Metall 37:1735–1745

78. Roy S, Suwas S (2017) Orientation dependent spheroidization response and macro-zone formation during sub β-transus processing of Ti-6Al-4V alloy. Acta Mater 134:283–301
79. Liu Y, Dunne FPE (2021) The mechanistic link between macrozones and dwell fatigue in titanium alloys. Inter J Fat 142:105971
80. Germain L, Gey N, Humbert M, Vo P, Jahazi M, Bocher P (2008) Texture heterogeneities induced by subtransus processing of near α titanium alloys. Acta Mater 56:4298–4308
81. Evans WJ, Gostelow CR (1979) The effect of hold time on the fatigue properties of a β-processed titanium alloy. Metall Trans A 10:1837–1846
82. Woodfield AP, Gorman MD, Corderman RR, Sutliff JF, Yamrom B (1996) Titanium'95. In: Proceeding of eighth world conference on titanium, IOM Birmingham, 1116
83. NTSBNSRA-t. l. (1998) Inspection recommendations, and discussion of areas with abnormal high-oxygen microstructure, in response to uncontained failure of GEAE CF6–50, -8OA, and -8OC2 HPC stage 3–9 spools. National Transportation Safety Board(NTSB), p 6975
84. Germain L, Gey N, Humbert M (2007) Reliability of reconstructed β-orientation maps in titanium alloys. Ultramicro 107:1129–1135
85. Paliwal D, Jha S, Raturi A, Ghosh A, Bandyopadhyay S, Pandey VK, Anoop CR, Narayanan PR, Mukherjee R, Gurao NP (2024) Processing-Microstructure-Mechanical Property Paradigm in Hot Rolled Ti-6Al-4V Alloy. Mater Perf Char 13:215–235
86. Dai Y, Xiao Y, Zeng W, Jia R, Jia W (2024) Mechanism of crystallographic orientation and texture evolution of Ti60 alloy during plane strain compression. Metal 14:359
87. Li Y, Gao P, Yu J, Jin S, Chen S, Zhan M (2022) Mesoscale deformation mechanisms in relation with slip and grain boundary sliding in TA15 titanium alloy during tensile deformation. J Mater Sci Tech 98:72–86
88. Wang L, Fan XG, Zhan M, Jiang XQ, Zeng X, Liang YF, Zheng HJ, Zhao AM (2020) The heterogeneous globularization related to crystal and geometrical orientation of two-phase titanium alloys with a colony microstructure. Mater Des 186:108338
89. Hashimoto S, Takebe H, Mori K, Miyahara M (2023) Effects of size of micro texture regions on the dwell fatigue properties of Ti-6Al-4V. ISIJ Inter 63:1774–1785
90. Xu Y, Joseph S, Karamched P, Fox K, Rugg D, Dunne FPE, Dye D (2020) Predicting dwell fatigue life in titanium alloys using modelling and experiment. Nat Commun 11:5868
91. Evans WJ (1987) The influence of microstructure on dwell sensitive fatigue in a near alpha titanium alloy. Scr Metall 21:469–474
92. Kechagias JD, Aslani KE, Fountas NA, Vaxevanidis NM, Manolakos DE (2020) A comparative investigation of Taguchi and full factorial design for machinability prediction in turning of a titanium alloy. Measure 151:107213
93. Pilchak AL (2014) A simple model to account for the role of microtexture on fatigue and dwell fatigue lifetimes of titanium alloys. Scr Mater 74:68–71
94. Collings EW (1984) The physical metallurgy of titanium alloys. Amer Soc Met Metals Park, OH
95. Adam P, Kate F, Eric P (2024) Cold dwell fatigue of titanium alloys: History, current state, and aviation industry perspective. United States Department of Transportation Federal Aviation Administration
96. Hasija V, Ghosh S, Mills MJ, Joseph DS (2003) Deformation and creep modeling in polycrystalline Ti–6Al alloys. Acta Mater 51:4533–4549
97. Zheng Z, Balint DS, Dunne FPE (2016) Rate sensitivity in discrete dislocation plasticity in hexagonal close-packed crystals. Acta Mater 107:17–26
98. Biavant KL, Pommier S, Prioul C (2002) Local texture and fatigue crack initiation in a Ti-6Al-4V titanium alloy. Fat Frac Eng Mater Struc 25:527–545
99. Briffod F, Shiraiwa T, Enoki M, Emura S (2022) Effect of macrozones on fatigue crack initiation and propagation mechanisms in a forged ti-6Al-4V alloy under fully-reversed condition. Mater 22:101401
100. Wu Z, Kou H, Li J, Hémery S, Li T, Franke C, Prima NCF, Sun F (2025) Preferential fatigue cracking at basal twist grain boundary (BTGB) in bimodal Ti-5Al-4V alloy: Dislocation activities and crack initiation. J Mater Sci Tech 217:281–295

101. Liu C, Xu X, Sun T, Thomas R, Fonseca JQ, Preuss M (2023) Microstructural effects on fatigue crack initiation mechanisms in a near-alpha titanium alloy. Acta Mater 253:118957
102. Jadhav S, Powar A, Patil S, Supare A, Farane B, Singh R (2017) Effect of volume fraction of alpha and transformed beta on the high cycle fatigue properties of bimodal Ti6Al4V alloy. IOP Conference Series: Mater Sci Eng 201:012035
103. Hémery S, Bertheau D, Hamon F (2024) Microtexture effects on fatigue and dwell-fatigue lifetimes of Ti-6Al-4V. Inter J Fat 179:108068
104. Li J, Davies HM, Fox K, Mulyadi M, Glavicic MG, Bache MR (2023) An appraisal of dwell sensitive fatigue in Ti-6Al-4V and the governing role of inhomogeneous micro-texture. Inter J Fa 171:107589
105. Hémery S, Stinville JC, Wang F, Charpagne MA, Emigh MG, Pollock TM, Valle V (2021) Strain localization and fatigue crack formation at (0001) twist boundaries in titanium alloys. Acta Mater 219:117227
106. Fang C, Qiu J, Zhang M, Zhang M, Ma Y, Lei J, Yang R (2024) Investigation on the dwell fatigue behavior of Ti-6.5Al-3.5Mo-1.5Zr-0.3Si alloy. Mater Sci Eng A 918:147430
107. Fan P, Fan J, Zhang W, Zhao J, Zhao R, Li B, Jiang P, Xue X (2024) Deciphering the role of macrozones in the microstructure globularization and tensile fracture: Insights from a new developed marine engineering titanium alloy. Mater Sci Eng A 917:147396
108. Wang B, Zeng W, Zhao Z, Jia R, Xu J, Wang Q (2024) Crack initiation mechanisms of micro-textured Ti60 alloys with different dwell sensitivities subjected to fatigue and dwell fatigue loading. Int J Plast 177:103986
109. Liu W, Huang J, Liu J, Wu X, Zhang K, Huang A (2021) Experimental and crystal plasticity modelling study on the crack initiation in micro-texture regions of Ti-6Al-4V during high cycle fatigue tests. Inter J Fat 148:106203
110. Chen X, Germain L, Berbenni S Hierarchy of the macrozone features in Ti-6Al-4V alloy inferred from massive polycrystal plasticity calculations. J Mater Res Tech 29:2468–2482
111. Silva BF, Jackson M, Foxb K, Wynne BP (2023) Tool for automatic macrozone characterization from EBSD data sets of titanium alloys. J App Cryst 56:737–749
112. Pilchak AL, Shank J, Tucker JC, Srivatsa S, Fagin PN, Lee Semiatin S A dataset for the development, verification, and validation of microstructure-sensitive process models for near-alpha titanium alloys. Integr Mater Manufac Innov 5:259–276
113. Shen J, Venkatesh V, Noraas R, Ghosh S (2023) Parametrically upscaled crack nucleation model (PUCNM) for fatigue nucleation in titanium alloys containing micro-texture regions (MTR). Acta Mater 252:118929
114. Tak TN, Nair KA, Venkatesh V, Pilchak A, Ghosh S (2024) Parametrically upscaled constitutive and crack nucleation models for investigating the effects of specimen geometry and microstructure on fatigue crack nucleation in Ti alloys containing micro-texture regions. Metall Mater Trans A 55:4834–4851

Part III
Application of EBSD to Unravel the Critical Problems in Materials Investigation

Chapter 10
Microstructure and Texture Development During Deformation and Recrystallization in Two-Phase Materials

Rajesh K. Khatirkar and Satyam Suwas

1 Introduction

In most applications, the alloys used consists of more than one phases. The proportion of these other phases in the microstructure could vary from as low as ~1% to as high as ~50% depending on the alloy being considered. The second phase could be in the form of dispersed hard particles in ductile matrix, in the form of islands of hard phase in ductile phase or with two ductile phases co-existing [1–6]. α/β brass [6, 7], α/β titanium alloys [5, 8], duplex stainless steels (DSSs) [1, 3], and dual-phase steel (DPS) [4, 9, 10] are the most often utilized two-phase materials. The second phase can affect almost all the properties of the material as well as its deformation/recrystallization behavior [11–13]. In today's world, steel is the most widely used engineering material, though other materials are gaining importance day-by-day. In the present work, DSSs and dual-phase steels are considered in which microstructure predominantly consists of two primary phases. In DSSs, the two phases of interest are austenite and ferrite. The proportion of austenite and ferrite varies in the range 40–50% each in DSSs. Of course, the phase balance in these types of steels/alloys is largely dependent on their prior thermomechanical processing history. In dual-phase steels, the two phases of interest are ferrite and martensite. The proportion of martensite varies from 5 to 40%. In order to tailor the mechanical/technological properties,

R. K. Khatirkar (✉)
Department of Metallurgical and Materials Engineering, Visvesvaraya National Institute of Technology (VNIT), Nagpur, Maharashtra, India
e-mail: rajeshk@mme.vnit.ac.in

S. Suwas
Department of Materials Engineering, Indian Institute of Science, Bangalore, India
e-mail: satyamsuwas@iisc.ac.in

S. Suwas and D. P. Field (eds.), *Advances in Texture, Microtexture, and Allied Techniques*, Materials Horizons: From Nature to Nanomaterials,
https://doi.org/10.1007/978-981-96-5346-1_10

it is of great practical and scientific importance to understand the microstructure and texture development during deformation and recrystallization in these steels.

All the DSSs consist of ferrite and austenite in nearly equal proportion in the solution-annealed condition [1–3, 9]. They are also called ternary ferrite and ternary austenite because of the substituted Cr and Ni together with interstitial C atoms in the cubic crystalline iron cells. Ferrite is harder than austenite in most DSSs (based on Fe–Cr–Ni system) in the solution-annealed condition [14, 15]. UNS S32205/S31803 is the most conventional DSS used for the industrial applications [1–3]. There are many other grades like lean DSSs, super DSSs, and hyper DSSs, which vary in the composition and stability of the austenite [1–3]. Super DSS possess higher strength and corrosion resistance, but are difficult to manufacture due to presence of larger amounts of Mo, Cr, and N. The lean DSS have corrosion resistance almost similar to austenitic stainless steels, but higher strength which makes them demanding in constructions of buildings, bridges, etc. Table 1 shows various grades of DSSs along with their chemical composition and typical properties in the solution-annealed condition [1, 16].

DPSs are advanced high strength steels typically used in automotive industries [4, 9, 17]. These steels consist of hard martensite islands dispersed in soft ferrite matrix. This specific microstructure gives these steels excellent combination of high strength, good formability, and high strain hardenability [9, 17, 18]. This combination depends on the fraction of martensite, grain size, carbon content of the steel, and so on. The carbon content ranges from 0.06 to 0.15wt% and Mn content from 1.5 to 3%. They typically have tensile strength from 500 to 900 MPa and elongation 15 to 30% [4, 17].

In order to produce sheets, cold rolling and annealing of these steels is inevitable. Since, these are two-phase materials, there are lot of changes in the microstructure during cold rolling and annealing which can significantly affect the mechanical properties of the final sheet. It is known that the two phases do not deform in a similar manner during deformation [1, 4] and hence needs to be studied in detail. Deformation results in changes in microstructure, increase in dislocation density, development of crystallographic texture, orientation/phase dependent strain partitioning to name a few [11, 13]. These changes directly affect the recrystallization behavior of the material. The differences in the behavior of each phase and the strain partitioning can be effectively studied by EBSD statistically. Image quality (IQ) maps/distribution or band contrast maps/distribution, which depends on the quality of Kikuchi pattern can be used as an estimate of stored deformation energy or dislocation density [13, 19, 20]. The quality of pattern also depends on the measurement parameters like detector gain, exposure, and sample preparation. However, if proper care is taken during the acquisition of EBSD patterns, normalized value of IQ or band contrast can be used quantitatively. Another important parameter which can be used in deformation studies is kernel average misorientation (KAM), which represents the average misorientation of a particular point/pixel with neighboring points/pixels within a particular threshold (usually taken as 5°) [11, 18, 19]. Similarly, depending on the degree of indexing and statistics, EBSD scans can be used to determine deformation or recrystallization texture. It is also possible to separate fraction recrystallized in

Table 1 The chemical composition, mechanical, and corrosion properties of different grades of duplex stainless steels [1, 16]

EN No/UNS	Composition (in weight%)								Mechanical properties				PREN
	C	Cr	Ni	Mo	N	Mn	Cu	W	YS	UTS	El	HB	
									MPa	MPa	%		
Lean duplex stainless steel													
S32304	0.03	21.5–24.5	3.0–5.5	–	0.05–0.20	2.5–0	0.05–0.60	–	400	600	25	290	25
S32101	0.04	21.0–22.0	1.35–1.70	0.1–0.8	0.20–0.25	4.0–6.0	0.10–0.80	–	450	700	30	270	26
Standard duplex stainless steel													
S31803/S32205	0.03	22.0–23.0	4.5–6.5	3.0–3.5	0.14–0.20	2.0–0	–	–	450	620	25	293	35
Super duplex stainless steel													
S32750	0.03	24.0–26.0	6.0–8.0	3.0–5.0	0.24–0.32	1.2–0	0.5	–	550	795	15	310	41
S32760	0.03	24.0–26.0	6.0–8.0	3.0–4.0	0.2–0.3	1.0	0.5–1.0	0.5–1.0	550	750	25	270	> 40
Hyper duplex stainless steel													
S32707	0.03	27.0–32.0	7.0	3.5–5.0	0.4–0.5	–	–	–	550	820	30	–	48–50

YS: Yield (Proof) Strength
UTS: Ultimate Tensile Strength
PREN: Pitting Resistance Equivalent Number = %Cr + 3.3(%Mo + 0 5%W) + 16%N

partially recrystallized samples by using a suitable criterion (grain size, KAM, grain orientation spread, IQ/band contrast, etc.). The following section(s) elucidate the developments in microstructure and crystallographic texture during deformation and recrystallization of DSSs and DPSs.

2 Microstructure

Plastic deformation in a polycrystalline material is expected to change the grain shape based on the imposed strain and strain mode. The "other" consequences can be generalized as follows:

(i) developments of deformation microstructure—often generalized in terms of dislocation substructures [21, 22].
(ii) misorientation, lattice strain developments and lattice re-orientations [23, 24].

Microstructures that develop during plastic deformation of polycrystalline metals essentially consists of dislocation structures [11, 13, 21, 22]. After moderate strain, plastic deformation is inhomogeneous and the features representative of deformation localize. The heterogeneities that occur during deformation depend on the studied length scale—see Fig. 1. Figure 1 generalizes the heterogeneities as deformation and shear bands. In a deformed microstructure, deformation bands are the regions inside the grains that deform homogenously, but the slip characteristics are different than the adjacent regions [11, 13]. Shear bands/strain localizations are the regions of higher-than-average misorientation and with strong localization of shear which can span over the individual grain or many grains. This localization of shear most frequently extends to macroscopic dimensions [11, 13]. Plastic deformation within a grain can be influenced by the local stress state variation resulting from the dislocation interactions [13]. Interactions between the grains, and the fact that grains can have different hardness to plastic deformation, may also add to such stress variations [13]. To maintain strain compatibility, there is variation of local strain within a grain also [13]. These results in heterogeneous distribution of plastic deformation/ slip across a grain. As deformation proceeds the strain heterogeneity at the grain scale can be amplified by misorientations [25, 26] that tend to develop between the deformation bands. There is formation of domains separated by boundaries usually denoted as cell blocks (CBs) or deformation bands [13, 27]. The term "deformation bands" is generally applied to the coarse bands seen in large grains in which there is significant lattice rotations. The adjacent bands rotate away from each other in opposite senses and are separated by micron sized zones of high orientation gradients termed transition bands [11].

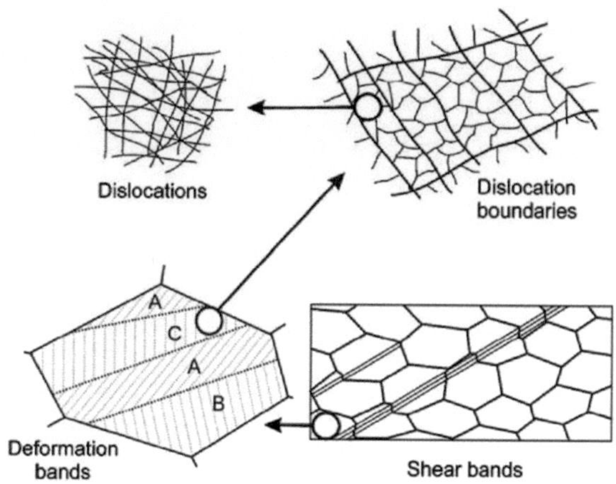

Fig. 1 Schematic of different types of deformation heterogeneities at different length scales [11, 13]

3 Duplex Stainless Steels

3.1 Microstructure

The microstructural changes that take place in individual phases of DSSs depend on the amount of strain, deformation process (rolling, compression, extrusion, forging etc.) and the temperature at which deformation is carried out. During cold rolling, both component phases of DSSs are plastically deformed resulting in a band-like morphology, a typical "pan-cake" structure. Further, the deformation in DSSs is complicated due to the presence of phase boundaries along with grain boundaries. In a hot-rolled sheet, the interfaces (δ/γ) are typically parallel to the rolling plane and the grain boundaries are perpendicular. There are contrary opinions on the primary deformation processes acting inside the regions of both phases. Some authors believe them to be identical to their single-phase counterparts, while some believe them to be significantly different due to presence of phase boundaries [28–31].

Figure 2 shows the secondary electron scanning electron microscope (SEM) image for SA, and 80% cold rolled UNS S32205 DSS samples [14], while Fig. 3 shows the electron back scattered diffraction (EBSD) phase map for SA and EBSD phase map superimposed with image quality (IQ) map for 40% cold rolled and 80% cold rolled UNS S32205 DSS samples [14].

For SA DSS sheet, with increased cold rolling reduction (deformation/strain), lamella thickness decreases in general with average band thicknesses being dependent on the amount of strain. For 80% cold rolling, shear bands are seen inside the austenite grains. These bands are formed preferentially inside the austenite grains. At highest cold rolling reduction, i.e., 80%, lenticular bulges are also seen in the austenite (Fig. 2c). In EBSD phase map superimposed with IQ map (Fig. 3) also, the lenticular bulges are visible in some austenite grains. These lenticular bulges

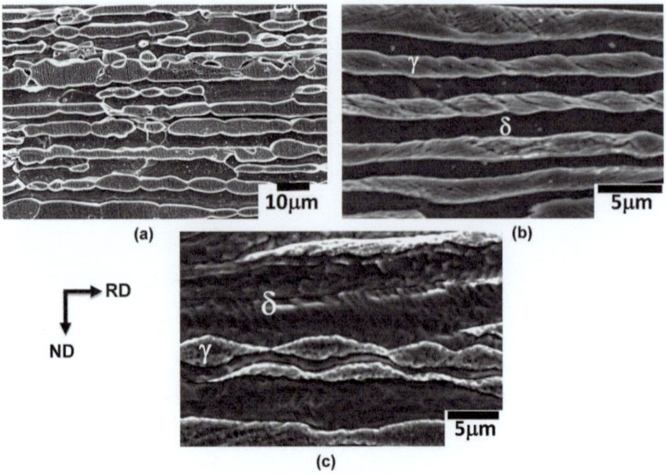

Fig. 2 Secondary electron SEM images of (**a**) SA, (**b**) 40% and (**c**) 80% cold rolled UNS S32205 DSS samples [14]

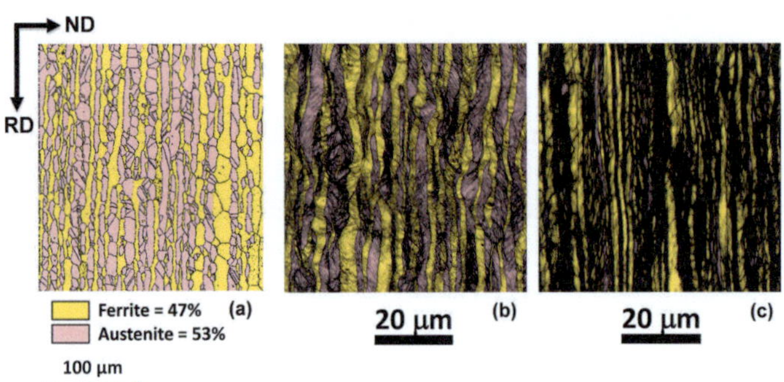

Fig. 3 (**a**) EBSD phase map for SA sample and EBSD phase map superimposed with IQ map for (**b**) 40% cold rolled and (**c**) 80% cold rolled UNS S32205 DSS samples [14]

are usually seen only in austenite and has been reported to be due to the formation of strain-induced martensite. The IQ deteriorated with increase in strain and is usually indicative of the strain partitioning. Austenite IQ is low indicating that more strain is partitioned in austenite during cold rolling [14, 32]. Wronski et al. [33] studied the deformation behavior of URN45N steel (composition close to UNS S32205 DSS) consisting of 50% austenite and 50% ferrite during tensile deformation. They did EBSD measurements for undeformed (initial), medium deformed (away from fractured region) and large/highly deformed (very near to fracture) region. The normalized IQ (($IQ_{max}-IQ_{min}$)/IQ_{max}) will be higher for undeformed regions and it will be lower for deformed regions. Figure 4 shows the normalized IQ distribution for initial, medium and large deformation for ferrite and austenite [33]. Increase in

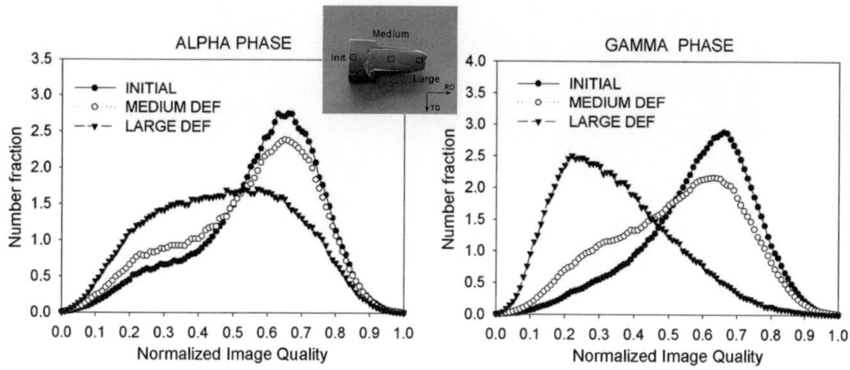

Fig. 4 Normalized IQ distribution for initial, medium and highly deformed regions for ferrite and austenite [33]

deformation results in shifting of the IQ distribution to lower side, i.e., towards the left [33].

The shift in the distribution is significantly less in ferrite than austenite. By calculation of the area under the curve, it was shown that the deformed fraction was higher in austenite than ferrite for both medium as well as large deformation. For medium deformation, the deformed fraction was 7% in ferrite and 15% in austenite. This increased to 15% in ferrite and 56% in austenite for large deformed sample.

Another significant feature of the microstructure in DSSs is the occurrence of shear bands during deformation, particularly in austenite [14, 32]. The development of shear bands in austenite is often related to its stacking fault energy (SFE), which depends on the chemical composition. The SFE also influences the stability of austenite. These shear bands in rolled DSSs are usually inclined at an angle of 25–35° with the sample rolling direction (RD) [11, 14, 15, 34, 35]. Austenitic stainless steels (ASSs) have SFE values of 50 mJm^{-2} (AISI 316) and 18 mJm-2 (AISI 304) [13, 34, 36, 37]. In ASSs, Ni and C tend to enhance SFE, whereas Cr, Si, Mn, and N tend to reduce SFE. The behavior of Ni has been linked with its effect on dislocation cross slip, i.e., cross slip is easy in high Ni alloys. SFE has a nearly linear relationship with Ni concentration. Dislocation arrays have been found to become more planar as Cr concentration increases. SFE changes very little at higher Cr values, while it changes extremely quickly at lower Cr concentrations [34–37]. N works in a similar way, reducing dislocation tangles and forming planar arrays. It has been reported, however, that it has little or no effect on SFE. Despite the fact that C is an interstitial element like N, it is reported to trigger dislocation arrays [35]. Because SFE is so sensitive to chemical composition, empirical formulae for calculating SFE of single-phase materials have been published.

1. SFE = (−53 + 6.2%Ni + 0.7%Cr + 3.2%Mn + 9.3%Mo) mJm^{-2} – Reed & Schramm [36]
2. SFE = (1.2 + 1.4%Ni + 0.6%Cr + 17.7%Mn – 44.7%Si) mJm^{-2} – Rhodes & Thompson [37]

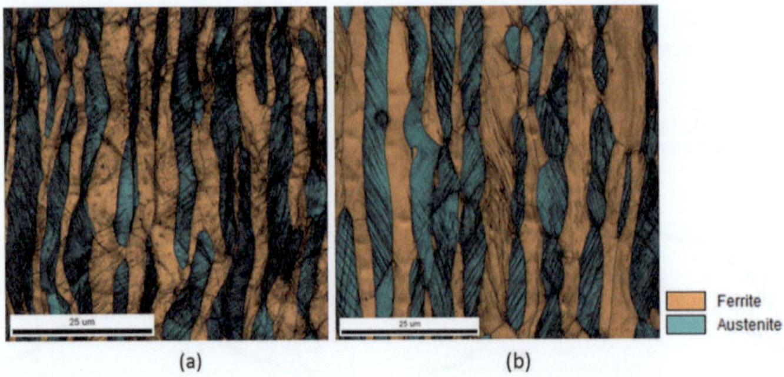

Fig. 5 EBSD IQ map of 40% (**a**) unidirectionally cold-rolled and (**b**) multi-step cross cold-rolled UNS S32205 DSS superimposed on its phase map. Vertical direction is RD and horizontal direction is normal direction (ND) [14, 15]

3. SFE = (16.7 + 2.1%Ni − 0.9%Cr + 26%C) mJm^{-2} – Brofman & Ansel [38, 39]

The SFE is a common measure of austenite stability and phenomenon such as strain-induced martensite (SIM) transformation and mechanical twinning predominantly occurs in materials with low (<20 mJm^{-2}) and medium (20–50 mJm^{-2}) SFE levels. During mechanical loading, ferrite primarily deforms by dislocation slip involving various slip systems, driven by high stacking fault energy (SFE) [23, 40], austenite, on the other hand, has multiple deformation modes, including dislocation slip, mechanical twinning, and martensitic transformation [23, 40]. Figure 5 shows the EBSD IQ map of 40% unidirectionally cold-rolled and multi-step cross cold rolled UNS S32205 DSS superimposed on its phase map. Shear bands are clearly visible in the austenite grains for both strain paths [14, 15].

The IQ of austenite is also lower indicating that more strain is partitioned in austenite than ferrite. In cross rolled sample, bands are visible in both positive and negative directions with respect to RD. The austenite in UNS S32205 and super DSSs is stable and less prone to strain SIM transformation. Lean DSSs has more tendency to form SIM during deformation due to lower alloying elements and less stability of austenite [32, 41]. In conventional and super DSSs, SIM forms after heavy cold deformation. During cold rolling, it forms after 60% thickness reduction and that too in very less amount. In heavily cold deformed conventional and super DSS, it is difficult to get good Kikuchi patterns due to high work hardening of austenite [1, 14, 42]. On the other hand, in lean DSSs, SIM has been reported to form even after 0.3 strain [32, 41]. Figure 6 shows the EBSD band contrast (equivalent to IQ) and phase maps for lean DSS deformed to a true strain of 0.30 (during tensile test at room temperature) [41]. The SIM transformation is clearly visible in the austenite (grain at the center). In lean DSSs, the formation of shear bands is less prominent than the conventional/super DSSs, while SIM transformation is more prominent. The formation of SIM was also associated with accumulation of higher strain in the adjoining region. Increase of N content also results in an increase in austenite content of lean DSSs and correspondingly SIM (deformation-induced) during subsequent

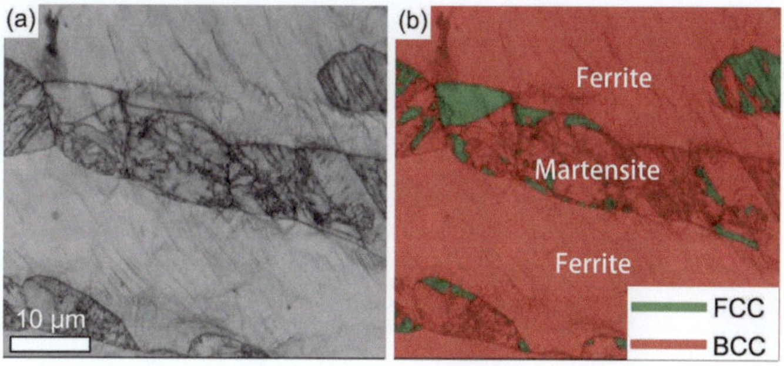

Fig. 6 EBSD band contrast (equivalent to IQ) and phase maps for lean DSS deformed to a true strain of 0.30 (during tensile test at room temperature) [41]

processing. In some DSSs like UNS S32304, ε-martensite forms first and then SIM forms at its intersection [41].

Higher strain partitioning in austenite is also evident in terms of higher reduction in band thicknesses along ND than ferrite for same deformation level (for example, during cold rolling). Ferrite is harder than austenite in the SA condition and austenite has higher work hardening and deforms more than ferrite during cold rolling. Typically, in a deformed microstructure, work hardening also manifests itself in the form of increased misorientation in the grains. KAM distribution becomes wider and average KAM increases in general with deformation. KAM is also higher in austenite than ferrite in deformed conditions [14, 42]. Figure 7 shows average KAM for URN45N steel after tensile deformation and lean DSS after cold rolling [33, 42]. The average KAM increases with increase in deformation and deformation becomes highly inhomogeneous with austenite having higher misorientation than ferrite. This further confirms that more strain is partitioned in austenite than ferrite. Figure 8 shows similar trend for strain path change [14, 15].

3.2 *Deformation Texture*

The processes that give rise to textures and texture changes are of three types, namely, deformation, recrystallization and phase transformation [11, 13, 43]. The orientation changes that take place during deformation are not random, except in case where there are large second phase particles. The zones near large non-shearable particles randomizes the texture. Deformation typically takes place due to favorably oriented slip and/or twinning systems, and as a consequence, the deformed metal acquires a texture known as deformation texture [11, 13]. If the deformed metal is subsequently recrystallized, there are three possibilities: (i) retaining the original deformation texture, (ii) texture randomization and (iii) a relatively minor component of deformation texture becoming significant through recrystallization. Cube recrystallization in FCC aluminum and the so-called γ-fiber recrystallization in BCC steels are the

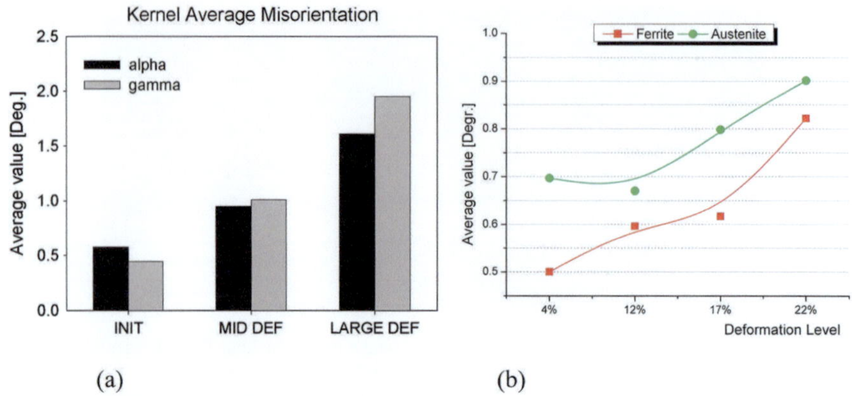

Fig. 7 Average KAM for (**a**) URN45N steel after tensile deformation and (**b**) lean DSS after cold rolling [33, 42]

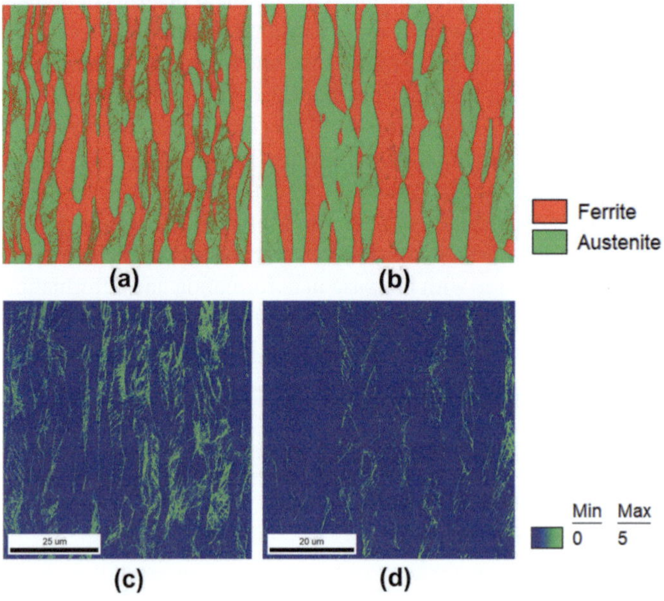

Fig. 8 Phase maps and KAM maps for unidirectionally cold-rolled (**a** and **c**) and cold cross rolled (**b** and **d**) UNS S32205 DSS [14, 15]

notable examples of (iii). In DSSs, the deformation texture may also be affected by the presence of interface boundaries (between austenite and ferrite grains/bands).

Textures in BCC materials have commonly been described in terms of certain orientation fibers in the Euler space. Usually, it is customary to plot orientation density along these fibers and/or to use the volume/area fraction of these fibers for comparing textures. Some of the important texture components/fibers that occur in

BCC and FCC materials during rolling are shown in Figs. 9 and 10, respectively. The specific orientation ranges for BCC materials are as follows [11–13, 44]:

(a) RD or α fiber running from {001} <110> to {110} <110> along RD// <110>
(b) ND or γ fiber running from {111} <110> to {111} <112> along ND// <111>
(c) TD or ε fiber running from {001} <110> to {110} <001> along TD// <110> and
(d) θ-fiber running from {001} < 100 > to {001} < 110 > along ND// <100> .

The specific orientation ranges for FCC materials are as follows:

(a) α-fiber which consists Goss (G) ({011} <100>) and the Brass (Bs) ({110} <112>) orientations.

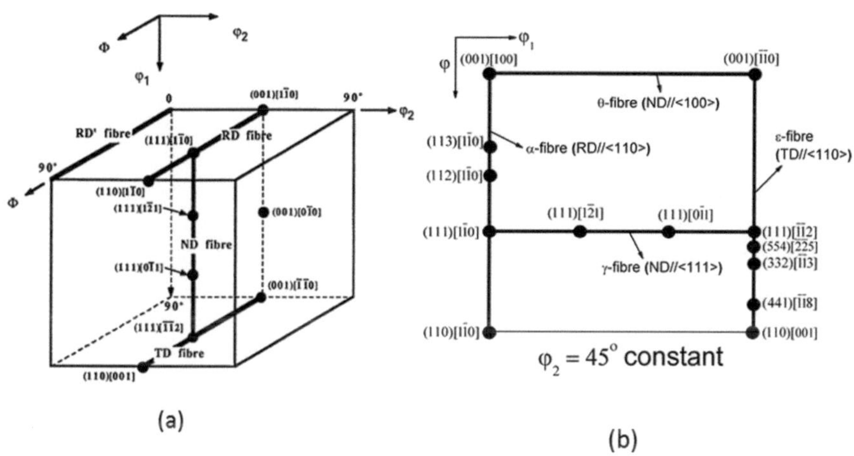

Fig. 9 (**a**) Three-dimensional view of Euler space with location of some important ideal orientations and fibers (Bunge notation) for BCC materials, and (**b**) $\varphi_2 = 45°$ constant section of (**a**) [11, 12, 23, 44]

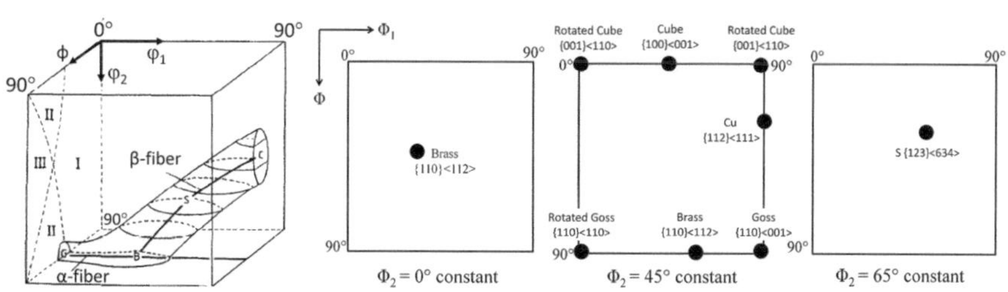

Fig. 10 (**a**) Three-dimensional view of Euler space with location of some important ideal orientations and fibers (Bunge notation) for FCC materials, and (**b**) $\varphi_2 = 0°, 45°, 65°$ constant section of (**a**) [11, 12, 23, 44]

(b) β-fiber which runs from Bs to Copper (Cu) ({112}<111>) with an intermediate S-orientation (({123} <634>).

It is well known that high SFE single-phase FCC materials develop copper-type texture, while low SFE single-phase FCC materials develop brass-type texture after cold rolling [11, 12]. On the other hand, BCC single-phase materials like low carbon (LC) steels, ferritic stainless steels, molybdenum, and tantalum develop γ- and α-fibers after cold rolling [12, 23, 45–47]. There are contradictory opinions about DSSs, which has both BCC and FCC phases. Some researchers believe that there is significant effect of interface boundaries on the deformation texture development and some believe that there is no effect. Further, during heavy cold rolling of UNS S32205 and super DSSs, formation of SIM is very less and hence its effect can be neglected [14, 15]. In lean DSSs like UNS S32304 and UNS S32101, significant amount of SIM is formed and the texture is of deformed ferrite and formed SIM [32, 41]. Figure 11 shows the ODF of austenite of UNS S32205 and UNS S32304 DSS for hot-rolled and solution-annealed condition [48]. The austenite texture of UNS S32205 and UNS S32304 hot-rolled steels showed strong S {123}⟨634⟩ and brass {110}⟨112⟩ components, brass being more intense in UNS S32205 steel. In lean DSS (UNS S32304) copper {112} ⟨111⟩ was stronger in the hot-rolled condition [48].

After hot rolling, solution annealing resulted in shifting the copper orientation by ~ 7° along ϕ which generated Dillamore {4 4 11}⟨11 11 8⟩ component (twin of Goss) in austenite. Apart from this change, the solution-annealed and hot-rolled texture of UNS S32304 remained same. The {110}⟨001⟩, i.e., Goss moved towards cube in UNS S32304 steel. In UNS S32205 steel, Goss formed after SA. Figure 12 shows the ODF of austenite after 75% cold rolling.

Figure 12 shows that cold rolling resulted in reduction of texture intensity for austenite than the hot-rolled and solution-annealed samples [48]. The Goss and brass were more intense in UNS S32304 alloy than in UNS S32205 alloy. The Goss component moved towards the cube along ⟨100⟩ direction for UNS S32205 cold rolled sample. A weak {111}⟨110⟩ component was observed with some spread around γ-fiber. UNS S32304 formed a brass-type texture which scattered towards Goss. In both steels, S-orientation had low intensity. Figure 13 shows the ODF of ferrite in hot-rolled and SA condition (A) and after 75% cold rolling (B) for UNS S32205 and UNS S32304 DSS samples. Strong partial α-fiber was observed after hot rolling, which significantly weakened after SA in UNS S32205 steel. After SA, weak γ-fiber components were also observed. In UNS S32304 steel, there was strong γ-fiber after hot rolling, which weakened after SA. After SA, the γ-fiber moved towards partial α-fiber. After 75% cold rolling, strong and uniform α-fiber was observed in both the steels. In UNS S32304 steel, a very weak but uniform γ-fiber was also formed.

The standard DSS showed both twinning and slip as deformation mechanisms, while in the lean steel, twinning was the main mechanism for austenite. Jia et al. [49] studied UNS S32507 DSS (25–85% cold rolled) and showed that at the highest reduction, very strong Goss and brass were formed, typical for deformed FCC materials with low SFE [49]. It has been reported that during deformation Cu is unstable and tends to rotate towards twinned Cu which then converts to Goss. This may be the

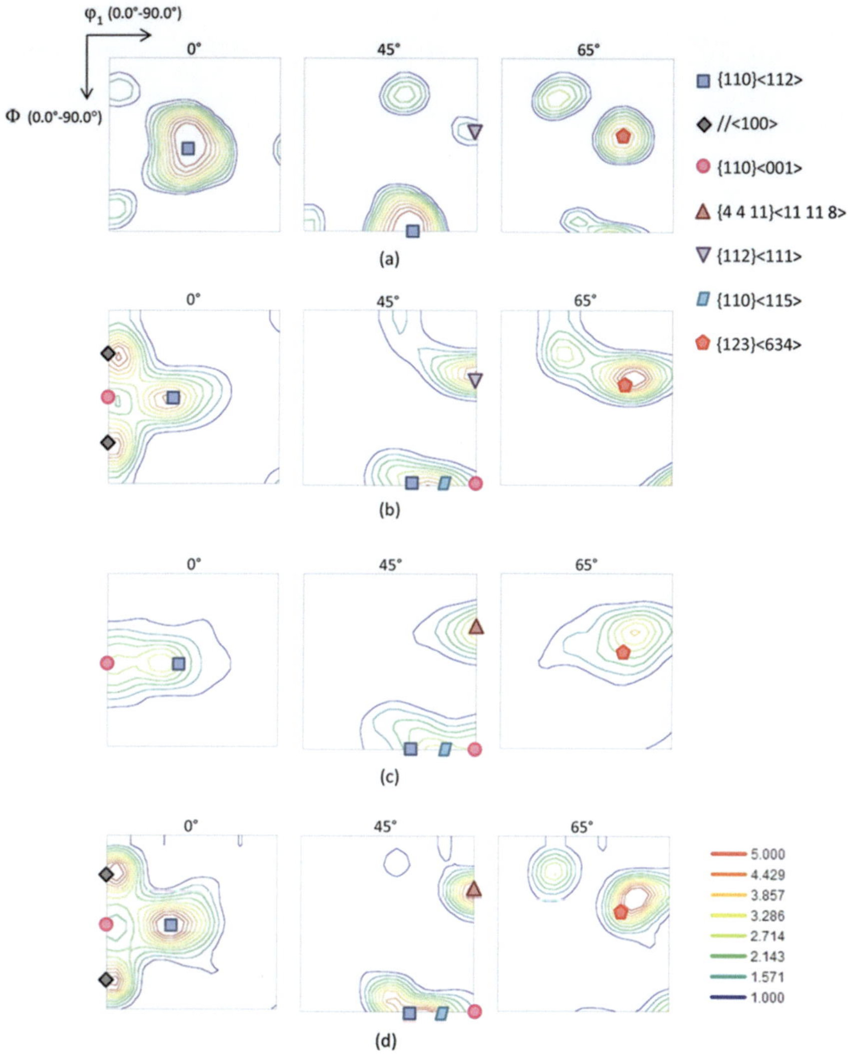

Fig. 11 ODF of austenite for hot-rolled (**a**, **b**) and SA (**c**, **d**) DSS samples. (**a**) and (**c**) are for UNS S32205 DSS, while (**b**) and (**d**) are for UNS S32304 DSS [48]

reason for increase in Goss after deformation. Brass continuously strengthens with increase in deformation. It has also been suggested that formation of shear bands in austenite also plays an important role in tailoring the texture. In lean DSSs, the formation of SIM and the corresponding orientation relationship may change the texture significantly. Ferrite in DSSs develops strong partial α-fiber, which is contrary to single-phase BCC materials which develop strong α- and γ-fibers after cold rolling. The interface boundaries may play a role in restricting the rotation of grains towards

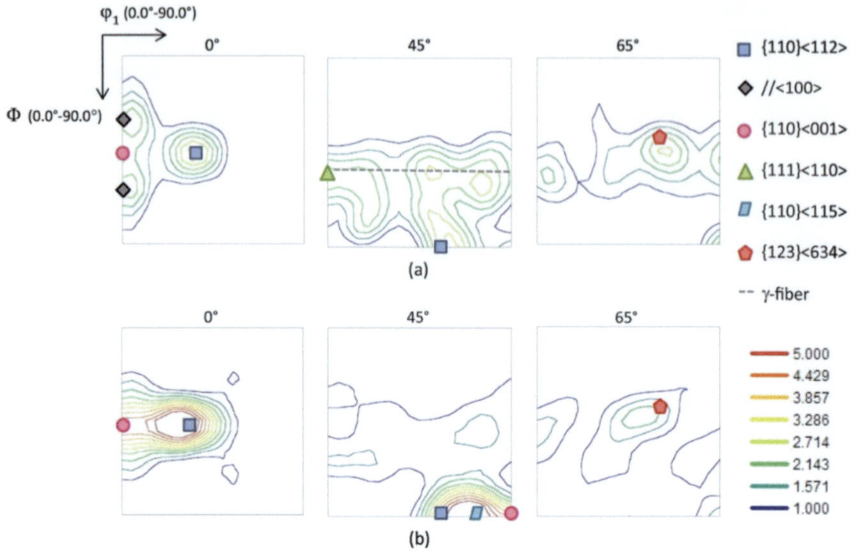

Fig. 12 ODF of austenite after 75% cold rolling. (**a**) UNS S32205 steel, and (**b**) UNS S32304 steel [48]

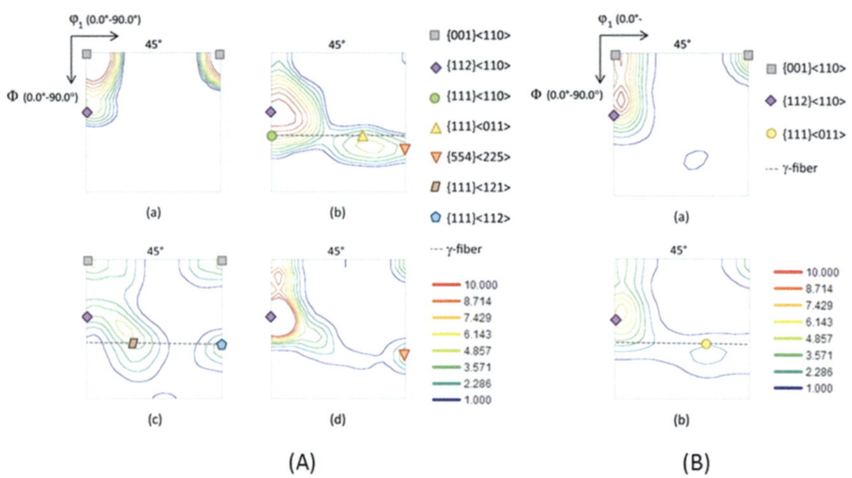

Fig. 13 ODF of ferrite in hot rolled (**a**, **c**) and SA (**b**, **d**) condition (**A**), and after 75% cold rolling (**B**) for UNS S32205 (top row) and UNS S32304 (bottom row) DSS samples [48]

γ-fiber orientations resulting in development of strong partial α-fiber texture. In addition, strain path during cold rolling also plays an important role. Austenite results in the development of ND rotated brass and cube in both standard and lean DSSs with increase in deformation. Ferrite, on the other hand develops strong rotated cube in

UNS S32205 DSS and strong rotated cube with γ-fiber in lean DSSs. It is important to study this, since recrystallization texture and forming properties can be largely affected by this. In tensile deformed UNS S32760 steel, in highly deformed region, strong brass, rotated cube and S was reported. Due to unconstrained rotation during tensile test, rotated cube texture might have developed. However, this remains to be tested by full field or mean field simulation of texture in DSS.

3.3 Recrystallization Texture

Deformation results in an increase in the dislocation density of the material (work hardening) and anisotropy of mechanical properties. It needs to be annealed to reduce the dislocation density and recover properties like ductility. During annealing, the deformed microstructure is replaced by new one. The final properties (r-bar and tensile) are highly dependent on the crystallographic texture of the cold rolled and annealed sheet [11, 13, 50]. It has been reported in single-phase materials like LC/ferritic/stainless steels that the formability (as measured by r-bar) can be tailored by suitable texture in the final cold rolled annealed sheet [11–13]. In DSSs, the situation is complicated due to the presence of two phases and precipitation of intermetallic phases (σ, χ and Laves) at low annealing temperatures. Therefore, DSSs are generally annealed at temperatures greater than 1000 °C and the recrystallization kinetics is also very fast due to high temperatures involved. Figure 14 shows the ODF of ferrite and austenite for initial, cold rolled and annealed condition for UNS S32101 (a) and UNS S32205 (b) DSS samples [50].

In general, the texture weakened after annealing for both austenite and ferrite. Weak rotated cube texture was formed in UNS S32101 steel, while strong partial α-fiber was observed in UNS S32205 steel in ferrite. Austenite, for both steels had retained rolling texture components (Cu), typically observed in austenitic stainless steels. The recrystallization in ferrite was continuous for both steels, while for austenite, recrystallization was discontinuous. Discontinuous recrystallization occurs when there is higher stored energy of deformation and lower SFE (austenite in DSSs), while continuous recrystallization occurs when the stored energy is lower and SFE is higher (ferrite in DSSs). The recrystallization in DSSs seems to be affected by the presence of interface boundaries, which can cause pinning effect during the growth of recrystallized nuclei/grain.

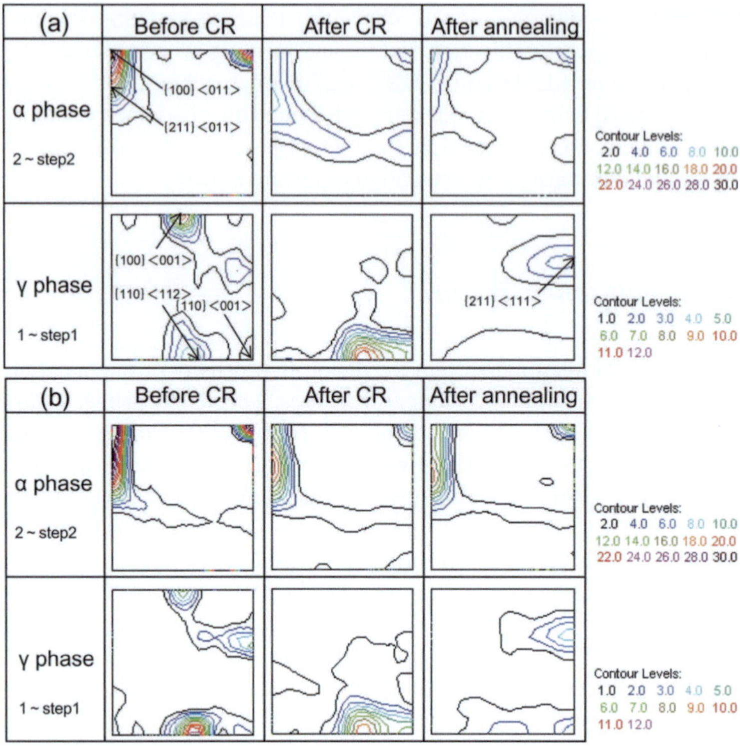

Fig. 14 ODF ($\varphi_2 = 45°$ constant section) of ferrite and austenite for initial, cold rolled and annealed condition for UNS S32101 (**a**) and UNS S32205 (**b**) DSS samples. Annealing was carried out at 1050 °C for 30 s [50]

4 Dual-Phase Steels

4.1 Microstructures

The microstructure of dual-phase steels consists of martensite (hard phase) dispersed in the ferrite matrix (soft and ductile). The properties like ductility are largely governed by the matrix ferrite and the distribution/size/shape of the dispersed martensite islands. During deformation, the strain portioning between ferrite and martensite plays a significant role in deciding the properties of the dual-phase steels. As the fraction of martensite increases, more and more strain get accumulates at the ferrite–martensite interfaces. In general, plastic deformation initiates in ferrite in regions relatively free of martensite and where grains are larger. Figure 15 [51] shows a typical EBSD IQ map of a dual-phase steel (DP600) in cold rolled and annealed condition along with distribution of IQ values.

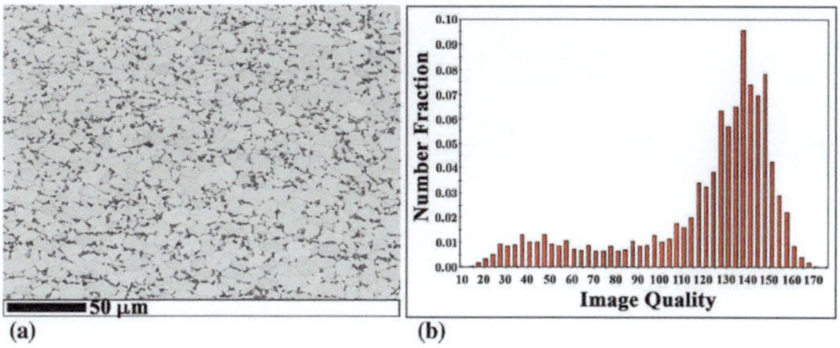

Fig. 15 (**a**) EBSD IQ map, and (**b**) IQ distribution for DP600 in cold rolled and annealed condition [51]

The IQ distribution shows two Gaussian peaks, one with lower average IQ values and one with higher average IQ values. The lower IQ value peak corresponds to martensite and higher IQ value peak corresponds to ferrite. In EBSD, both martensite and ferrite are indexed as bcc and hence it is very difficult to distinguish between them based on phase maps. IQ map on the other hand, provides a reliable method for its identification in dual-phase steels. IQ has been used by many authors to distinguish martensite from ferrite in dual-phase steels. Ashrafi et al. [51] reported that KAM values at the martensite–ferrite interface were higher (>1°) than those in ferrite (<1°) in the initial condition. They attributed this increase in KAM to the volume expansion associated with austenite to martensite transformation. Das et al. [52] studied the influence of strain rate during tensile testing on the strain partitioning between ferrite and martensite in dual-phase steels (DP600 and DP800). They showed that the evolution of misorientation in ferrite is primarily dependent on the martensite fraction of the steel. During deformation, more strain accumulation occurred in the ferrite regions close to martensite than the interior regions. They showed that the rotation of ferrite grains was dependent on the martensite fraction, higher martensite in DP800 grade resulted in lower rotation of ferrite grains than the DP600 grade. The restricted rotation of ferrite in DP800 did not change the crystallographic orientation significantly. For DP600, higher rotation of ferrite grain resulted in preferential orientation of grains towards ND// <111> . Yokoi et al. [53] and Nakada et al. [54] studied the deformation behavior of dual-phase steel during punching operation. They tried to quantify the changes in microstructure by a combination of EBSD with digital image correlation (DIC) technique. The amount of plastic deformation can be quantified by KAM value. They showed that average KAM can be considered to be an effective index for quantitatively characterizing the damage introduced by the hole punching operation. They went on to correlate the average KAM with mechanical properties (tensile strength and hardness, measured with micro samples near the punched hole). Figure 16 shows the correlation of KAM with the damaged region and its mechanical properties [53]. KAM correlates with the geometrically necessary dislocations (GNDs) and usually corresponds to the equivalent plastic strain. The micro-tensile strength and ferrite nano-hardness increased with increase in KAM. The hard martensite dispersed in dual-phase steel promote the heterogeneity of strain

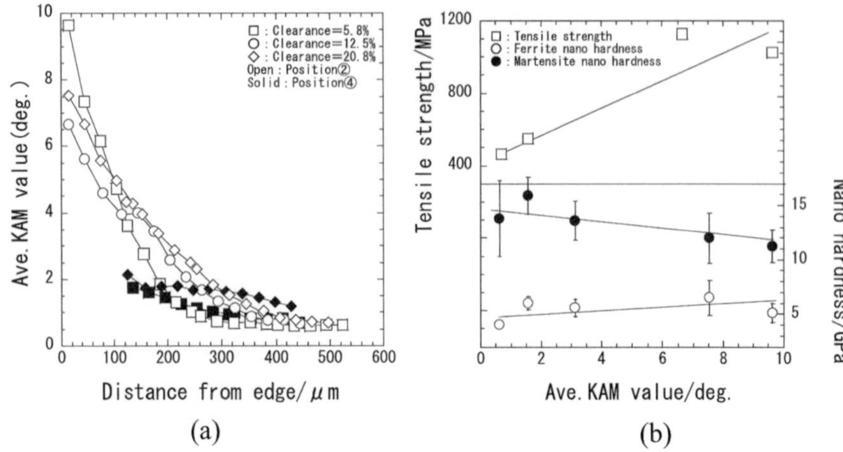

Fig. 16 (**a**) Relationship between distance from fractured edge and average KAM value, and (**b**) Relationship between average KAM and nano-hardness, tensile strength [53]

distribution even when the ferrite grain is significantly away from the punched hole edge.

Calcagnatto et al. [55] studied two dual-phase steels with different martensite volume fraction, one with 24% martensite + 2% cementite and other with 38% martensite. The increase in martensite lead to lowering of the elastic limit. Figure 17 shows the IQ and KAM maps for the two dual-phase steels [55]. The martensite has low IQ and the KAM map shows large orientation gradients spreading from martensite–ferrite interface into the ferrite grain interior. Orientation gradients are present around each martensite island, small as well as big.

In above, the dual-phase steel was not deformed. The average misorientation and GND density were observed to be higher in the vicinity of the ferrite–martensite interface than the ferrite–ferrite interface. These strains were pure because of the phase transformation during the manufacturing of the dual-phase steels. High strength dual-phase steel sheets with a low yield ratio, high work hardening rate, non-aging property, bake hardenability are a good choice for automotive industry. These steels need to be cold rolled to obtain sheet form and their deformation behavior becomes important due to presence of two-phase microstructure. In dual-phase steels, the presence of martensite islands in the ferrite matrix causes changes in the plastic deformation behavior of the ferrite resulting in different substructural developments than purely ferritic steel like interstitial free steels. Dillien et al. [56] studied the microstructural developments of DP600 steel during cold rolling (50% thickness reduction) and tensile deformation (logarithmic strain of 0.16). They compared the cold rolling of DP600 steel with IF steel, while the tensile deformation was studied for dual-phase steel only. In order to enhance the contrast of the substructure, they used maximized misorientation contrast (MMC) plots. Figure 18 shows the MMC plots for undeformed and tensile deformed (strain of 0.16) DP600 steel samples [56].

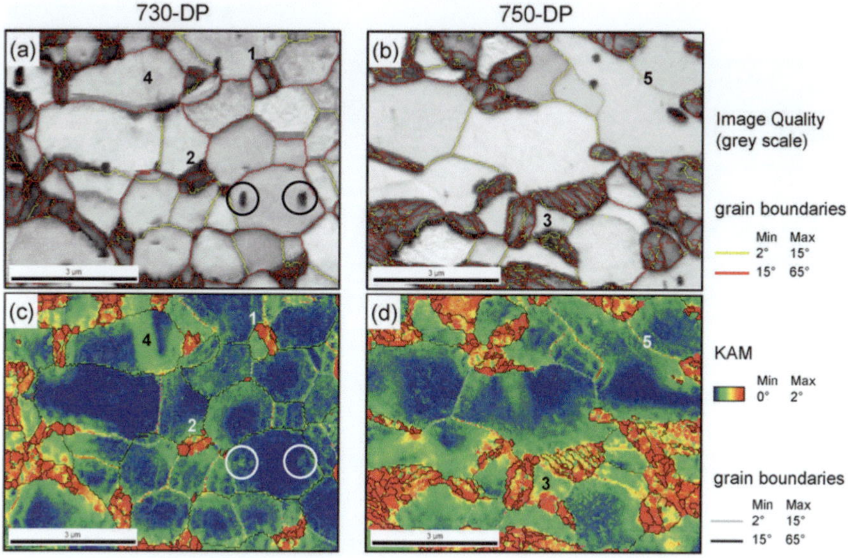

Fig. 17 Orientation gradients near ferrite–ferrite boundaries, ferrite-cementite boundaries and ferrite–martensite boundaries in dual-phase steel containing 24% martensite (730-DP) and 38% martensite (750-DP). (**a**) and (**b**) are IQ maps, while (**c**) and (**d**) are KAM maps [55]

For undeformed DP600 sample, the MMC plots indicate a larger variation in lattice orientation close to martensite islands and regions farther from the martensite islands had negligible gradient. No excess dislocations are present at the interior or between ferrite boundaries in regions farther from the martensite islands. In tensile deformed sample, significant curvature was present in the ferrite grains. The hard martensite islands make flow path of ferrite heterogenous causing pile up of dislocations. The excess dislocations give rise to curvature and result in high hardening exponent in dual-phase steels. After cold rolling, it was not possible to distinguish between ferrite and martensite in dual-phase steel. The ferrite and martensite both were elongated along the rolling direction. Ferrite grains has a smooth transition of orientation, i.e., gradual curvature. IF steel on the other hand, also showed gradual transition of orientation but with the presence of shear bands. Li et al. [57] studied the deformation of DP600 grade using in situ uniaxial tensile stage. The EBSD scans were taken after interrupting the test at regular intervals up to a strain of 0.20. They showed an increase in low angle grain boundary fraction with increase in strain. They also found more strain accumulation near martensite–ferrite interface than ferrite–ferrite interface after deformation.

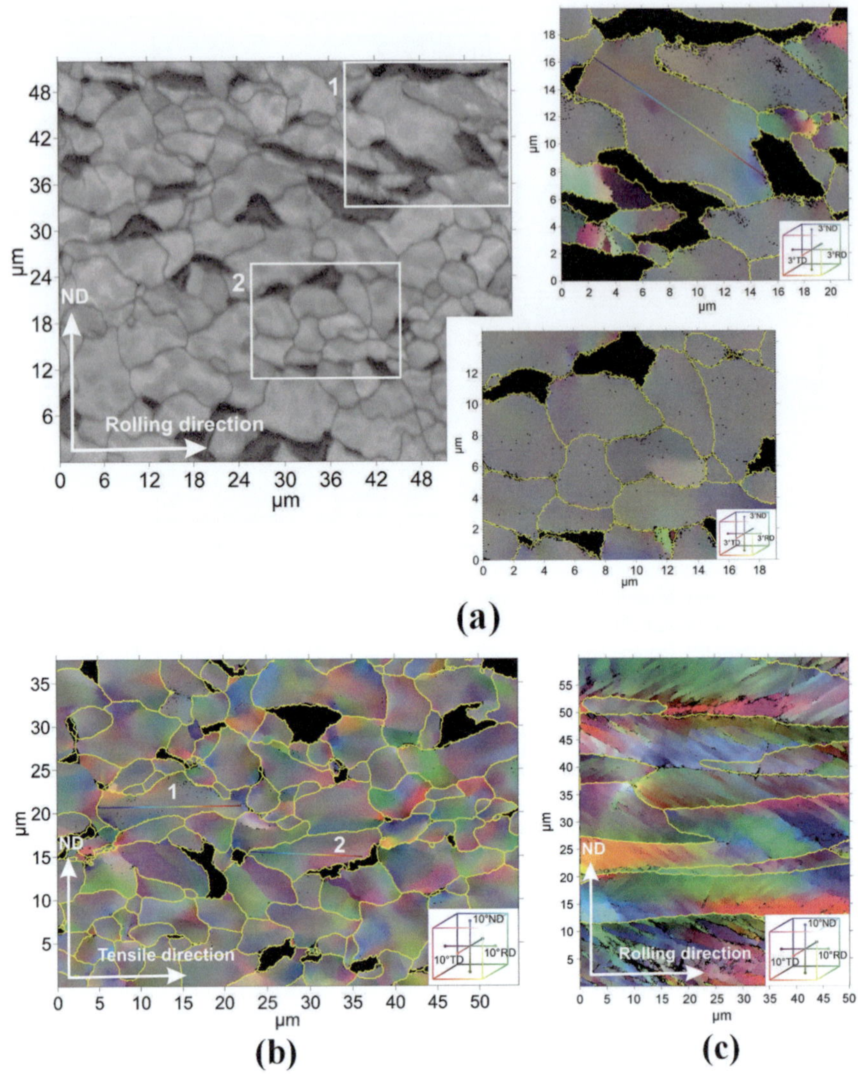

Fig. 18 (**a**) IQ map and MMC plots for undeformed dual-phase steel, (**b**) MMC plot for tensile deformed (strain of 0.16) DP600 steel and (**c**) MMC plot for 50% cold rolled IF steel sample [56]

4.2 *Texture*

The volume fraction of ferrite/martensite, ferrite texture and their morphologies are usually dependent on the recovery, recrystallization and phase transformation during inter-critical annealing and prior deformation. During the production of dual-phase steel, the sheet is subjected to cold rolling before being inter-critically annealed to form martensite. The prior cold rolling of the sheet has been reported to develop strong rotated cube and {112} <110> texture, i.e., α-fiber component, while the γ-fiber was significantly weaker. After annealing, the α-fiber becomes weaker and γ-fiber strengthens slightly. The high solubility of carbon in the ferrite matrix results in the weakening of the γ-fiber. Han et al. [58] studied the microstructure and texture evolution in hot-rolled, cold-rolled, and annealed dual-phase steel. They obtained 3.5% (DP-1) and 5% (DP-2) martensite volume fraction by varying the thermomechanical processing. DP-1 was finished rolled at 900 °C, soaked at 700 °C for 1 h and then furnace cooled. DP-2 was finish rolled at 780 °C, soaked at 550 °C for 1 h and then furnace cooled. The final sheet thickness for both was 3.2 mm. Subsequently, these steels with varying martensite content were cold rolled to 0.8 mm final thickness. They were then heated in an infrared furnace to simulate continuous annealing. The hot-rolled and cold-rolled sheets exhibited a distinct texture gradient in the thickness direction. Figure 19 shows the $\varphi_2 = 45°$ constant section of ODF of ferrite after hot rolling and cold rolling for different sections [58]. After hot rolling, the texture at surface was dominated by Cu, Bs and Goss for both the processing conditions. The texture at mid-thickness of the sheet consisted of α-fiber and {332} <113> component. After cold rolling, the texture was similar for both DP-1 and DP-2 consisting of α-fiber and weak γ-fiber, α-fiber being significantly stronger in DP-2 due to low temperature processing. The volume fraction of α- and γ-fibers was higher in DP-2, particularly in the mid-thickness of the sheet.

Calcagnotto et al. [59] optimized the thermomechanical processing to obtain cementite free dual-phase steels. For a plain C-Mn steel, they suggested a holding temperature of 730 °C and a holding time of 1 min to obtain ferrite–martensite microstructure free of cementite. They used EBSD IQ maps to quantify the martensite fraction.

5 Summary

EBSD is a very effective tool to study strain partitioning, deformation texture developments and recrystallization phenomenon, particularly in two-phase materials. Deformation can be studied as long as good indexed patterns can be obtained from the deformed material.

(1) Following is the summary of different features studied for different types of DSSs:

- Plastic deformation causes more strain (as measured by IQ, KAM, and shear banding tendency) to be partitioned in austenite than ferrite. Austenite in standard and super DSSs deform by slipping, twinning, and shear banding, followed by formation of SIM. Lean DSSs, on the other hand, preferentially form SIM. The amount of SIM was significantly higher in lean DSSs than standard and super DSSs for same level of deformation.

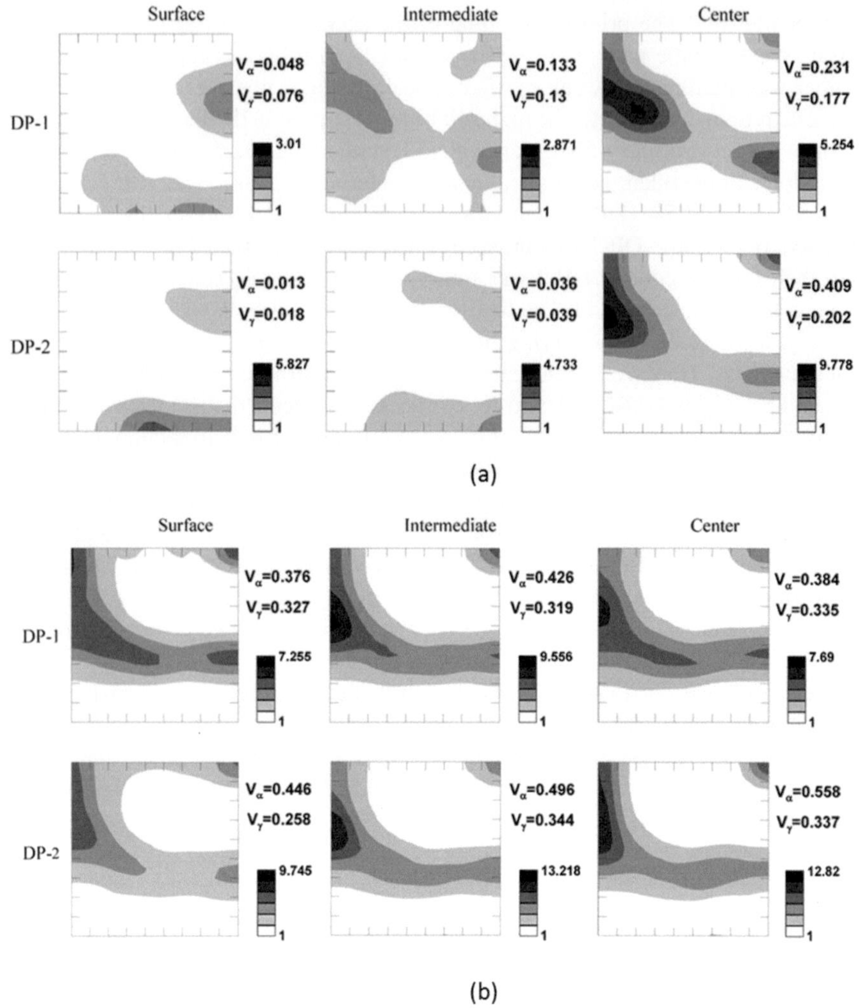

Fig. 19 $\varphi_2 = 45°$ constant section of ODF of ferrite after (**a**) hot rolling and (**b**) cold rolling for different sections [58]

- The development of deformation texture is strongly dependent on the alloy chemistry and strain path. For unidirectional cold rolling and tensile deformation, standard and super DSSs form strong partial α-fiber in ferrite and strong brass-type texture in austenite. During unidirectional rolling, lean DSSs form strong α- and γ-fibers in ferrite and strong brass-type texture in austenite. For multi-step cross cold rolling, DSSs form a very strong rotated cube in ferrite and ND rotated brass in austenite.
- After recrystallization, textures in all DSSs weakens. Ferrite forms a weak rotated cube texture and austenite retains most of deformation texture components.

(2) Following is the summary of different features studied for dual-phase steels:

- In all the dual-phase steels, ferrite–martensite interfaces had more strain accumulations and more orientation gradients in the ferrite grains adjacent to martensite islands. Ferrite grains away from the martensite islands had near zero orientation gradients. IQ maps can be used to partition the martensite islands from the ferrite grains. Martensite, due to high dislocation density and more strain localizations results in very low IQ, which can be effectively used to identify and quantify it. Plastic deformation (tensile deformation or cold rolling) causes more orientation gradients in ferrite initially and after large deformation, martensite and ferrite becomes indistinguishable from each other due to work hardening in ferrite.
- Ferrite in dual-phase steels develop α- and γ-fibers after cold rolling or tensile deformation. α-fiber is stronger than the γ-fiber in most cases. Recrystallization after cold rolling leads to strengthening of the γ-fiber slightly. Through thickness texture gradient is also observed in hot-rolled, cold-rolled, and recrystallized dual-phase steels.

References

1. Gunn R (1997) Duplex stainless steels: microstructure, properties and applications. Woodhead Publishing
2. Armas IA (2013) Low-cycle fatigue at room temperature, in: Duplex Stainl Steels
3. Armas IA (2008) Duplex stainless steels: brief history and some recent alloys. Recent Patents Mech Eng 1:51–57
4. Peranio N, Roters F, Raabe D (2012) Microstructure evolution during recrystallization in dual-phase steels. In: Mater Sci Forum 13–22
5. Semiatin SL (2020) An overview of the thermomechanical processing of α/β titanium alloys: current status and future research opportunities. Metal. Mater Trans A 51:2593–2625
6. Garg R, Ranganathan S, Suwas S (2010) Effect of mode of rolling on development of texture and microstructure in two-phase (α+ β) brass. Mater Sci Eng A 527:4582–4592
7. Metals Handbook (1998): Properties and Selection: Nonferrous Alloys and special purpose materials
8. Leyens C, Peters M (2003) Titanium and titanium alloys: fundamentals and applications. John Wiley & Sons
9. Metals Handbook (1990) Volume 1: Properties and selection: Irons, steel, and high-performance alloys
10. Masimov M, Peranio N, Springub B, Roters F, Raabe D (2010) EBSD study of substructure and texture formation in dual-phase steel sheets for semi-finished products 160:251–256. https://doi.org/10.4028/www.scientific.net/SSP.160.251
11. Verlinden B, Driver J, Samajdar I, Doherty RD (2007) Thermo-mechanical processing of metallic materials. Elsevier
12. Suwas S, Ray RK (2014) Crystallographic texture of materials. Springer
13. Humphreys FJ, Hatherly M (2012) Recrystallization and related annealing phenomena. Elsevier
14. Kumar A, Khatirkar RK, Chalapathi D, Kumar G, Suwas S (2017) Microstructure and texture development during cold rolling in UNS S32205 and UNS S32760 Duplex Stainless Steels, Metall. Mater. Trans. A. 48:2349–2362

15. Kumar A, Khatirkar RK, Gupta A, Shekhawat SK, Suwas S (2018) Deciphering the Possible Role of Strain Path on the Evolution of Microstructure, Texture and Magnetic Properties in a Fe-Cr-Ni Alloy, Metall. Mater. Trans. A. 49:3402–3418
16. https://www.outokumpu.com/en/products/stainless-steel-types/duplex-stainless-steel
17. Peranio N, Li YJ, Roters F, Raabe D (2010) Microstructure and texture evolution in dual-phase steels : Competition between recovery, recrystallization, and phase transformation. Mater Sci Eng A 527:4161–4168
18. Pinard PT, Schwedt A, Ramazani A, Prahl U, Richter S (2013) Characterization of Dual-Phase Steel Microstructure by Combined Submicrometer EBSD and EPMA Carbon Measurements, Micro. Microan. 19:996–1006
19. Masimov M, Peranio N, Springub B, Roters F, Raabe D (2010) EBSD study of substructure and texture formation in dual-phase steel sheets for semi-finished products, in: Solid State Phenom. Trans Tech Publications Ltd, 251–256
20. Kang JY, Park SJ, Moon MB (2013) Phase analysis on dual-phase steel using band slope of electron backscatter diffraction pattern. Microsc Microanal 19:13–16
21. Wert JA, Liu Q, Hansen N (1995) Dislocation boundaries and active slip systems. Acta Metall Mater 43:4153–4163
22. Kuhlmann-Wilsdorf D, Van der Merwe JH (1982) Theory of dislocation cell sizes in deformed metals. Mater Sci Eng 55:79–83
23. Bunge H (1991) The basic concepts of texture investigation in polycrystalline materials. Steel Res 62:530–541
24. Gairola BKD (1993) Gauge theory of dislocations. Mater. Trans Tech Publ, Sci. Forum, pp 579–590
25. Hughes DA, Liu Q, Chrzan DC, Hansen N (1997) Scaling of microstructural parameters: misorientations of deformation induced boundaries. Acta Mater 45:105–112
26. Hughes DA, Chrzan DC, Liu Q, Hansen N (1998) Scaling of misorientation angle distributions. Phys Rev Lett 81:4664
27. Hughes DA (2001) Microstructure evolution, slip patterns and flow stress. Mater Sci Eng A 319:46–54
28. Hutchinson WB (1984) Development and control of annealing textures in low-carbon steels. Int Met Rev 29:25–42
29. Tavares SSM, da Silva MR, Pardal JM, Abreu HFG, Gomes AM (2006) Microstructural changes produced by plastic deformation in the UNS S31803 duplex stainless steel. J Mater Process Technol 180:318–322
30. Keichel J, Foct J, Gottstein G (2003) Deformation and annealing behavior of nitrogen alloyed duplex stainless steels. Part I: Rolling, ISIJ Int 43:1781–1787
31. Ryś J, Witkowska M (2010) Influence of band-like morphology on microstructure and texture evolution in rolled super-duplex steel. Arch Metall Mater 55:733–747
32. Ramdas Dandekar T, Kisni Khatirkar R, Kumar A, Bibhanshu N, Suwas S (2022) Unidirectional cold rolling of Fe-21Cr-5Mn-1.5Ni alloy – Microstructure, texture and magnetic properties. J Magn Magn Mater 549:169040
33. Wroǹski S, Tarasiuk J, Bacroix B, Baczmaǹski A, Braham C (2012) Investigation of plastic deformation heterogeneities in duplex steel by EBSD, Mater. Charact. 3:1–9
34. Duckham A, Knutsen RD, Engler O (2001) Influence of deformation variables on the formation of copper-type shear bands in Al–1Mg. Acta Mater 49:2739–2749
35. Donadille C, Valle R, Dervin P, Penelle R (1989) Development of texture and microstructure during cold-rolling and annealing of FCC alloys: Example of an austenitic stainless steel. Acta Metall 37:1547–1571
36. Schramm RE, Reed RP (1976) Stacking fault energies of fcc fe-Ni alloys by X-ray diffraction line profile analysis, Metall. Mater. Trans. A. 7:359–363
37. Rhodes CG, Thompson AW (1977) The composition dependence of stacking fault energy in austenitic stainless steels. Metall Trans A 8:1901–1906
38. Talonen J, Hänninen H (2007) Formation of shear bands and strain-induced martensite during plastic deformation of metastable austenitic stainless steels. Acta Mater 55:6108–6118

39. Brofman PJ, Ansell GS (1978) On the effect of carbon on the stacking fault energy of austenitic stainless steels. Metall Trans A 9:879–880
40. Dieter GE, Bacon D (1976) Mechanical metallurgy, McGraw-hill New York
41. Chen L, Jia Q, Hao S, Wang Y, Peng C, Ma X (2021) The effect of strain-induced martensite transformation on strain partitioning and damage evolution in a duplex stainless steel with metastable austenite. Mater Sci Eng A 814:141173
42. Rodrigues DG, Gianini G, Maria B, Aparecida N, Viana L (2019) Materials Characterization Effect of low cold-rolling strain on microstructure, texture, phase transformation, and mechanical properties of 2304 lean duplex stainless steel. Mater Charact 150:138–149
43. Barbatti C, Triguero PR, Van Bohemen S, Celotto S, Hanlone D (2012) Study of recrystallization and phase transformation in a cold-rolled dual-phase steel: Influence of temperature and heating rate during first annealing stages. Mater. Trans Tech Publications Ltd., Sci. Forum, pp 818–821
44. Ray RK, Jonas JJ (1990) Transformation textures in steels. Int Mater Rev 35:1–36
45. Raabe D, Schlenkert G, Weisshaupt H, Lücke K (1994) Texture and microstructure of rolled and annealed tantalum. Mater Sci Technol 10:299–305
46. Park S, Kim K, Lee Y, Park C (2002) Evolution of microstructure and texture associated with ridging in ferritic stainless steels. ISIJ Int 42:100–105
47. Lobanov ML, Danilov SV, Pastukhov VI, Averin SA, Khrunyk YY, Popov AA (2016) The crystallographic relationship of molybdenum textures after hot rolling and recrystallization. Mater Des 109:251–255
48. Malta PO, Dias FL, de Souza ACM, Santos DB (2018) Microstructure and texture evolution of duplex stainless steels with different molybdenum contents. Mater Charact 142:406–421
49. Jia N, Peng RL, Wang YD, Johansson S, Liaw PK (2008) Micromechanical behavior and texture evolution of duplex stainless steel studied by neutron diffraction and self-consistent modeling. Acta Mater 56:782–793
50. Hamada J, Ishimaru E, Inoue H (2016) Texture and planar anisotropy in lean duplex stainless steel sheet, ISIJ Int. 56:1831–1839
51. Ashrafi H, Shamanian M, Emadi R, Sanayei M, Farhadi F, Szpunar JA (2021) Characterization of Microstructure and Microtexture in a Cold-Rolled and Intercritically Annealed Dual-Phase. J Mater Eng Perform 30:7306–7313
52. Das A, Tarafder S, Sivaprasad S, Chakrabarti D (2019) Influence of microstructure and strain rate on the strain partitioning behaviour of dual phase steels. Mater Sci Eng A 754:348–360
53. Yokoi T, Shuto H, Ikeda K, Nakad N, Tsuchiyama T (2016) Quantification of large deformation with punching in dual phase steel and change of its microstructure—Part I : Proposal of the quantification technique of the punching damage of the dual phase steel, ISIJ Int. 56: 2068–2076
54. Nakada N, Ikeda K, Shuto H, Yokoi T, Tsuchiyama T (2016) Quantification of large deformation with punching in dual phase steel and change of its microstructure—Part II : Local strain mapping of dual phase steel by a combination technique of electron backscatter diffraction and digital image correlation method, ISIJ Int. 56: 2077–2083
55. Calcagnotto M, Ponge D, Demir E, Raabe D (2010) Orientation gradients and geometrically necessary dislocations in ultrafine grained dual-phase steels studied by 2D and 3D EBSD. Mater Sci Eng A 527:2738–2746
56. Dillien S, Seefeldt M, Allain S, Bouaziz O, Van Houtte P (2010) EBSD study of the substructure development with cold deformation of dual phase steel, Mater. Sci. Eng. A, 527:947–953
57. Li S, Guo C, Hao L, Kang Y, An Y (2019) In-situ EBSD study of deformation behaviour of 600 MPa grade dual phase steel during uniaxial tensile tests. Mater Sci Eng A 759:624–632
58. Han S, Choi S, Choi J, Seong H, Kim I (2010) Effect of hot-rolling processing on texture and r -value of annealed dual-phase steels. Mater Sci Eng A 527:1686–1694
59. Calcagnotto M, Ponge D, Raabe D (2012) Microstructure control during fabrication of ultrafine grained dual-phase Steel: characterization and effect of intercritical annealing parameters. ISIJ Inter 52:874–883

Chapter 11
Role of Microtexture on Cyclic Deformation of 304 LN Austenitic Stainless Steel

Amrita Kundu, David P. Field, and Pravash Chandra Chakraborti

1 Introduction

The local orientation distribution of the grains in polycrystalline microstructures influences initiation of fatigue cracks which consists of the formation of persistent slip bands and evolution of local dislocation density [1]. The engineering importance of fatigue has resulted in comprehensive works on mechanistic modeling of fatigue using molecular dynamics, discrete dislocation dynamics, crystal plasticity, and conventional continuum plasticity which incorporate dependence on microstructures at different length scales [2–31]. The von Mises criterion in continuum plasticity and crystal plasticity can include the strain heterogeneity and geometrically necessary dislocation densities. Prediction of slip localization in the intergranular length scale can be challenging using crystal plasticity modeling. The prominence of fatigue damage characterization is associated with the formation of persistent slip bands, essential for the fatigue crack initiation. Planar dislocation structures involving dislocation pile-ups along grain boundaries, twins, formation of dislocation walls or channels triggering multiple slip and cross slips can produce persistent slip bands [15–21]. Strain localization occurs in regions of specific microstructures, and deformation gradients between grains can be responsible for fatigue crack initiation, and the energy requirements for fatigue crack nucleation can be calculated by molecular dynamics [25–31]. Knowledge about grain orientation distributions can be obtained from analyzing the crystallographic texture using X-ray diffraction or electron diffraction methods, and it is associated with the average value of texture

A. Kundu (✉) · P. C. Chakraborti
Metallurgical and Material Engineering Department, Jadavpur University, Kolkata, India
e-mail: akundu05@gmail.com

D. P. Field
Institute of Materials Research, School of Mechanical and Materials Engineering, Washington State University, Pullman, WA, USA

S. Suwas and D. P. Field (eds.), *Advances in Texture, Microtexture, and Allied Techniques*, Materials Horizons: From Nature to Nanomaterials,
https://doi.org/10.1007/978-981-96-5346-1_11

of the grain population in the microstructure. Microtexture offers information about the orientation topography [32]. Concerted efforts to generate information about local microstructure and crystal structure including crystal orientation distributions, conventionally happened in parallel, which will be discussed in the next section.

2 Characterization of Microtexture

Microstructure information like grain size, shape, presence of second phases can be obtained at a different length scales using an optical microscope, scanning electron microscope (SEM) and transmission electron microscope (TEM). Macrotexture can be determined using X-ray, EBSD, diffraction in TEM, Kossel diffraction, selected area channeling (SAC) in SEM [32]. The primary drawback of bulk analysis techniques is the difficulty to establish a direct connection between microstructural information and the average texture from X-rays. Applying TEM, microstructure information can be linked with the crystallographic texture in a very small imaging area as TEM examination suffers from the lack of generating statistically relevant datasets and difficulty in specimen preparation. Additionally, it is generally tedious to obtain information about grain boundary character and grain boundary triple junctions in the TEM [32]. In order to solve this difficulty, EBSD, which is performed in the SEM by focusing a stationary electron beam on a crystalline specimen, can be applied. The technique was first referred in 1928 when Kikuchi [33] revealed 'thin' lines of contrast in mica in the TEM [33]. The first backscatter diffraction pattern was captured by Alam, Pashley, and Nicholson in 1954 [34]. In 1970, a film camera was used by Venables et al. [35] to measure crystal orientation. Dingley [36] and Dingley et al. [37] in 1984 used a low light video camera to generate the first computerized crystal orientation program merging live camera images with a computer algorithm that gave individual orientation results. This computer-enabled system that allowed users to interactively analyze individual EBSD patterns online was marketed by Link Analytical (Oxford Instruments) starting in 1986. Further development in EBSD was marked by Wright et al. (1990) who used edge detection to identify Kikuchi bands, and the application of the Hough transformation in 1990 by N. C Krieger-Lassen. In the 1990, HKL Technology (Schmidt) was founded, followed by TexSEM Laboratories (TSL–Dingley, Adams) in 1993. Wright's PhD thesis in 1992 [38], followed by "Orientation Imaging Microscopy: The Emergence of a New Microscopy" in 1993 by Adams et al. [39] showed the first automated orientation maps with information about orientation, governed by the processing or specimen geometry and misorientation, or orientation difference between crystal lattices.

EBSD offers ready measurement of the crystal orientations. The specimens are tilted 70 degrees with respect to the horizontal. A phosphor screen is used to capture the diffraction pattern or Kikuchi pattern and the image is recorded using a CCD camera or other imaging/detecting device. The bands seen in the Kikuchi patterns signify reflecting planes present in the crystal volume undergoing diffraction. Therefore, the geometrical arrangement of the bands in the Kikuchi pattern is connected with the crystal structure and orientation of the crystal lattice undergoing electron diffraction. While scanning the electron beam from grain to grain, the Kikuchi pattern will change owing to the change in the crystal orientation in the diffracting volume. The relevant information is the orientation of a reference frame in the crystal lattice with respect to a reference frame in the specimen. As a result, mathematical description of the orientation of a known coordinate system in terms of an identified reference coordinate frame is needed. A multiple step process is involved with the indexing of the Kikuchi pattern concerning peak identification, band indexing, and crystal orientation determination. Straight lines are required to locate for peak identification. Hough transformation results in conversion of lines to points in the Hough space. Points are integrated normal to a line identified with an angle θ. Peaks are described using a mask. Angles between the planes are independent of crystal orientation. The angle between two bands can be compared with an interplanar angle look up table, the hkl pair connected with the band pair can be recognized. The indices (hkl) of three independent bands in the pattern can result in the determination of the crystal orientation.

In the past decade, many variations of Kikuchi pattern indexing and EBSD imaging strategy have been developed. Dictionary indexing, popularized by De Graef and co-workers [40], for example, does not require the identification of individual Kikuchi bands. High-angular resolution EBSD (HR-EBSD) uses auto-correlation of neighboring Kikuchi patterns to identify small changes in EBSD patterns, allowing for determination of residual stresses [41]. Transmission EBSD is performed using electron transparent specimens in the SEM that improves spatial resolution of the EBSD technique [42]. Each of these has become standard practice in laboratories throughout the world.

The measurement of the orientation of individual grains in the specimen should allow merging of crystallography and morphology of the microstructure generating information about local properties (like slip in preferred crystallographic orientations and propagation of slip, determined by individual crystal orientation, or in clusters resulting in deformation gradients between varying regions of the specimen). In addition, interfacial parameters, size, shape and aspect ratio of the grains, orientation gradients in individual grains, phase data, and direct measurement of the orientation distribution function (ODF) by generating full grain orientations are all possible.

Quantitative characterization of grain boundary complexions [43] allows the researcher to relate properties to local grain boundary structure. The microstructure of engineering polycrystals mainly consists of high-angle boundaries. However, low-angle boundaries can also be found in the microstructure. Physical property differences exist between high- and low-angle boundaries present in the microstructure influencing the intergranular transport properties by changing the diffusion path and performance of the materials. In a nutshell, microtexture refers to several preferred orientations and misorientations present in the microstructure. The spatial distributions of preferred orientations are generally non-random, leading to anisotropic properties. Thus, microtexture is linked with local determination of texture in the grain scale connecting crystallographic texture with other features of the microstructure. The primary goal of microtextural analysis is associated with the revelation of an entire range of morphology and crystallography attached with the microstructures.

Polarized light microscopy can also be used to image optically anisotropic surface regions [44, 45]. The surface wave microscopy techniques such as acoustic microscopy or acoustic spectroscopy can also be applied for characterization of orientation as determined from elastic properties. These techniques are dependent of the surface finish and require specialized instrumentation [45–47]. Nondestructive techniques such as eddy current or the ultrasound can also be utilized for determination of the microtexture regions [44–48]. The challenges of using nondestructive techniques are the presence of false signals due to gradients in the microtexture.

3 Importance of Microtexture on Cyclic Plastic Deformation

The cyclic deformation characteristics of engineering materials are strongly influenced by the microstructures, grain size distribution, amount and distribution of phases, macro and microtextures [1–31, 49, 50]. The negative effect of microtexture gradients is most apparent in reducing the continuous and dwell fatigue life [51, 52]. Literature reveals the presence of hard and soft grain boundary regions acting as fatigue crack initiation sites due to formation of slip gradients. Generation of slip gradients is associated with the difference in grain orientation resulting in uneven load distribution by the favorably oriented soft grains. Facets on fracture surfaces are not always formed by cleavage in polycrystals that fail due to cyclic loading. There may be signs of plasticity [2–14, 18–22]. The initiation of cyclic plasticity is associated with the generation of dislocations. In-grain generation and migration of dislocations can result in pile-ups of dislocations at the grain boundaries [15–17]. Grain boundary slip can happen due to the presence of dislocation pile-ups [15–17]. Residual dislocations can be left at the grain boundaries and dislocations may move backward in the pile-up [15–17]. The grain scale plastic deformation process is connected with the dynamics of dislocation motion and grain boundary slip geometry [8, 9]. There are three conditions for slip transfer across the grain boundaries [15–17, 49, 50]:

The angle between the intersecting slip planes and grain boundary plane should be minimized [49].

Maximum critical resolve shear stress should be present on the slip systems of the adjacent grains [50].

The Burgers vector of the grain boundary dislocation has to be lowest [52].

Different stress relaxation mechanisms can also be operative at the grain boundaries when slip systems with low stresses are not available [2, 49, 50, 52]. During cyclic deformation, generation of stress magnification is linked with the lack of slip transfer across the grain boundaries, resulting in generation of fatigue cracks [2, 4, 16, 18, 49, 50]. The influence of the microtexture on initiating the fatigue crack including dwell fatigue was investigated by several researchers [5, 26, 31, 51–59] using high resolution EBSD. The characterization of microtexture is especially important in Titanium alloys because of the dominance of the c-axis with respect to the loading axis and crack growth [2–7, 14, 18, 20, 22, 24, 27, 44, 51, 53, 54]. The microtexture data can also be utilized in the progress of the crystal plasticity models to predict the microstructure heterogeneity and evolution of the microtexture [55]. The application of this model is in the prediction of elastic anisotropy [60, 61]. The Schmid factor can be ambiguous in predicting the local stress level and requires the use of the microtexture data for accurate prediction of the local stress level [62, 63]. Hexagonal close-pacekd (HCP) materials were investigated to correlate the grain orientation with the elastic anisotropy [62, 63]. Speckle interferometry was used to measure the grain scale strain concentration in coarse grain Ti which could be revealed by crystal plasticity modeling [55] and could foretell the crack initiation sites. In FCC materials the hard and soft grain orientation combination can be unfavorable for initiation of the fatigue damage [8, 15–17, 57]. In polycrystalline directionally solidified Nickel, microtexture is used to understand the fatigue crack nucleation sites. In ferritic steel (BCC crystal structure), the grain size, grain morphology and microtexture are important in nucleation of fatigue cracks [64]. Accumulation of slip and slip localization rates are important in fatigue damage and can be illustrated using crystal plasticity models incorporating the microtexture. In fretting fatigue, local grain orientations are used to predict the fatigue life and wear [65]. The presence of twins leading to strain concentrations due to dislocation pile-ups is also responsible for generation of the fatigue crack [1, 15–17, 46, 49, 50]. Man et al. [19] and Kundu et al. [17] examined 316 and 304LN austenitic stainless steel, respectively, to characterize the strain localization and dislocation distribution at the grain boundaries and twin boundaries. Fatigue crack nucleation takes place at the twin boundaries mostly in Ni alloys [46] and twin thickness stimulates fatigue crack initiation. On the contrary, favorable slip transfer without generation of interface debris and defects is the pre-requisite of homogeneous cyclic plastic deformation. Thus, macro and microtexture will affect fatigue fracture mechanisms, specially influencing the initiation of the fatigue crack. The microtexture adjacent to the fatigue crack can contain scatter in intensity and distribution resulting in scatter in the fatigue life prediction. Scatter in fatigue life data due to variation in the microtexture is reported extensively in the literature, especially in Waspaloy, and both alpha and beta Ti alloys [51, 66]. The lower limit or the minimum fatigue lives do not depend strictly on the microtexture. The second phase precipitates have strong influence in controlling the minimum fatigue lives. The processing, such as forging of the alloys has been designed to eliminate the local variation in microtexture [51]. The presence of any residual microtextural gradients can form a weak link and accelerate fatigue failure, especially when scaling from the laboratory specimens to the engineering components. The random texture or the random grain sizes produced by thermomechanical processing in alpha and beta Ti alloys by thermomechanical treatments in several temperatures can result in larger scatter in the fatigue life data with the largest average fatigue life due to greater resistance against fatigue crack propagation. The effect of the processing history on the fatigue life can be insignificant when fatigue crack initiation takes place at the beginning of the fatigue life and crack propagation is prolonged, thus reducing the microstructure dependence. The minimum fatigue life time cannot be changed by thermomechanical processing. The average fatigue life can be expanded by increasing the resistance against fatigue crack propagation in the microstructure containing the random texture. The research in this area is still minor.

Fatigue deformation mechanisms are also governed by the micromechanical loading state, stiffness compatibility, grain boundary plasticity and formation of stress gradients due to the presence of discontinuities in microstructures such as grain boundaries, interface boundaries and second phase regions [49]. In the presence of hard second phases, plastic deformation is partitioned to the soft primary matrix phase. There is an additional microstructural feature, known as a macrozone with common orientations which deform as a large structural unit at the same time [3, 49, 52, 67–77]. There is evidence in the literature where fatigue crack initiation took place at the interface between hard and soft macrozones [3, 49, 52, 67–77]. Cyclic deformation is also associated with grain boundary migration, as each cycle could result in a new position in the grain boundary [69]. Uniform distribution of grain boundary migration does not take place [3, 49, 52, 67–77]. The boundaries which are oriented 45° with respect to the loading axis, move faster than the others [3, 49, 52, 67–77] due to the presence of maximum shear stress, showing the importance of the microtexture. The grain boundary migration also changes with time during the course of cyclic deformation [3, 49, 52, 67–77]. When grain boundary migration occurs in two phase materials, it can lead to formation of cavities [3, 49, 52, 67–77] and shortening of cyclic lifetimes of the materials. Grain boundary migration results in a change in the microtexture and misorientation between neighboring grains [3, 49, 52, 67–77]. Stacking fault energy significantly affects the grain boundary degradation of the materials [3, 49, 52, 67–77]. In low stacking fault energy materials, dynamic recrystallization can initiate when subjected to cyclic deformation at low strain amplitude, i.e., 0.2% [3, 49, 52, 67–77]. Dynamic recrystallization can lead to the formation of new strain free grains, presenting new grain boundaries into the system [3, 49, 52, 67–77] and thus changing the grain boundary migration [3, 49, 52, 67–77] and orientation of the grains with respect to the loading axis [3, 49, 52, 67–77]. This was first shown in high purity lead [78]. Grain boundary motion and dynamic recrystallization during LCF can lead to a change in the grain size distribution, often grain coarsening [31, 49, 52, 59, 67–86]. This can be represented using the orientation distribution function (ODF).

4 Effect of Cyclic Deformation on ODF

Brodesser and Gottstein [68] characterized the ODF in Euler space in high purity Nickel after interrupted low cycle fatigue (LCF) experiments at 600 °C. The ODF (estimated with 10° Gaussian orientation spread) of undeformed Nickel was compared with the cyclically deformed Nickel after 300 cycles. This is presented using constant φ_2 sections in the ($\varphi_1 \Phi \varphi_2$) Bunge convention for Euler angles [87] in Fig. 1. Macrotexture in the undeformed specimen consisted of a component at (88°, 35°, 42°), first order twin (T_1) and a weak incomplete < 111 > fiber texture. The significant increase in the fiber component and a simultaneous decrease in the component at (88°, 35°, 42°) of the macrotexture was observed with an increase in the number of cycles during LCF. At 300 cycles, the fiber component of the macrotexture died out and a cube orientation (0°, 0°, 5°) emerged. The ODFs were mostly in agreement with the pole figures obtained using X-ray diffraction. They showed that dynamic recrystallization, grain boundary migration and change of orientation of the grain boundaries with respect to the loading axis took place during the LCF. The orientation distribution changed from incomplete < 111 > fiber texture to incomplete < 100 > fiber texture leading to an increase in the Schmid factor after 300 cycles. In their studies, a large fraction of grain boundaries in undeformed and in deformed Nickel were low-Σ CSL boundaries. With the progress of cyclic deformation, the twin fraction decreased and the grains showed < 111 > rotations.

Weib et al. [70] investigated microtexture changes in high purity Nickel following the same fatigue test condition using a quasi in-situ technique. They observed grain growth, grain boundary migration and grain boundary sliding which was discussed earlier. They quantified that about 40% of the grain boundaries had a grain boundary angle of 45° with respect to the loading axis. The low-angle and twin boundaries are stable against rotation. The preferred crystallographic planes for the boundaries exhibits a difference of orientation of 10° from {211} and {111} planes (twins). The ODF after 100 cycles showed the dominance of < 100 > orientations (Fig. 2). The texture of growing grains was characterized by the strong cube orientation. The newly formed grains showed a near cube orientation at (5°, 0°, 0°). The shrinking grains contained an α fiber texture with components from (0°, 45°, 0°) via (45°.

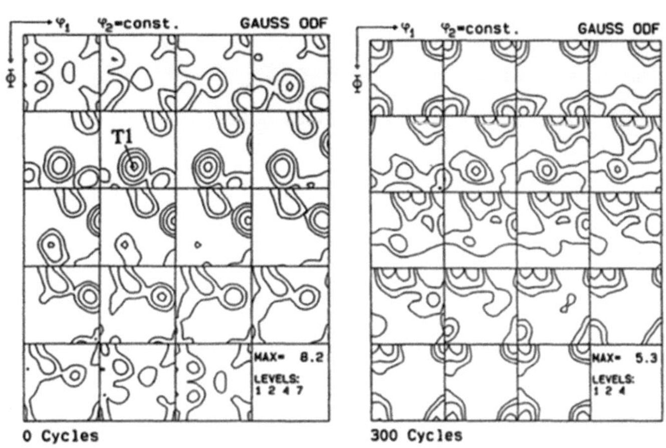

Fig. 1 ODF estimated using the single point orientation data with an estimated Gaussian scatter band of 10° [68]

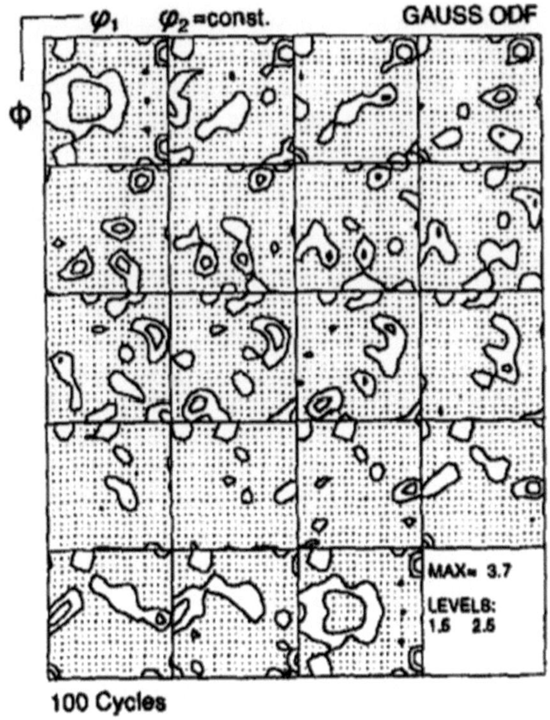

Fig. 2 The ODF of Nickel after 100 LCF cycles at 600 °C [70]

45°, 0°) to (90°, 45°, 0°). The development of microtexture was dependent upon the formation of new grains and the growth of new grains that influenced the grain boundary character.

5 Development of Grain Structures During Cyclic Plastic Deformation: EBSD Characterization

Weib and Gottstein [72] examined the grain boundary motion during cyclic deformation of Aluminum bicrystals by employing quasi in-situ experiments in the scanning electron microscope (SEM). Their experiments facilitated the observation of the same area on the specimen before and after LCF deformation without additional surface polishing [72] allowing to trace the grain boundary movement of the specimens at a free surface under known experimental settings. The bicrystal experiments allowed complete characterization of the orientations of the individual grains revealing orientation relationships and spatial arrangements of the grain boundaries [72], ignoring the interference of neighboring grain boundaries and triple junctions. They applied EBSD to measure the orientation of the adjacent grains near grain boundaries [72]. They studied grain boundaries with misorientations relationships close to low- ΣCSL misorientations Σ5, Σ7, Σ15, Σ19, and Σ21 a during cyclic deformation. Their experiments included characterization of grain boundary morphologies [72]. The morphology of the Σ5 grain boundary showed many slip traces and straight grain boundaries and a wavy grain boundary surface [67]. Adjacent to the grain boundaries, there was no change of orientation. At 4.5 mm from the grain boundary, 15–20° orientation differences developed [72]. Σ7 boundaries showed the presence of tooth protrusions on the grain boundaries parallel to the slip traces [72]. There was development of the orientation of 5.0° to 14.5° adjacent to grain boundaries and band structure with orientation gradients between 5.0° and 9.0° at 4.5 mm from the grain boundary [72]. Σ15 and Σ19 were similar to Σ7 boundaries containing tooth protrusions [72]. However, the orientation change near grain boundaries was only 3.0° to 4.0° [72]. The microstructure did not change at 4.5 mm away from the grain boundaries, but showed band structure, similar to what occurs in Σ7 boundaries [72]. Σ21 a boundaries had small indents at some parts of the boundaries and strong slip traces and the orientation angle near grain boundaries was 2.8–3.0° [72]. Orientations at 4.5 mm away from the grain boundaries were 15° to 20°, similar to that of Σ5 grain boundaries [72]. Weib and Gottstein [72] reported that microtexture involving orientation change did not occur near grain boundaries after 175 cycles. On the contrary, at 4.5 mm away from the grain boundary, 15° to 20° orientation gradients were recorded. The Σ5 grain boundary demonstrated stable characteristics. The bicrystals did not deform homogeneously, resulting in the development of dislocation-rich and dislocation-poor regions. The large gradients in the dislocation density generated a force on the boundary allowing the boundary to move in the direction of high dislocation density. The boundary displacement took place in the lower dislocation density regions of the bands having different orientations, resulting in the creation of an acute angle between the initial and displaced boundary positions. This stress magnification resulted in generation of the excess dislocations. The excess dislocations could be annihilated by straightening of the boundaries. Further grain boundary migration could lead to alignment of the grain boundaries with the primary, and less favored slip systems, leading to the formation of saw tooth shapes at the boundaries. Migration of grain boundaries was stabilized by grain boundary sliding and generation of new dislocations. The boundaries could further expand the sliding sections, changing the grain boundary connectivity and triple junction topology. Weiss and Gottstein [73] showed the development of microstructure and microtexture during high temperature LCF of oligocrystals of a specimen containing 15 grains in the region of deformation. They employed in-situ optical microscopy to track each grain boundary migration step without resorting to additional surface preparation after carrying out tension compression LCF experiments at 300 °C and a strain amplitude of 1% and using a frequency of 0.00013 Hz. Optical microstructural examination showed that coherent twin boundaries were stable in high stacking fault energy Aluminum in contrast to high purity nickel as discussed earlier (although the LCF conditions are different). There was a disappearance of small grains (having random orientation) due to movement of other grain boundaries and straightening of curved grain boundaries as discussed earlier. Within 100 cycles, there was change of grain orientation up to 12°. The frequency of {111} planes oriented 45° to the stress axis increased with an increase in the number of cycles. The misorientation of stable grain boundaries was primarily low-angle boundaries and twins, i.e., Σ3 boundaries. The grain boundary misorientation was more common in grain boundaries with misorientation in excess of 10°. In this study, the position of the grain boundary triple junction influences grain boundary motion due to exertion of a force on a rather immobile boundary [73] leading to activation of the less favorable slip system. The reasons could be incompatibility of primary slip systems of the adjacent grains or grain boundary sliding [73]. The in-situ experiments revealed that the horizontal grain boundaries remained stable in comparison to other boundaries. There was a misfit at the grain boundary triple junction due to movement of two of the three boundaries. Grain boundary sliding gathered at the horizontal grain boundaries in a plane normal to the surface to balance the misfit. Triple junctions could stimulate new slip activities and grain boundary sliding. Therefore, stability of the microstructure against high temperature could be achieved by increasing the population of low-angle boundaries and twin fractions [73].

6 Effect of Microtexture on Fatigue Lives

Microtextured regions (macrozones) influence fatigue lives. Bantounas et al. [74] performed TEM and EBSD investigations on bimodal Ti–6Al–4V plate after performing the fatigue experiments using displacement-controlled loading and the stress ratio of 0.3 (in the high cycle fatigue regime). This was done until fracture of the specimens to understand the effect of microtexture or macrozones on the formation of fatigue facets. It was known that in Ti–6Al–4V, the initiation of cracks occurs in the grains showing basal slip orientation [3, 4, 27, 31, 77–82]. In Ti–6Al–4V alloy, the orientation of the c-axis with reference to the loading direction, affects strongly the fatigue crack initiation [82]. The preferred site for fatigue crack initiation was regions of non-basal faceting of α grains [77–82]. The regions of microtexture with less favorable slip orientation were the potential sites of crack initiation and coalescence of cracks could result in the formation of larger cracks [77–82]. The facets were formed within microtexture regions with 40° c-axis misorientation with reference to the loading direction [77–82]. Fractographic investigation along with EBSD substantiated that the non-favorably oriented basal planes were the preferred sites for crack initiation. Facets formed along < 11–20 > (0001) slip systems. Unhindered crack propagation took place along macrozones whose c-axis was inclined along the loading direction [77–82]. Macrozones with c-axes perpendicular to the loading direction offer resistance to the faceted crack propagation [2–7, 14, 18, 20, 22, 24, 27, 44, 51, 53, 54, 77–84]. Variant selection during β to secondary α phase transformation governed the facet size and therefore resistance against fatigue crack propagation. Joseph et al. [52] reported that in near-α Ti6242Si alloy cyclic deformation occurred by planar slip in slip bands in primary α, after characterizing dislocation structure using TEM and microtexture using EBSD. In soft oriented grains in regions of varying microtexture, or macrozones, accumulation of slip bands was responsible for straight transfer of slip in the adjacent grains of similar orientation [52], facilitating slip alignment and multiple cross slip [52]. No strain gradient was visible through the boundaries between soft grains [52]. Slip takes place by basal, pyramidal or prismatic slip mechanisms [52]. Dislocation storage in soft oriented grains that neighbor hard grains resulted in activation of few dislocations in the hard grains [52] instead of propagation of dislocations through the grain boundaries. As a result, the hard grains offered resistance against cyclic plastic deformation [52]. The dislocation source might be present either at the grain boundaries or at the hard grains. The various types of nucleation of dislocations resulted in the strain transfer rather than slip transfer. After nucleation, dislocations multiply by cross slip and result in patterning [52]. No dislocation glide was observed. Multiplication of dislocations happened by moving from one plane to another by cross slip in the hard oriented grains. Dislocation interaction resulted in the formation of dislocation loops and tangles and was responsible for formation of cavities in the β phase boundaries owing to the interaction between retained β and secondary α [52]. Cross slip usually initiated under high stresses on the hard grains originated by the dislocation dense soft grains. Dislocation velocity was different on the cross slip planes and on the slip planes. Bantounas et al. [74] characterized the fatigue cracking under high cycle fatigue (HCF) ($R = 0.3$) in Ti–6Al–4 V alloy and established correspondence between the fatigue fracture surface and macrozones. Fatigue cracks were arrested in the grains poorly oriented for prismatic {10–10} < 11–20 > slip [59]. Crack propagation was found to happen along {0002} planes undergoing < c + a > slip [74]. The size of regions of striations at varied angles were connected with the microtextured regions.

Similar to Ti–6Al–4V alloy, facet creation and fatigue crack growth by striations were the responsible fracture mechanism in Ti6242Si alloy (near α Ti alloy) [52]. The fatigue experiments were performed using a trapezoidal waveform with ramp up/down duration of 1 s and 1 s hold at maximum stress which is 95% of yield stress and a stress ratio of 0.05 until failure of the specimens. Some dislocations could be accommodated at the primary α boundary resulting in the change of the geometry of the grain boundaries and eventually leading to initiation of the transgranular cracking in the boundary [52]. Thus, it was required to learn the development of the dislocation structure during LCF for better life prediction.

7 Micromechanism of Cyclic Plastic Deformation: Influence of Microtexture

The contribution of geometrically necessary dislocation (GND) density on micromechanisms of fatigue has been investigated by many researchers [17, 31, 59, 84–86]. Liu et al. [84] studied the HCF and LCF of Nickel-based superalloy IN713C at 600 °C and the microstructural heterogeneity because of carbide precipitates in the dendritic network was characterized by GND distribution. Higher GND density at the oxidized carbide precipitates were responsible for the initiation of the fatigue cracks determining the micromechanism of fatigue fracture. Carbide morphology influenced the oxidation process and hence crack growth. Microtexture and grain orientation contributed to the HCF fracture by forming facets on the grains with Goss (<110 > || LD) orientations and the slip systems with the highest Schmid factors were prone to crack initiation.

Kundu et al. [17] characterized the effect of strain amplitude on the evolution of microstructure and dislocation structure, especially GND density by applying EBSD and ECCI in solution annealed austenitic stainless steel. The strain amplitudes were varied between 0.25% and 1.2% under completely reversed strain cycling and the tests were carried out to complete fracture of the specimens. GND density increased with the increase in the strain amplitudes and this explains the cyclic hardening character of the austenitic stainless steel at room temperature. This was also consistent with the results of Kamaya [85] in the SS 316 steel. The GND density distribution was uneven and smaller grains were found to accumulate larger GND content than the coarser grains. The GND density pile-up was observed near grain boundaries than grain centers. Increase in strain amplitudes resulted in reducing the heterogeneity of GND structure distribution. EBSD was used to quantify the orientation dependent GND density distribution. The austenite grains with orientation {111} < 0–11 > contained highest GND density. The difference in the grain orientation dependent GND density distribution increases with the increase in the cyclic strain amplitudes. The fidelity of the dislocation structure was captured using ECCI, showing the development of planar dislocation structure. At lower strain amplitude, the dislocation structure consisted of loosely packed dislocations and tangles. Low energy arrangements of dislocations in cellular form were seen at intermediate strain amplitude. At higher strain amplitude, dislocation-rich and dislocation-depleted areas were visible in the microstructure forming dislocation vein structures. The spacings between the veins decreased due to thickening of the veins with the increase in the strain amplitudes resulting in the retardation of the mobility of the dislocations. The reduced mobility of the dislocations was the precursor of increased cyclic hardening with the strain amplitudes.

Sinha et al. [86] examined the development of the misorientation and GND density with strain amplitudes in interstitial free (IF) steel and found that the local average misorientation and GND density increases with the strain amplitude of 0.4% and then decreases. Their investigation showed that the finer grains were responsible for significant storage of dislocation density in IF steel. Littlewood and Wilkinson [31, 59] characterized the cyclic deformation heterogeneity and grain size based GND distribution in Ti–6Al–4V and reported that the grains with substructure retain more GNDs.

The above review unveils the importance of microtexture in the cyclic deformation in generating deformation heterogeneities, damage in the form of grain boundary sliding, grain boundary migration, cavity formation and crack initiation, and shortening of fatigue lives in various metals and alloys. The alteration of grain structure is the key in controlling the fatigue damage mechanisms. The next section of this chapter aims to characterize the microtexture in the form of the change of grain boundary character, misorientation, and grain scale deformation (in terms of grain reference orientation deviation (GROD)) in austenitic stainless steel subjected to different cyclic lives. The heterogeneity in GND density evolution with the cyclic lives is discussed. The present investigation also quantifies the influence of grain size and orientation of austenite grains on the buildup of cyclic deformation heterogeneities and GND density and relating these to the cyclic lifetimes.

8 Experimental Details

Commercial variety 304 LN austenitic stainless steel was used in the form of 12 mm diameter rod as the test material in the solution annealed condition. The chemistry of the steel in weight percent is C—0.027, Mn—1.64, Si—0.374, S—0.028, P—0.015, Ni—10.06, Cr—16.85, Mo—2.11, Ti—0.0008, N—0.057. The steel was solution annealed at 1100 °C for 45 min trailed by water quenching at room temperature (~28 °C).

LCF tests of the steel in solution annealed condition were done at room temperature under completely reversed total strain-control mode ($R\varepsilon = -1$) at $\pm$ 0.5% strain amplitude using a total strain rate of 5×10^{-3} s^{-1} in INSTRON 8501R of $\pm$ 100 kN capacity. A triangular wave form was employed. The fatigue tests were carried out under strain-control mode using an extensometer of 25.00 mm gauge length. Tests were conducted at room temperature for different life fractions (545 cycles, 1089 cycles, 2178 cycles and 3267 cycles).

The LCF tested specimens were characterized using EBSD to understand the evolution of GND density during cyclic loading. For this purpose, specimens were extracted from the gauge sections. The extracted samples were polished through different stages of polishing, and final polishing was done using dilute 0.05 micron colloidal silica suspension.

EBSD scans were executed at 20 keV in a field emission scanning electron microscope (FEI, Quanta 450) equipped with an EDAX EDS/EBSD system at a working distance of 15 mm. Each scan area was equivalent to 600 to 800 grains. Approximately 2500 grains were scanned in both the deformed and solution annealed specimens using 0.1 micron step size. The raw EBSD scan data were analyzed using TSL-OIM™ software with no data manipulation or "cleanup." Characterization using ECCI of the cyclically strained samples was performed using a Quanta 450 FEG-SEM using 20 keV operating voltage.

9 Results and Discussion

9.1 *Orientation Maps with Varying Cyclic Lives of 304 LN Austenitic Stainless Steel*

The orientation maps, (or inverse pole figure (IPF) maps) of the undeformed and the fatigue tested samples using strain amplitudes of 0.5% with different numbers of cycles are shown in Fig. 3a–e. The orientation color code signifies the crystallographic poles aligned with the surface normal orientation of the samples. The LCF deformation gives rise to the crystal rotation, seen from the orientation maps. With increasing number of cycles, there is an accompanying formation of orientation differences (represented by the color gradient) within the individual grains, which indicates heterogeneous deformation and generation of matrix dislocations leading to cyclic hardening of the steel. The cyclic hardening behavior of this steel was characterized in detail in a separate work [17]. Grain boundaries are associated with the discontinuous change in color in the orientation map. A wide range of in-grain crystal orientations are generated with an increasing number of cycles. The formation of orientation gradients are restricted within the small sized grains as compared with the larger grains.

The grain size distribution of the solution treated specimen is shown in Fig. 4. The grain size distribution is broad, containing 4.5 to 126 micron grains. The smaller austenite grains undergo straining with constraint by the surrounding structure, and generation of excess dislocation density in these small austenite grains is more rapid, making the smaller austenite grains harder. As a result, more orientation changes develop in small austenite grains than in larger grains.

The evolution of crystal orientations can be represented by inverse pole figures, presented in Fig. 5 a–c, corresponding to the undeformed specimen and specimens with 1089 cycles and 2178 cycles of LCF. The IPFs indicate the orientation distribution aligned with the deformation axis in the steel. The crystal orientation distribution accumulates around individual grains in the undeformed solution treated specimen. On the contrary, in the cyclically strained specimens, the points are spread out. The progressive cyclic deformation resulted in the change in crystal orientations. This is consistent with the development of the color gradients within a grain in the orientation maps presented in Fig. 3.

9.2 *Evolution of Grain Size Distribution and Grain Boundary Character*

In order to determine the microstructural changes that happen during the progress of cyclic deformation, the grain size distributions are plotted in Fig. 6. The solution annealed steel consists of equiaxed austenite grains with twins present in the

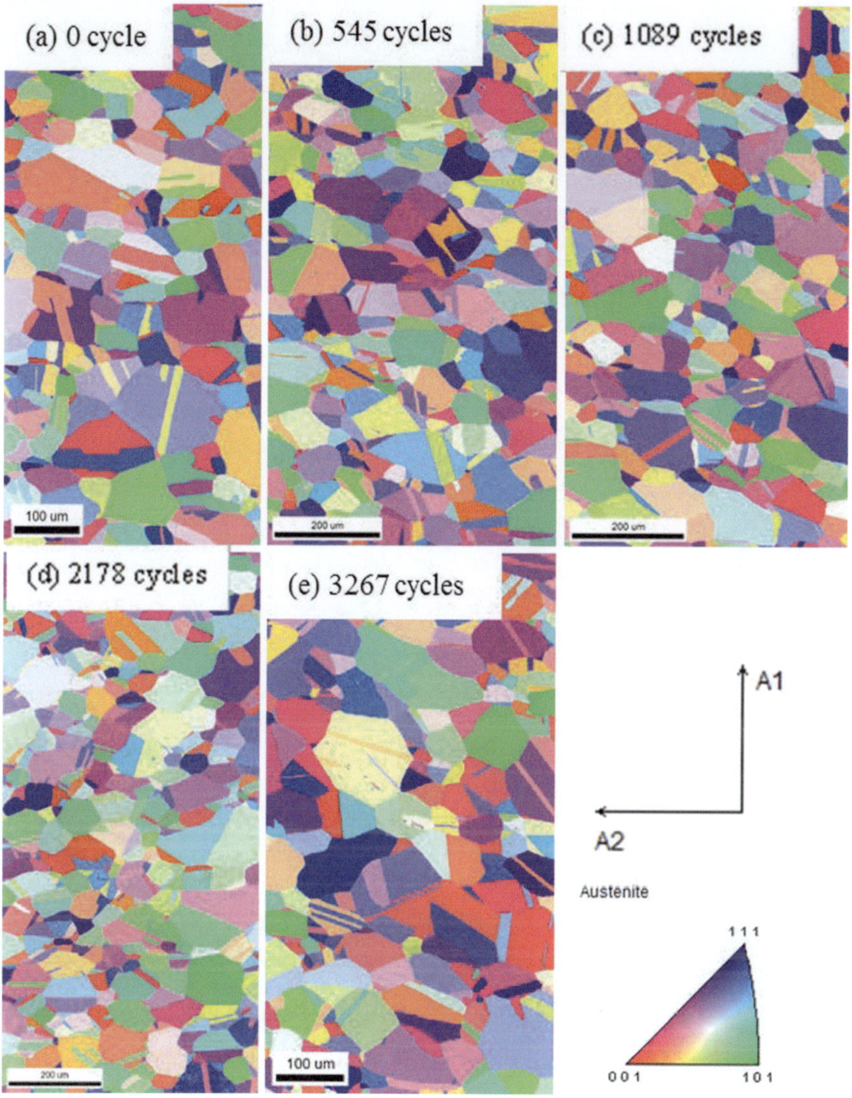

Fig. 3 Orientation or inverse pole figure (IPF) maps (pole orientation along normal direction) of the specimens deformed at various number of cycles at strain amplitude 0.5%

grains. The average austenite grain size in the solution treated condition is 50 micron (determined after measuring grain size of 2000 grains). When the steel is subjected to strain cycling to 545 cycles, the average austenite grain size decreases to 42.5 microns because of fragmentation of larger grains to smaller grains. After 1089 cycles, grain subdivision further continues and the average austenite grain size is 37 microns. When the number of cycles increases to 2178, there is a small increase in the

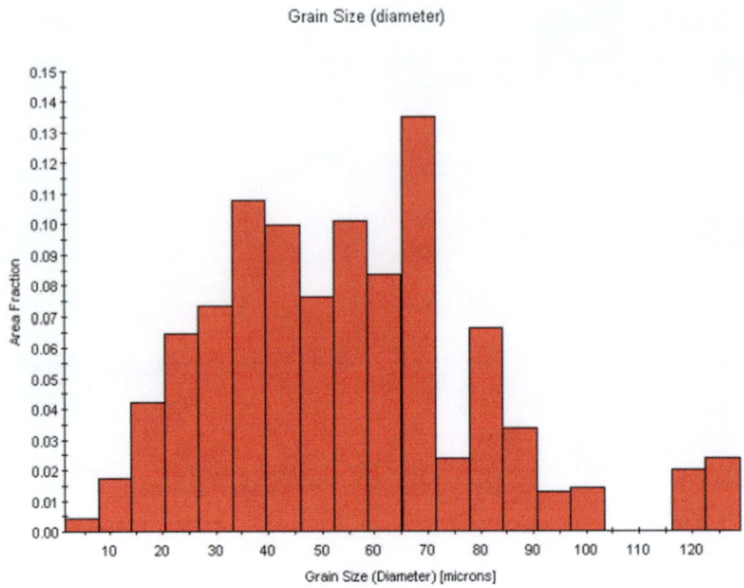

Fig. 4 Grain size distribution of solution treated austenitic stainless steel

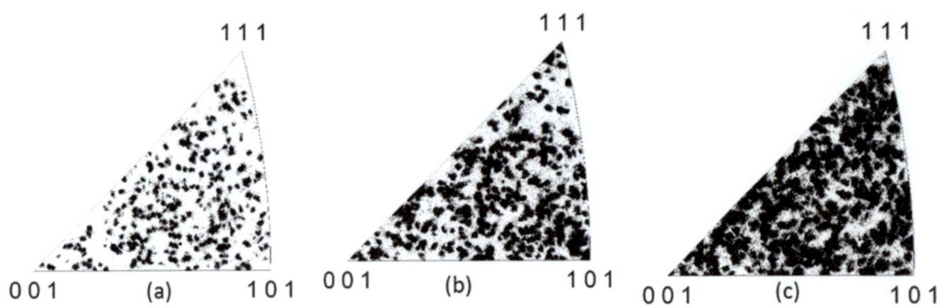

Fig. 5 Inverse pole figures showing the orientation of the axis of deformation of **a** the solution treated specimen and after cyclic deformation at **b** 1089 cycles and at **c** 2178 cycles

average grain size to 40 microns. The austenite grain size is 42.5 micron after 3267 cycles. Similar variation of grain structure with number of cycles was mentioned by Brodesser and Gottstein [68] in Nickel subjected to cyclic plastic deformation at 600 °C.

There is also change in the twin boundary fraction with number of cycles during the progress of the deformation. There is a slight 6% increase in the twin boundary fraction as opposed to the solution treated specimen after deformation to 1089 cycles. This may be associated with the grain fragmentation during cyclic deformation and

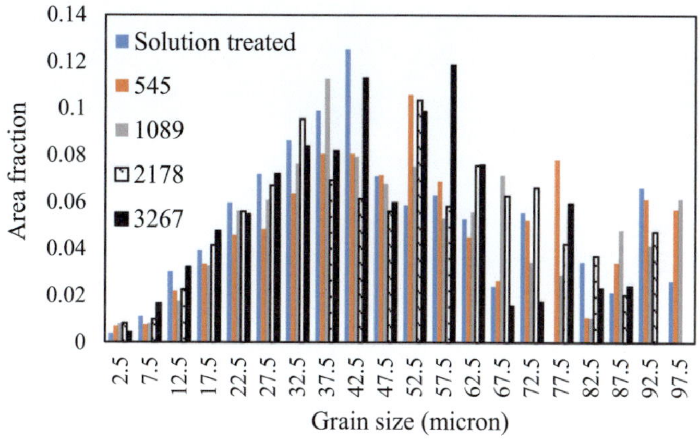

Fig. 6 Evolution of grain size distribution after cyclic deformation at various numbers of cycles at strain amplitude of 0.5%

cyclic hardening. Thereafter, the amount of twin boundaries decreases with the higher number of deformation cycles. The change of twin boundaries with the number of cycles is presented in Fig. 7. In total, the twin boundaries covered 13–26% of total boundaries. Kundu and Field [88] and Kundu et al. [89] quantified the change in the twin fraction in 304 LN austenitic stainless steel. Brodesser and Gottstein [68] in Nickel showed that first and second order twin boundaries decreases with increase in the number of cycles during LCF. With advancing number of deformation cycles, the decrease in twin density can be explained by the rotation of the grains, resulting in the exhaustion of the twin relationship. This also relates to the generation of the deformation twins which do not increase the 60° < 111 > misorientation component. Similar observations are available in the literature of the variation of twin density with cyclic deformation in low stacking fault energy materials [73, 74].

The change of character of grain boundaries is reflected in the evolution of the grain boundary misorientation angle. The grain boundaries are classified in to three groups, namely a. low-angle grain boundaries (LAGBs) with misorientation 2–5°, b. medium-angle grain boundaries (MAGBs) with misorientation 5–15° and HAGBs with misorientation 15–180°. The grain boundary fraction of the specimens deformed using varying strain cycles are presented in Fig. 8. The amount of LAGBs increases with increase in the strain cycles. Significant increase in LAGBs starts from 2178 cycles. The presence of LAGBs in the microstructure means development of dislocation cell structure separated by low misorientation angles. Dislocation rearrangements during LCF can involve two different mechanisms, namely recovery and recrystallization which will predominantly be governed by the deformation history, i.e., strain amplitudes, strain rates, amount of deformation and austenite grain sizes [49, 52, 68]. The characteristic features of the dynamic recovery are the presence of elongated grains with in-grain misorientations as seen in Fig. 3 in the specimen which was deformed cyclically using the cycles of 2178 and 3267. The MAGB

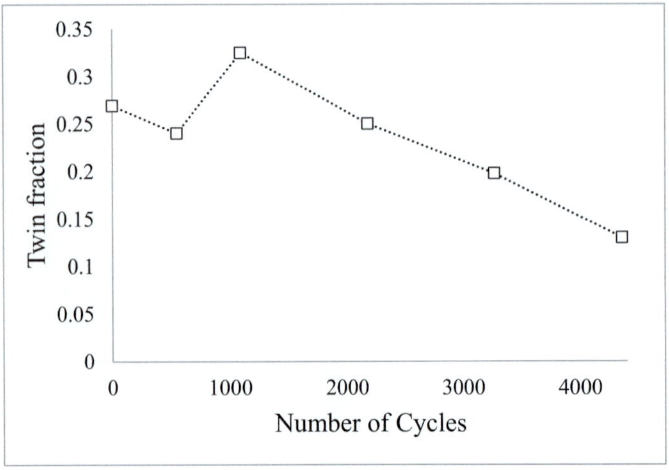

Fig. 7 Change of twin fraction at various numbers of cycles at strain amplitude 0.5%

amount changes mildly in comparison with the amount of LAGBs and HAGBs. The fraction of HAGB decreases with increasing the number of cycles which is also consistent with the increase in the average austenite grain size at this two levels of cycles.

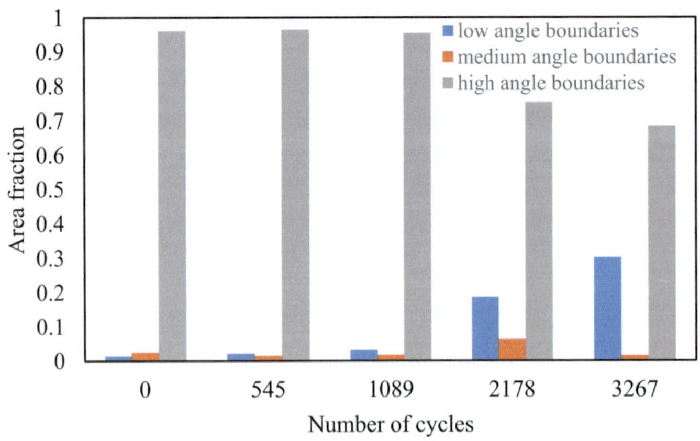

Fig. 8 Variation of fraction of low-, medium-, and high-angle boundaries at various number of cycles at strain amplitude 0.5%

9.3 Variation of Schmid Factor with the Progress of Cyclic Deformation

In order to determine the orientation of grains with respect to the stress axis as function of the number of fatigue cycles, i.e., alignment of the grains parallel to the direction of maximum stress axis, Schmid factor distribution is determined. The distribution of Schmid factor with number of cycles is shown in Fig. 9. With the raise in the number of cycles, the Schmid factor undergoes a slight increase. The maximum value of Schmid factor after 3267 cycles is 0.45. The alteration of Schmid factor can be considered as the trigger for the texture change. The steel is having fairly large grain size resulting in excess grain boundary deformation and setting up deformation heterogeneities. Brodesser and Gottstein [68] in Nickel described similar variation of Schmid factor with number of cycles. They stated that orientation of individual grains may change as the stress axis shifts further from < 111 > direction. The orientation develops along < 100 > specifying buildup of the < 100 > fiber texture. Brodesser and Gottstein [68] applied Taylor model to predict the grain orientation in the same material after interrupted cyclic deformation using strain amplitude of 0.5%. They predicted development of < 121 > orientations and < 111 > orientation stabilization. Brodesser and Gottstein [68] found the Schmid factor is very close to 0.5, for a given stress axis near to < 100 > , the {111} slip planes with the highest Schmid factor are at an angle 45° with the stress axis. They used Rodrigues plot to identify < 111 > rotations between the neighboring grains. When the rotation axis is primary slip plane normal, then the boundaries at an angle 45° are close to < 111 > twist boundaries. Twist boundaries are very less mobile than the < 111 > tilt boundaries. The analysis of Brodesser and Gottstein [68] showed that 50% of grain boundaries are twist boundaries and controlled the grain rotation and grain boundary mobility during deformation. The gradient in Schmid factor can influence the dislocation storage [49, 52, 68]. The change of microstructure and texture are the outcome of grain boundary motion due to anisotropy of the driving force [68].

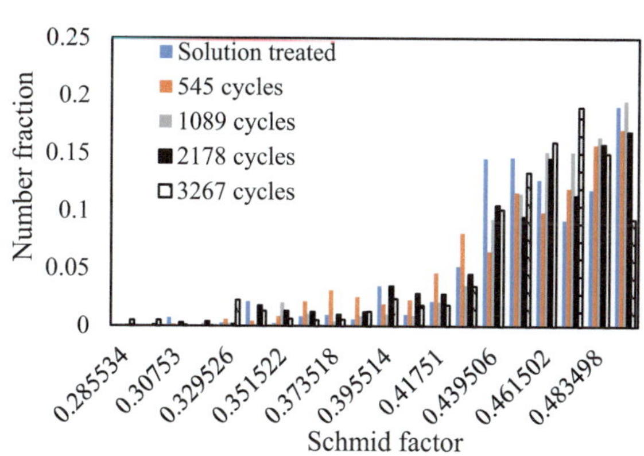

Fig. 9 Distribution of Schmid factor with different cyclic lives

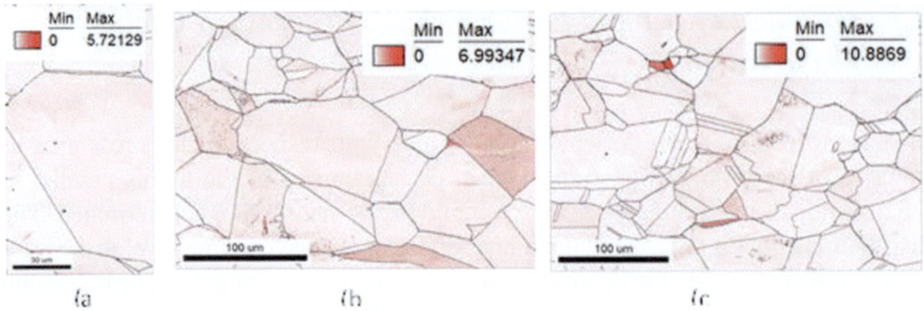

Fig. 10 The maps of variation of grain reference orientation deviation (GROD) **a** solution annealed austenitic stainless steel, **b** after 545 cycles and **c** at 3267 cycles

9.4 Grain Reference Orientation Deviation with Progress of LCF

Another descriptor of microstructural heterogeneity that evolves during progressive cyclic deformation is the grain reference orientation deviation (GROD) [88]. The GROD maps of the austenitic stainless steel in the undeformed condition and after deformation to 545 cycles and 2178 cycles are presented in Fig. 10a through (c). The larger the evolution of deformation heterogeneity, the higher will be the magnitude of the GROD distribution. The higher GROD regions are near grain boundaries and twin boundaries. The requirement of grain scale compatibility changes along the grain boundaries resulting in the accumulation of higher GROD along the grain boundaries [88]. The formation of GROD distribution also validates the change in grain size distribution after exposure to different number of cycles in solution treated austenitic stainless steel (Fig. 4).

9.5 Kernel Average Misorientation

The definition of Kernel Average Misorientation (KAM) is an average misorientation centering a pixel with respect to a definite set of adjacent neighboring points [17, 88, 90, 91]. The application of KAM is in describing a deformation state of a material. In this study, KAM values are estimated using a 0.5 micron step size of measurement and a grain boundary angle definition of 5 degrees. The kernel is herein defined as nearest neighbor orientations only. The KAM maps of the solution treated specimen and the specimens which experienced 545 and 3267 strain cycles, are presented in Fig. 11a–c. In the solution annealed specimen, the KAM map is relatively uniform across the region of observation with the exception of a small angle boundary that is seen just above the micron bar (Fig. 11a). With cyclic deformation, a heterogeneous distribution of KAM is visible as seen in Fig. 11b. The higher KAM regions are near austenite grain boundaries and twin boundaries showcasing the requirement

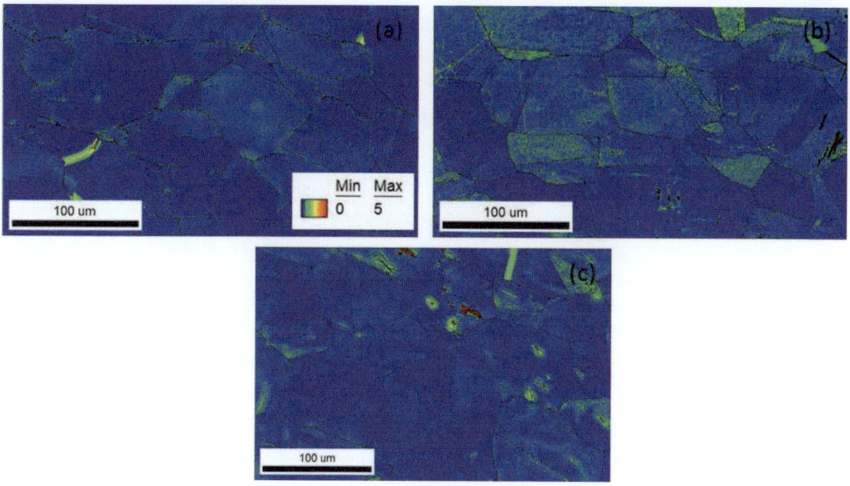

Fig. 11 The maps of Kernel Average Misorientation (KAM) **a** solution annealed austenitic stainless steel, **b** after 545 cycles, and **c** after 3267 cycles. The legend is from 0° to 5°

of maintaining compatibility in the microstructure. The presence of higher KAM regions are consistent with the higher GROD regions, presented in Fig. 10. The KAM distribution becomes nonuniform in the specimen which experienced higher strain cycles. This is indicative of damage accumulation and the restoration process that sets in simultaneously and keeps the KAM distribution uneven.

9.6 Development of the GND Structure with LCF and Influence of Grain Size and Grain Orientation

The characterization of dislocation structure using EBSD is based on Nye's theory [92] associating GND density with the lattice curvature. Nye's dislocation density tensor can be estimated from the determination of orientations over a regular set of points. Nye's dislocation density tensor results in a lower bound estimation of the total GND density values at each point in the structure. The data can be generated in 2D or 3D space. The methodology for estimation of GND density for the present study is available in the literature [92–99]. A brief depiction of the methodology is described here. Equation (1) yields the components of the curvature tensor:

$$\alpha_{\mathrm{ij}} = \mathrm{e}_{\mathrm{ikl}}\left(\varepsilon^{\mathrm{e}}_{\mathrm{jl,k}} + \mathrm{g}_{\mathrm{jl,k}}\right) \tag{1}$$

where α_{ij} is the dislocation density tensor, e_{ikl} is the permutation tensor, $\varepsilon^{e}_{jl,k}$ is the gradient of the elastic strain, and $g_{jl,k}$ is the lattice curvature as measured by EBSD (g_{jl} is the orientation matrix obtained from EBSD scans) [92–99]. Consequently, the dislocation density tensor can be found straight using the EBSD data (albeit, only five components of the tensor can be computed when scans are done on a plane section). Nye's tensor can also be described by the Burgers vectors and line directions for all dislocation types that are available in the given crystal system [92–99]:

$$\alpha_{ij} = \sum_{k=1}^{k} \rho^k \left(b_i^j z_j^k \right) \tag{2}$$

In Eq. (2), ρ^k is the density of the kth type of dislocation that is defined by a Burgers vector of b^k and a line direction of z^k. The crystal structure decides the number of dislocation types, k. Screw and edge dislocations are assumed to exist on the twelve unique {111} < 1–10 > slip systems for the present discussion.

A unique solution of the set of equations described in 2 is not available. Several solution techniques can be applied to estimate the lower level of dislocation density resulting in the measured crystal lattice curvature. In this work, an L^2 minimization approach was adopted to generate the values of ρ^k [92–99]. L^2 minimization does not result in measures of the various types of dislocation densities nor does it take into account distinct line energy minimization conditions. It is employed solely as a mathematical tool in generating the GND content. The procedure does not estimate the elastic strain. Lattice curvatures normal to the plane of measurement are not estimated [92–99] in the current work (2D measurements). Statistically stored dislocation (SSD) density cannot be estimated using this procedure. The GND estimation is controlled by the step size of the scan and the resolution of the orientation measurement. The physics of the dislocation structure does not influence the GND measurement procedure. This methodology gives a measure of the GND density which can result in generation of the estimated misorientation or lattice curvature [92–99].

EBSD-generated GND density maps of these specimens are presented in Fig. 12a–e. The maps were made using 0.5 μm step size and a 5° misorientation criterion for grain boundary definition. The average GND density in the solution treated 0% strained material is 5.75×10^{12} m^{-2}. This establishes a base value for the GND density measurement in this material at the given step size and scan conditions, and the value is consistent with reported expectations [89–96].

The variation of average GND density with the number of cycles is shown in Fig. 13. There is a correlation between the KAM maps shown in Fig. 11 and the GND maps in Fig. 12 as higher KAM regions are coincided with the higher GND regions, for example austenite grain boundaries. The overall GND density shows slight change after cyclic deformation to 545 cycles. However with further progress of cyclic deformation there is slight increase in the GND density. It should be noted that these samples are in the stable cyclic state. In case of cyclic deformation, the contribution of GND density to the total dislocation density is limited, only about 12% of the total dislocation density [17]. Littlewood and Wilkinson [31, 59] and Jiang et al. [100] reported similar observation of the evolution of GND density with

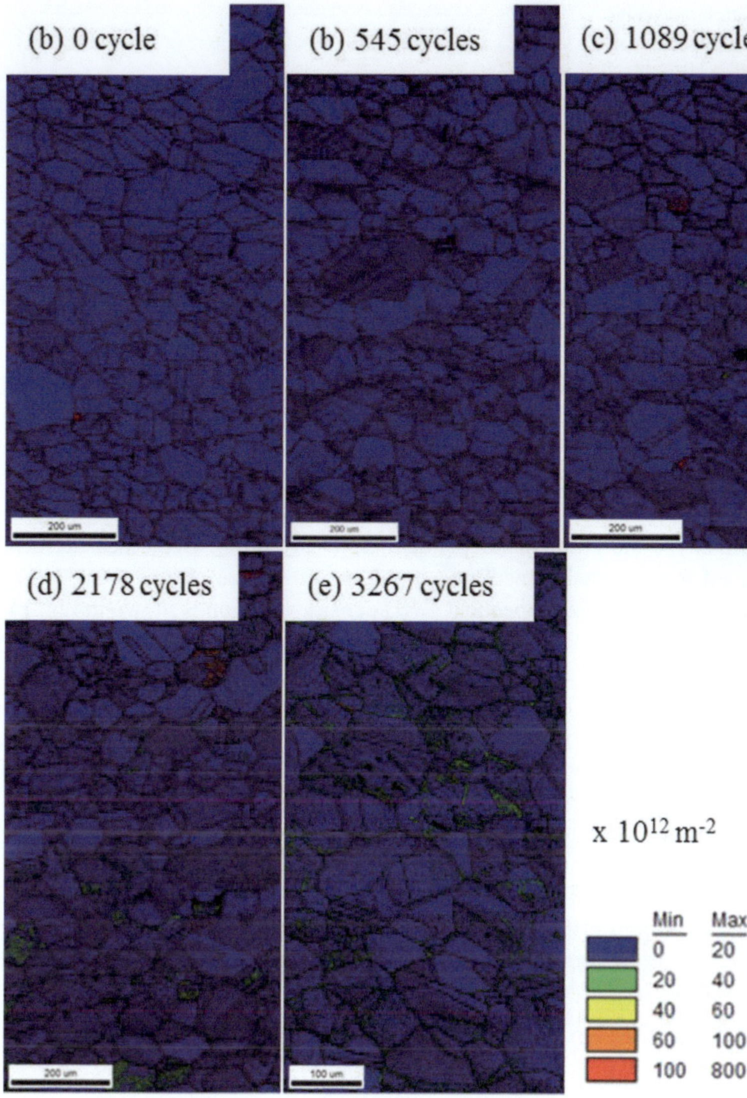

Fig. 12 **a–e** GND density maps of the specimens deformed at various number of cycles at strain amplitude 0.5%

the progress of cyclic deformation in case of Ti–6Al–4 V alloy and polycrystalline copper, respectively.

The effect of austenite grain size on the generation of GND content has been studied quantitatively. The small grains are associated with the larger storage of GND density. The GND distribution is uniform in the smaller grains. On the contrary, the gradient in GND distribution is visible in the coarse austenite grains as presented in

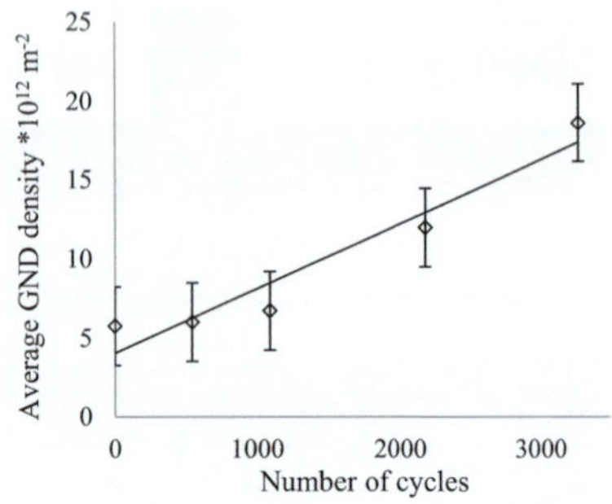

Fig. 13 Variation of average GND density with the number of cycles at strain amplitude of 0.5%

the Fig. 12b. The literature shows evidence of heterogeneous arrangements of GNDs depending on the grain size in austenitic stainless steel [17, 85], subjected to LCF using varying strain amplitudes.

The austenite grain size distribution was divided into three groups: small: 4.5–42 μm, (area %: 40), medium: 44–86 μm (area %: 48.6), and large: 87–126 μm (area %: 11.4) to analyze the grain size dependent GND evolution. GND measurement was performed for the small-, medium-, and large-grain size classes using a step size of 0.5 μm. The grain size dependent GND density variation is shown in Fig. 14. Smaller grains store higher GND content than medium and larger grains. The change in GND density in three different grain size classes increases with increasing number of cycles of deformation. Kundu et al. [17, 89], Jiang et al. [100], Kundu and Field [97], and Littlewood and Wilkinson [31, 59] have reported an identical variation in grain averaged GND density, which was larger for smaller grains in solution annealed 304 LN austenitic stainless steel, OFHC Copper, dual phase steel, and Ti–6Al–4 V samples strained to moderate strain in fatigue or tension.

In addition to grain size, orientation of austenite grains with respect to the strain axis controls the buildup of the GND density. Kundu and Field [97] reported that GND storage is favored in {111} type of ferrite grains in tensile tested dual phase steel with strains varying from 2.1% to 11.2%. In austenitic stainless steel, TEM and

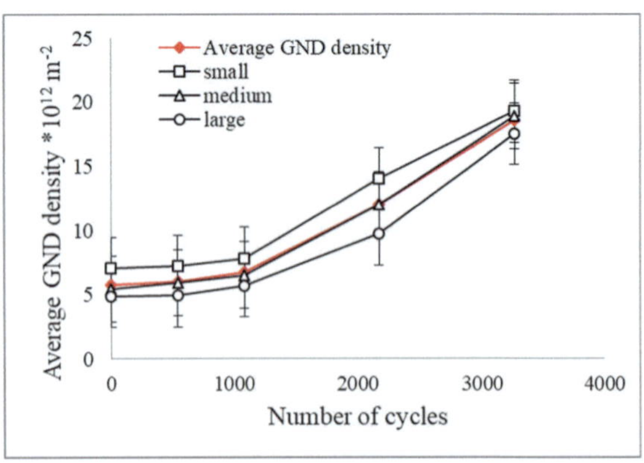

Fig. 14 Effect of austenite grain size on the change of GND density with the number of cycles

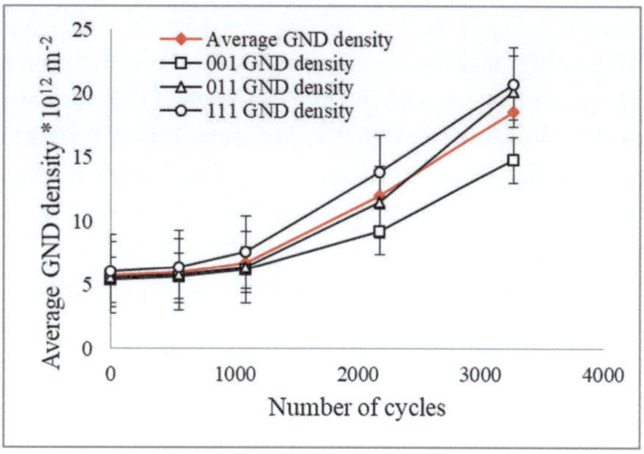

Fig. 15 Effect of austenite grain orientation on the change of GND density with the number of cycles

ECCI are used to characterize orientation dependent dislocation structure development following different types of loading schemes [17, 101, 102]. Merriman et al. [103] showed that in ultrafine grained aluminum, at the beginning of the monotonic straining GND storage begins in {001} type of grains as these grains deform at lower stress than highest Taylor factor grains. With the progress of the deformation, after plastic strain of 0.025, the GND density was larger in {111} and {011} types of grains. In the current work, evolution of the GND density in the austenite grains with {001}, {011} and {111} poles with respect to the deformation axis was estimated. 0.5 μm scan step size was used in this measurement. These grains have been characterized in Fig. 15 as {001}, {011}, and {111} Austenite grains. The slip systems for the FCC metals, i.e., {111} < 1–10 > are used to estimate the Taylor factors. The Taylor factor of {001} < 100 > grains is 2.53, that of {011} < 0–11 > grains is 3.29 and for {111} < 0–11 > grains is 3.35. GND density rises in the {111} < 0–11 > grains with maximum Taylor factor. In FCC metals, plastic deformation occurs on the slip system {111} < 0–11 > having the maximum slip activity (on the basis of Taylor's assumption of strain compatibility). Higher slip activity for a given strain can lead to greater development of GND content with the increasing number of strain cycles, as seen in Fig. 15.

9.7 Characterization of the Evolution of Substructure at Different LCF Life Fraction Using ECCI

After 545 cycles, the substructure of the steel consists of dislocation-rich and depleted regions eliminating the intergranular strain incompatibilities (Fig. 16a). The substructure suggests the activation of secondary slips including cross slip promoting dislocation interactions, annihilation of dislocations, and generation of low energy configuration of dislocations. With further increase in the number of cycles to 2178, the refinement of low energy arrangement of the dislocation structure takes place as shown in Fig. 16b. The spacing between low density dislocation regions decreases suggesting packing of more dislocations in the planar structure representing veins. This is consistent with the increase in the GND density with higher number of fatigue cycles, presented in Fig. 13. The outcome of formation of this dislocation structure is the relief of nonuniform strain distribution in the microstructure. At 3267 LCF cycles, the vein structure is sustained signifying the stability of the dislocation structures at this fatigue loading condition (Fig. 16c). The progressive alteration of the substructure is systematic with the change in the GND density, discussed in the previous section (Fig. 12). Pham and Holdsworth [16] reported a similar variation of dislocation substructure with number of cycles in 316L austenitic stainless steel subjected to LCF at room temperature using strain amplitude of 0.7%. The change in substructure during the span of LCF is responsible for the change in the long range and short range interactions of dislocations resulting in the change in the back stress and effective stress and subsequent cyclic hardening behavior. Larger dislocation pile-ups can induce larger back stress in the microstructure. The measurement of GND density (Fig. 13) in the present study shows an increase with increasing number of cycles, validating the increase in the back stress with the progress of the cyclic deformation maintaining the microstructural continuity of the material.

10 Summary

The significance of microtexture on cyclic plastic deformation was presented, especially controlling the fatigue crack initiation, the fatigue lives and the fracture mechanism in engineering metals and alloys such as nickel, aluminum, Ti–6Al–4 V, and austenitic stainless steel. The use of microtexture in the fatigue crack nucleation model using crystal plasticity was described. The literature review in the first part of the chapter revealed the importance of microtexture in governing the nucleation of the fatigue crack and the characterization strategies. The difference in the macrotexture and the microtexture is also discussed. The influence of microtexture in the slip transfer, slip localization, and the development of the ladder structure and persistent slip bands in FCC, BCC, and HCP alloys are studied. Particularly, the influence of number of cycles on the development of microstructure and microtexture in the austenitic stainless steel was assessed. The significance of grain boundary character and twin boundaries in nucleating fatigue crack initiation after intrusion and extrusion was systematically characterized. The steel exhibited change in the grain size distribution, amount of low- and high-angle boundaries, and twin fraction. The deformation heterogeneity developed during the cyclic deformation in the austenite grains was evaluated with the help of GROD. The wide distribution of GROD with the number of fatigue cycles validated the strain gradient developed in the microstructures generating the slip localization. The interaction with the slip build ups and the grain boundaries and twin boundaries result in enlargement of the slipped regions. With the increase of the number of cycles during LCF, the evolution of KAM increases

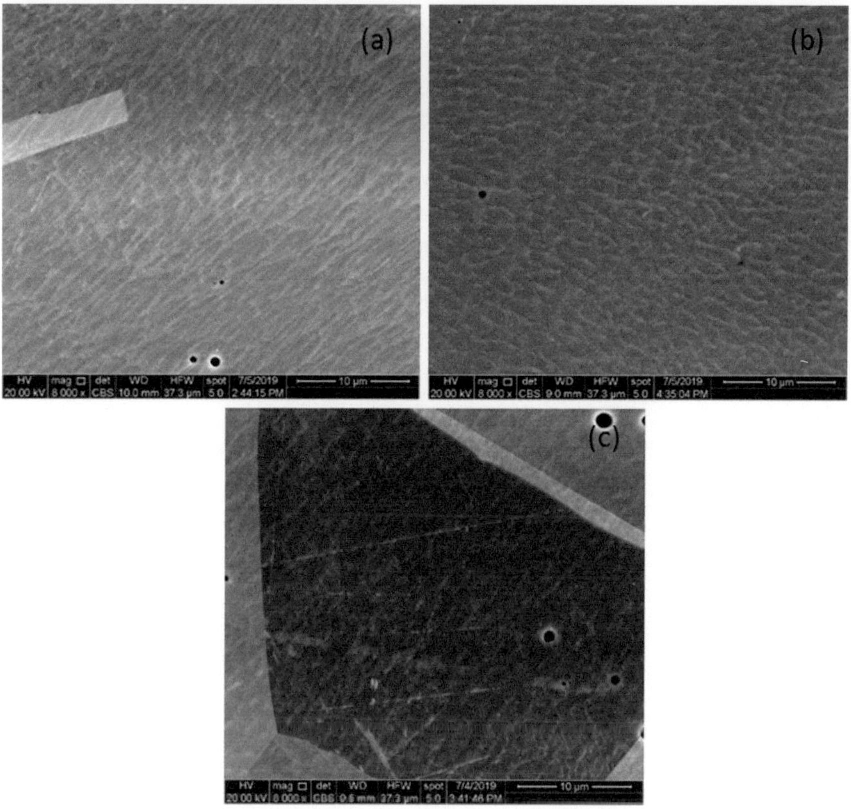

Fig. 16 Substructure or the dislocation structure revealed using ECCI at **a** 547 cycles **b** 2178 cycles and **c** 3267 cycles

and demonstrates nonuniform distribution throughout the grains. For a strain amplitude of 0.5%, the evolution of GND density increases with the number of cycles of deformation. However, the rate of increase of GND density is limited with the number of cycles of deformation. It is inferred that this is due to formation of stable dislocation structures as the specimens used for characterization were obtained from the cyclically stable zone. The spread of GND is not homogeneous throughout the microstructure. The higher GND density areas are close to austenite grain boundaries and twin boundaries, potential sites of crack initiation. The finer the austenite grains, the more uniform is the distribution of the GND density. The development of the GND density is governed by the orientation of the austenite grains. The austenite grains with orientation {111} < 0–11 > store the highest GND density. The fidelity of the stable dislocation structure was validated by ECCI affecting the development of back stress and effective stress by changing the long range and short range interaction of the dislocations, and by maintaining the continuity in the microstructural length scales. Therefore, for prediction of the cyclic lives, it is necessary to consider the in-grain deformation heterogeneity characterized by microtexture, change of grain boundary character and distribution of misorientation and GND density. The 3D characterization of microtexture or deformation heterogeneity is a challenge and therefore 3D models of fatigue crack nucleation are more difficult. The slip localization and slip transfer can be characterized using EBSD and ECCI. The modeling should utilize discrete dislocations and molecular dynamics. The temperature dependent constitutive modeling of fatigue damage is still prohibitive as it is not possible to capture diffusion controlled recovery. All these modeling techniques require the detailed knowledge of development of the dislocation structures which can be characterized using microtexture giving the influence of the grain size and grain orientation.

Acknowledgements The financial support of the work has been received from Science and Engineering Research Board, Department of Science and technology, Government of India (file no.: ECR/2016/000260) under early career research award. The OIM^{TM} research facility at Metallurgical and Material Engineering Department, Jadavpur University was used in the work.

References

1. Stroh AN (1954) The formation of cracks as a result of plastic flow, Proc. Royal Soc. Lon Math Phy Eng Sci 223:404–414
2. Dunne FPE (2014) Fatigue crack nucleation: mechanistic modelling across the length scales. Curr Opi Sol Sta Mat Sci 18:170–179
3. Cappola J, Stinville J, Charpagne M, Callahan PG, Echlin MP, Pollock TM, Pilchak A, Kasemera M (2021) On the Localization of Plastic Strain in Microtextured Regions of Ti-6Al-4V. Acta Mater 204:116492
4. Lindemann J, Wagner L (1999) Microtextural effects on mechanical properties of duplex microstructures in (α+β) titanium alloys. Mater Sci Eng A 263:137–141
5. Pilchak AL (2014) A simple model to account for the role of microtexture on fatigue and dwell fatigue lifetimes of titanium alloys. Scr Mater 74:68–71
6. Jha SK, Szczepanski CJ, John R, Larsen JM (2015) Deformation heterogeneities and their role in life-limiting fatigue failures in a two-phase titanium alloy. Acta Mater 82:378–395
7. Ozturk D, Shahba A, Ghosh S (2016) Crystal plasticity FE study of the effect of thermo-mechanical loading on fatigue crack nucleation in titanium alloys. Fat Fract Engg Mater and struct 39:752–769
8. Wen W, Ngan AHW, Zhang Y, Xu B, Zhai T (2013) A study of the effects of particle 3-dimensional geometry and micro-texture on fatigue crack initiation behaviors in an Al–Cu alloy using focused ion beam and electron backscatter diffraction. Mater Sci Eng A 564:97–101
9. Chan KS (2010) Roles of Microstructure in Fatigue Crack Initiation. Int J Fat 32:1428–1447
10. McDowell D, Dunne FPE (2010) Microstructure-sensitive computational modeling of fatigue crack formation. Int J Fat 32:1521–1542
11. Mughrabi H (2013) Microstructural fatigue mechanisms: cyclic slip irreversibility, crack initiation, non-linear elastic damage analysis. Int J Fat 57:2–8
12. Sangid MD (2013) The physics of fatigue crack initiation. Int J Fat 57:58–72
13. Zhai T, Jiang XP, Li JX, Garratt MD, Bray GH (2005) The grain boundary geometry for optimum resistance to growth of short fatigue cracks in high strength Al-alloys. Int J Fat 27:1202–1209

14. Hémery, S Stinville JC (2022) Microstructural and load hold effects on small fatigue crack growth in α+β dual phase Ti alloys. Int J Fat 156
15. Pham M, Solenthaler C, Janssens K, Holdsworth S (2011) Dislocation structure evolution and its effects on cyclic deformation response of AISI 316L stainless steel. Mater Sci Eng A 528:3261–3269
16. Pham M, Holdsworth S (2014) Evolution of relationships between dislocation microstructures and internal stresses of AISI 316L during cyclic loading at 293 K and 573 K (20°C and 300°C) Metall. Mater Trans A 45A:738–751
17. Kundu A, Field DP, Chakraborti PC (2019) Influence of strain amplitude on the development of dislocation structure during cyclic plastic deformation of 304 LN austenitic stainless steel. Mater Sci Eng A 762:138090
18. Briffod F, Blueset A, Shiraiwa T, Enoki M (2019) Effect of crystallographic orientation and geometrical compatibility on fatigue crack initiation and propagation in rolled Ti-6Al-4V alloy Acta Mater 177:56–67
19. Man J, Vystavel T, Weidner A, Kubena I, Petrenec M, Kruml T, Polak J (2012) Study of cyclic strain localization and fatigue crack initiation using FIB technique. Int J Fat 39:44–53
20. Briffod F, Shiraiwa T, Enoki M, Emura S (2022) Effect of macrozones on fatigue crack initiation and propagation mechanisms in a forged ti-6Al-4V alloy under fully-reversed condition. Materia 101401
21. Polak J, Man J, Vystavel T, Petrenec M (2009) The shape of extrusions and intrusions and initiation of stage I fatigue cracks. Mater Sci Eng A 517:204–211
22. Ozturk D, Kotha S, Pilchak A, Ghosh S (2019) Two-way multi-scaling for predicting fatigue crack nucleation in titanium alloys using parametrically homogenized constitutive models. J Mech Phys Sol 128:181–207
23. Polak J, Man J (2014) Mechanisms of extrusion and intrusion formation in fatigued crystalline materials. Mater Sci Eng A 596:15–24
24. Liu W, Huang J, Liu J, Wu X, Zhang K, Huang A (2021) Experimental and crystal plasticity modelling study on the crack initiation in micro-texture regions of Ti-6Al-4V during high cycle fatigue tests. Int J Fat 148:106203
25. Ma L, Xiao S, Deng H, Hu W (2014) Molecular dynamics simulation of fatigue crack propagation in bcc iron under cyclic loading. Int J Fat 68:253–259
26. Sangid MD, Maier HJ, Sehitoglu H (2011) An energy-based microstructure model to account for fatigue scatter in polycrystals. J Mech Phys Sol 59:595–609
27. Mohtadi-Bonab MA, Eskandari M, Sanayei M, Das S (2018) Microstructural aspects of intergranular and transgranular crack propagation in an API X65 steel pipeline related to fatigue failure. Eng Fail Analy 94:214–224
28. Sangid MD, Maier HJ, Sehitoglu H (2011) The role of grain boundaries on fatigue crack initiation–an energy approach. Int J Plast 27:801–821
29. Sangid MD, Ezaz T, Sehitoglu H, Robertson IM (2011) Energy of slip transmission and nucleation at grain boundaries. Acta Mater 59:283–296
30. Sangid MD, Rotella J, Naragani D, Park JS, Kenesei P, Shade PA (2020) A complete grain-level assessment of the stress-strain evolution and associated deformation response in polycrystalline alloys. Acta Mater 201:36–54
31. Littlewood PD, Wilkinson AJ (2012) Local deformation patterns in Ti–6Al–4V under tensile, fatigue and dwell fatigue loading. Int J Fat 43:111–119
32. Engler O, Randle V (2010) Introduction to Texture Analysis Macrotexture, Microtexture, and Orientation Mapping, CRC Press. Taylor & Francis Group, Boca Raton
33. Nishikawa S, Kikuchi S (1928) The diffraction of cathode rays by calcite. Proc Imperial Academy (of Japan) 4:475–477
34. Alam MN, Blackman M, Pashley DW (1954) High-angle Kikuchi patterns. Proc Royal Society of London A221:224–242
35. Venables JA, Harland CJ (1973) Electron back-scattering patterns—A new technique for obtaining crystallographic information in the scanning electron microscope. Phil Mag 2:1193–1200

36. Dingley DJ (1984) Diffraction from sub-micron areas using electron backscattering in a scanning electron microscope. Scan Elec Micro 11:569–575
37. Dingley DJ, Alderman J, Longdon M, Wienbren J (1987) On-line analysis of electron back scatter diffraction patterns. I. Texture analysis of zone refined polysilicon. Scan Elect Micro 11:451–456
38. Wright SI (1992) Individual lattice orientation measurements: development and application of a fully automatic technique, PhD thesis, Yale University
39. Kunze K, Wright SI, Adams BL, Dingley DJ (1993) Advances in automatic EBSP single orientation measurements. Text Str Micro 20:41–54
40. De Graef M (2020) A dictionary indexing approach for EBSD. IOP Conf Ser Mater Sci Eng 891:012009
41. Abdolvand H, Wilkinson AJ (2016) Assessment of residual stress fields at deformation twin tips and the surrounding environments. Acta Mater 105:219–231
42. Garner A, Gholinia A, Frankel P, Gass M, MacLaren I, Preuss M (2014) The microstructure and microtexture of zirconium oxide films studied by transmission electron backscatter diffraction and automated crystal orientation mapping with transmission electron microscope. Acta Mater 80:159–171
43. Cantwell PR, Tang M, Dillon SJ, Luo J, Rohrer GS, Harmer MP (2014) Grain boundary complexions. Acta Mater 62:1–48
44. Cherry MR, Sparkman D, Pilchak A, Homa L (2019) Characterization of microtexture regions in Ti alloys using eddy current testing. AIP Conf Proc 2102:080008
45. Briggs GAD, Kolosov OV (2010) Acoustic Microscopy. Oxford University Press
46. Sharples SD, Clark M, Somekh MG (2006) Spatially resolved acoustic spectroscopy for fast noncontact imaging of material microstructure. Opt Exp 14:10435–10440
47. Blodgett M, Hassan W, Nagy PB (2000) Theoretical and experimental investigations of the lateral resolution of eddy current imaging. Mat Eval 58:647–654
48. Pilchak A, Li J, Rokhlin SI (2014) Quantitative comparison of microtexture in near-alpha titanium measured by ultrasonic scattering and electron backscatter diffraction. Metall Mater Trans A 45A:4679–4697
49. Suresh S (1998) Fatigue of Materials, Second edition, Cambridge University Press, ISBN, 0521578477
50. Dieter GE (1988) Mechanical metallurgy, McGraw Hill Book Company, UK, ISBN 007–084187
51. Szczepanski CJ, Semiatin SL, Larsen JM (2011) The role of microtexture on fatigue lifetime variability and crack initiation mechanisms. In: Proceeding 12th World Conference Tita, pp 1281–1284
52. Joseph S, Bantounas I, Lindley TC, Dye D (2018) Dislocation interactions and crack nucleation in a fatigued near-alpha titanium alloy. Int J Plast 100:90–103
53. Sinha V, Spowart JE, Mills MJ, Williams JC (2006) Observations on the faceted initiation site in the dwell-fatigue tested ti-6242 alloy: Crystallographic orientation and size effects. Metall Mater Trans A 37A:1507–1518
54. Szczepanski CJ, Jha SK, Larsen JM, Jones JW (2008) Microstructural influences on very-high-cycle fatigue-crack initiation in Ti-6246. Metall Mater Trans A 39A:2841–2851
55. Bache MR, Dunne FPE, Madrigal C (2010) Experimental and crystal plasticity studies of deformation and crack nucleation in a titanium alloy. J Strain Anal 45:391–399
56. Dunne FPE, Rugg D (2008) On the mechanisms of fatigue facet nucleation in titanium alloys. Fat Fract Engg Mater Struct 31:949–958
57. Guilhem Y, Basseville B, Curtit F, Stephan JM, Callitaud G (2010) Investigation of the effect of grain clusters on fatigue crack initiation in polycrystals. Int J Fat 32:1748–1763
58. Dunne FPE, Wilkinson AJ, Allen R (2007) Experimental and computational studies of low cycle fatigue crack nucleation in a polycrystal. Int J Plast 23:273–295
59. Littlewood PD, Wilkinson AJ (2012) Geometrically necessary dislocation density distributions in cyclically deformed Ti–6Al–4V. Acta Mater 60:5516–5525

60. Lebensohn RA, Brenner R, Castelnau O, Rollet AD (2008) Orientation image-based micromechanical modelling of subgrain texture evolution in polycrystalline copper. Acta Mater 56:3914–3926
61. Sauzay M, Jourdan T (2006) Polycrystalline microstructure, cubic elasticity, and nucleation of high-cycle fatigue cracks. Int J Fract 141:431–446
62. Dunne FPE, Rugg D, Walker A (2007) Lengthscale-dependent, elastically anisotropic, physically-based hcp crystal plasticity: application to cold-dwell fatigue in Ti alloys. Int J Plast 23:1061–1083
63. Shanthraj P, Zikry MA (2013) Microstructurally induced fracture nucleation and propagation in martensitic steels. J Mech Phys Sol 61:1091–1105
64. Sweeney CA, Vorster W, Leen SB, Sakurada E, Dunne FPE (2013) The role of elastic anisotropy, length scale and crystallographic slip in fatigue crack nucleation. J Mech Phys Solids 61:1224–1240
65. McCarthy OJ, McGarry JP, Leen SB (2014) The effect of grain orientation on fretting fatigue plasticity and life prediction. Tribo Int 76:100–115
66. Brogdon MI, Rosenberger AH (2008) Evaluation of the influence of grain structure on the fatigue variability of Waspaloy. Superalloy 583–588
67. Yavari P, Langdon TG (1983) Cyclic grain boundary migration during high temperature fatigue-I. Microstruct Obs, Acta Metal. 31:1595–1603
68. Brodesser S, Gottstein G (1993) Microstructure and microtexture evolution during high temperature low cycle fatigue of nickel. Text Str Micro 20:179–193
69. Brodesser S, Bruckner G, Gottstein G (1994) Influence of the initial texture on the microstructure- and microtexture development during high temperature low cycle fatigue. Mat Sci Forum 157:1153–1160
70. Weiß S, Ponge D, Gottstein G (1995) Grain boundary control during high temperature low cycle fatige. Canad Metall Quart 34:237–241
71. Weiß S, Bruckner G, Gottstein G (1994) Microstructure- and microtexture-development during high temperature low cycle fatigue, Oikawa H (ed) Strength of Materials, Jap Inst Met 591:591–594
72. Weiß S, Gottstein G (1998) Grain boundary motion during high temperature cyclic deformation of high purity aluminium bicrystals. Mater Sci Eng A 256:8–17
73. Weiss S, Gottstein G (1998) Evolution of microstructure during high temperature low cycle fatigue of high purity aluminium oligocrystals. Mat Sci Tech 14:1169–1174
74. Bantounas I, Dye D, Lindley TC (2010) The role of microtexture on the faceted fracture morphology in Ti–6Al–4V subjected to high-cycle fatigue. Acta Mater 58:3908–3918
75. Snowden KU (1960) The distribution of deformation in lead fatigued in vacuo. Phil Mag 6:321
76. Lutjcring G (1998) Influence of processing on microstructure and mechanical properties of (α+β) titanium alloys. Mater Sci Engg A 243:32–45
77. Evans WJ (1999) Microstructure and the development of fatigue cracks at notches. Mater Sci Engg A 263:160–175
78. Bridier F, Villechaise P, Mendez J (2005) Analysis of the different slip systems activated by tension in a α/β titanium alloy in relation with local crystallographic orientation. Acta Mater 53:555–567
79. Bridier F, Villechaise P, Mendez J (2008) Slip and fatigue crack formation processes in an α/β titanium alloy in relation to crystallographic texture on different scales. Acta Mater 56:3951–3962
80. Neal DF, Blenkinsop PA (1976) Internal fatigue origins in α-β titanium alloys. Acta Metall 24:59–63
81. Evans WJ, Bache MR (1994) Dwell-sensitive fatigue under biaxial loads in the near-alpha titanium alloy IMI685. Int J Fat 16:443–452
82. Bantounas I, Lindley TC, Rugg D, Dye D (2007) Effect of microtexture on fatigue cracking in Ti–6Al–4V. Acta Mater 55:5655–5665

83. Peters M, Gysler A, Lutjering G (1984) Influence of texture on fatigue properties of Ti-6Al-4V. Metall Trans A 15:1597–1605
84. Liu G, Salvat Canto J, Winwood S, Rhodes K, Birosca S (2018) The effects of microstructure and microtexture generated during solidification on deformation micromechanism in IN713C nickel-based superalloy. Acta Mater 148:391–406
85. Kamaya M (2009) Characterization of microstructural damage due to low-cycle fatigue by EBSD observation. Mater Char 60:1454–1462
86. Sinha S, De PS, Mukherjee S, Kundu A, Chakraborti PC, Bhattacharya B, Shome M, Bhattacharjee D (2016) Local deformation heterogeneity in cyclically deformed interstitialfree steel. Fat Fract Engg. Mater Struct 39:110–119
87. Bunge HJ (1982) Texture analysis in materials science: mathematical methods. Butterwoth & Co
88. Kundu A, Field DP (2020) Influence of microstructural heterogeneity and plastic strain on geometrically necessary dislocation structure evolution in single-phase and two-phase alloys. Mater Char 170:110690
89. Kundu A, Field DP, Chakraborti PC, Raut P (2022) Development of microstructural heterogeneity in 304 LN austenitic stainless steel: Effect of temperature and strain. J Mater Engg Perform 31:9050–9059
90. Field DP (1997) Recent advances in the application of orientation imaging. Ultramic 67:1–9
91. Field DP, Merriman CC, Allain-Bonasso N, Wagner F (2012) Quantification of dislocation structure heterogeneity in deformed polycrystals by EBSD. Model Simul Mater Sci Eng 20:024007
92. Nye JF (1953) Some geometrical relations in dislocated crystals. Acta Metall 1:153–162
93. Ashby MF (1970) The deformation of plastically non-homogeneous materials. Phil Mag 21:399–424
94. El-Dasher BS, Adams BL, Rollett AD (2003) Viewpoint: experimental recovery of geometrically necessary dislocation density in polycrystals. Scr Mater 48:141–145
95. Pantleon W (2008) Resolving the geometrically necessary dislocation content by conventional electron backscattering diffraction. Scr Mater 58:994–997
96. Konijnenberg PJ, Zaefferer S, Raabe D (2015) Assessment of geometrically necessary dislocation levels derived by 3D EBSD. Acta Mater 99:402–414
97. Kundu A, Field DP (2016) Influence of plastic deformation heterogeneity on development of geometrically necessary dislocation density in dual phase steel. Mater Sci Eng A 667:435–443
98. Kundu A, Field DP (2018) Geometrically necessary dislocation density evolution in interstitial free steel at small plastic strains. Metall Mater Trans A 49A:3274–3282
99. Kundu A, Field DP, Chakraborti PC (2020) (2020), Effect of strain and strain rate on the development of deformation heterogeneity during tensile deformation of a solution annealed 304 LN austenitic stainless steel: An EBSD study. Mater Sci Eng A 773:138854
100. Jiang J, Britton TB, Wilkinson AJ (2015) Evolution of intragranular stresses and dislocation densities during cyclic deformation of polycrystalline copper. Acta Mater 94:193–204
101. Winther G, Huang X (2007) Dislocation structures. Part II. Slip system dependence. Phil Mag 87:5215–5235
102. Nellessen J, Sandlobes S, Raabe D (2015) Effects of strain amplitude, cycle number and orientation on low cycle fatigue microstructures in austenitic stainless steel studied by electron channelling contrast imaging. Acta Mater 87:86–99
103. Merriman CC, Field DP, Trivedi PB (2008) Orientation dependence of dislocation structure evolution during cold rolling of aluminum. Mater Sci Eng A 494:28–35

Chapter 12
Dependence of Environmental Sensitivity on Microtexture

Darothi Bairagi, Pallabi Bhuyan, Rahul Rakshit, and Sumantra Mandal

1 Introduction

Corrosion or environmental degradation is regarded as a crucial factor/phenomenon in many practical applications as it leads to gradual deterioration of material, resulting in loss of material's function. Among the various distinct forms of corrosion, the localized ones (e.g., pitting corrosion, intergranular corrosion (IGC), galvanic corrosion, and stress corrosion cracking (SCC)) are considered to be more insidious in nature as compared to the uniform dissolution. It is due to this fact that the prediction of localized corrosion and identification of the exact sites of localized attacks are often very difficult in practice [1]. Hence, in many such cases, the corrosion process ultimately leads to complete collapse of the structure, costing billions to the economy. Therefore, the corrosion engineers are constantly trying to enhance the corrosion resistance of materials by regulating different factors that exert pronounced influence on materials degradation [2, 3]. Among the various metallurgical factors associated to corrosion, the grain orientation is an important aspect [4–6]. As the corrosion phenomenon is a strong function of surface energy, which, in turn, is directly dependent upon the orientation of crystallographic planes, the corrosion process is significantly affected by the orientation of grains within the specimen or, in other words, the microtexture of the material [7–10]. The microtexture commonly originates during various stages of material processing [11] and hence, the development of microtexture in materials is quite inevitable. Additionally, the evolution of microtexture has also been observed during different surface modification techniques [12–14]. Depending upon the type or extent of microtexture developed in a

Darothi Bairagi, Pallabi Bhuyan and Rahul Rakshit have contributed equally to this Chapter.

D. Bairagi · P. Bhuyan · R. Rakshit · S. Mandal (✉)
Department of Metallurgical and Materials Engineering, Indian Institute of Technology Kharagpur, Kharagpur, India
e-mail: sumantra.mandal@metal.iitkgp.ac.in

S. Suwas and D. P. Field (eds.), *Advances in Texture, Microtexture, and Allied Techniques*, Materials Horizons: From Nature to Nanomaterials,
https://doi.org/10.1007/978-981-96-5346-1_12

material, its electrochemical degradation property varies widely. Therefore, in order to regulate the degradation rate as per the application requirement, it is important to understand the one-to-one correlation that exist between microtexture and the extent of corrosion.

Knowing the individual orientations of grains, the orientation difference or misorientation between abutting grains can be calculated which could provide information about the grain boundary character distribution (GBCD) in a given microstructure [15]. This is typically denoted as mesotexture (i.e., the texture between grains) and considered to be an important aspect of microtexture to elucidate the grain boundary structure and property relationships [16]. The susceptibility to corrosion of any grain boundary directly depends upon its interfacial energy, which is essentially governed by the atomic structure of the concerned boundary [17, 18]. The low Σ ($\Sigma \leq 29$) coincidence site lattice (CSL) boundaries may possess low interfacial energy and help to disrupt the interconnectivity of susceptible random high-angle grain boundary (RHAGB) network within the microstructure, thus making the alloy more resistant to corrosion [15, 17–21]. Clearly, tailoring microstructure by modifying GBCD employing grain boundary engineering (GBE) approach is also a useful way to impart control over the corrosion susceptibility of materials [15, 17, 18, 20].

Although, a large number of research works have been reported in literature that deal with evolution of microtexture in materials [11, 22–26], the implications of developed microtexture on electrochemical degradation property of the material is comprehensively discussed only in a very few articles [4–6, 27–29]. Additionally, the literatures associated to this topic are not always consistent and some of them holds a complete contradictory opinion with respect to others [30–32]. Therefore, this chapter aims to elucidate the influence of both microtexture and GBCD (i.e., mesotexture) on corrosion of materials in a coherent and concise manner. The first part of this chapter elaborates the influence of microtexture on the corrosion behavior of different alloy systems. The second part of this chapter, on the other hand, briefly explains the effect of GBCD and grain boundary connectivity on the electrochemical degradation of materials. This chapter is expected to help the readers to acquire some valuable insights on tailoring a material in terms of either microtexture or GBCD in order to achieve a desired rate of degradation.

2 Effect of Microtexture on Corrosion

The electrochemical properties of a material are related to the microtexture by virtue of its surface energy [7, 33, 34]. Since surface energy varies with change in crystallographic texture of a surface [7, 8], it is expected that texture will influence the corrosion properties of a material. It should be clarified at this point of time that although a microtexture developed on the surface is defined by a set of planes and directions, the planes are of more importance as far as the corrosion properties of the material are concerned. This is because of the reason that irrespective of the grain directions, the electrochemical degradation rate of a material will be directly related to the crystallographic planes exposed to the surface and its associated surface energy [8, 10]. In general, the densely packed crystallographic planes with high atomic packing density have strong atomic bonding, and high coordination number of atoms. Therefore, these planes possess lower surface energy, needing higher amount of activation energy to dissolve [7, 8, 10]. However, the packing fraction is not the sole factor that influence surface energy of planes. There are situations where the surface energy of the planes is predominantly influenced by some other factors (for e.g., the solution characteristics) rather than the packing fraction of planes [35, 36]. Therefore, contradictory results are also present in literature where the lowest corrosion rate is not associated with the most densely packed plane.

In general, the dissolution rate of a material (I_a) and the surface energy of different crystallographic planes can be correlated through the following equation [7, 8]:

$$I_{\mathrm{a}} = nFk \exp\left(\frac{Q + \alpha nFE}{RT}\right) \quad (1)$$

where n denotes the number of electrons associated with the electrochemical reaction, F stands for the Faraday constant, k is the reaction constant, R denotes the universal gas constant, T denotes the absolute temperature in K, E is the electron potential, α is the transit coefficient calculated based on surface energy of a particular plane, and Q accounts for the activation energy required for a metallic ion to escape and dissolve in electrolyte [7, 8]. The equation clearly denotes that the material dissolution rate is exponentially related to Q for a particular value of α (i.e., for a particular crystallographic plane), assuming that n and k would be same for different planes [7, 8]. Using this equation, the theoretical dissolution rate has been calculated with respect to different crystallographic planes of a material [7, 8, 10, 36]. Since the miller indices of "highly packed crystallographic plane" vary from one crystal structure to another, it is evident that the texture evolution will also vary with respect to crystal structure of materials [22, 37]. Besides, the different processing stages of a material namely heat treatment, thermomechanical processing (e.g., rolling, forging, and extrusion) and different severe plastic deformation techniques (e.g., equal channel angular pressing, and high pressure torsion) can modify the grain orientations and lead to evolution of different microtextures [11, 24, 25, 38]. In this regard, a generalized correlation between microtexture and corrosion resistance has been put forward in the following sections. The first section discusses the dependency of room temperature corrosion on microtexture of a material with respect to three commonly encountered crystal structures, namely face centered cubic (FCC), body-centered cubic (BCC), and hexagonal closed pack (HCP) systems. The second section focuses on the relationship between microtexture of substrate surface and high-temperature corrosion process (i.e., oxidation phenomenon). Finally, the effect of microtexture, developed though different surface modification techniques or coatings, on electrochemical degradation behavior of different materials are discussed in the last section.

2.1 *Effect of Microtexture on Corrosion in Terms of Crystal Structures*

2.1.1 FCC

The important texture components found in FCC material include Brass {110} < 112 >, Cube {001} < 100 >, Rotated Cube {001} < 110 >, Copper {112} < 111 >, Goss {110} < 001 >, Rotated Brass {111} < 112 > and S {123} < 634 > [11, 39, 40]. In addition, the presence of {110} < 110 > and {111} < 110 > orientations have also been reported in some FCC materials [11]. It is worth mentioning here that in FCC crystals, the {111} planes are the most closely packed planes, followed by the {100}, {110}, and {210} planes, respectively [41]. As discussed in the previous section, the densely packed planes possess the lowest surface energy. Taking the above factors into account, the surface energy of the planes in the FCC crystals can be arranged in the following order: {111} < {100} < {110} < {210} [42, 43]. This implies that the dissolution rate of {111} planes is considerably lower than that of {100} and {110} planes as enormous amount of energy is required for removing an atom from a stack of densely packed atoms in the crystallographic planes. So, it can be inferred that the development of textural components with {111} oriented grains parallel to the rolling plane (RP) could significantly improve the corrosion resistance of the material. A large number of investigations have revealed that the evolution of R-Brass {111} < 112 > and {111} < 110 > textural components, associated with {111} densely packed crystallographic planes parallel to the surface, are beneficial in improving the corrosion resistance of austenitic stainless steels and Al alloys [41, 44].

Interestingly, when rolled Al alloy was annealed at very high temperature (673 K and 773 K), its corrosion rate was accelerated irrespective of the presence of higher fraction of {111} oriented grains [31]. High temperature annealing promoted relatively easier movement of dislocations in rolled aluminum alloy and led to formation of sub-grain boundaries. The dislocations at grain boundaries increased energy and subsequently decreased corrosion resistance of the alloy [31]. Other contradictory results have also been observed in literatures where the corrosion rate does not follow the trend with respect to packing fraction of the planes [30]. Apart from packing fraction, the surface activity is also found to be dependent upon the ability of the grain to adsorb ionic species from solution. It is known that the surface energy of a material can be reduced by the adsorption of site-blocking species from solution on its surface. Such adsorbed layer would reduce the tendency of dissolution of the grains, impeding further surface-solution interactions [32, 45]. In this case, the corrosion rates of the {111} oriented grains are observed to be higher since they barely adsorb any ions on surface due to their low surface energy. On the other hand, {100} and {110} planes show higher affinity towards adsorption of ions, owing to higher surface energy associated with them. This, in turn, would form a uniform adsorbent layer over the surface, leading to significant reduction in electrochemical activity of the planes [32].

Another interesting study, carried out by Lapeire et al. [30], has shown that in addition to the texture of an individual grain, the orientation relationship existing between two neighboring grains is also an important factor that influences the corrosion properties of an alloy. When two differently oriented grains are present adjacent to each other, they form a galvanic couple. Based on their potential difference, the severity of preferential dissolution of anode varies [8, 30]. Since a large potential difference exists between the {001} and {111} type of grains, their presence adjacent to one another would induce severe differences in dissolution between the two types of orientations [30]. The dissolution rates in the study were qualitatively measured by the height difference between neighboring grains as a result of preferential dissolution (as shown in Fig. 1a–b), employing atomic force microscopy (AFM). It is evident from Fig. 1c that significant height differences were estimated for a pair of < 111 > // ND and < 001 > // ND oriented grains as compared to other orientations (Fig. 1d). Most importantly, they have reported that the < 111 > // ND orientation are severely attacked when submerged in corrosive environment (0.1 M NaCl), provided that these were adjacent to < 001 > // ND oriented grains. It is important to point out here that although the {100} planes are more active compared to the {111} planes, the preferential dissolution has been observed for the latter. However, no concrete explanations behind such observation have been made in this study and hence need further investigation.

Besides uniform dissolution, a few studies have revealed that the crystallographic orientations also have a profound effect on crack initiation and propagation during SCC of FCC materials [46–49]. It is well known that the cracks are induced into the material due to a simultaneous effect of tensile stress and corrosive environment [50, 51]. To be precise, SCC is a consequence of dislocation pileups, generated during the application of tensile stress. It is well established that dislocations propagate along {111} crystal planes and intersect to produce sessile dislocations which restrict further dislocation motion and create dislocations pileup [46]. It has been reported that the dislocation pileups act as preferential sites for localized dissolution, thereby causing discontinuous micro-cleavage on {100} planes and shearing on {111} planes in 316 austenitic stainless steel [46]. On the contrary, a study by Meletis et al. [47] have suggested primary cracking on {100} planes and secondary cracking on {110} planes in 304L austenitic stainless steel. Other studies have reported that among the {100} and {110}, the {110} oriented grains are more effective in resisting crack propagation [48, 49]. This observation could be explained on the basis of the twist and tilt angles between the above-mentioned orientations and their adjacent grains. A larger twist angle is more likely to promote crack deflection and resist crack propagation whereas a higher tilt angle provides the necessary driving force for the crack growth [52]. The twist and tilt angles between grains close to Brass orientation {110} < 112 > and adjacent grains were estimated to be 18.2° and 8.8°, respectively. As the twist angle is higher than the tilt angle, the crack is deflected and consequently arrested on encountering {110} orientated grains such as Brass {110} < 112 > and Goss {110} < 001 > . On the contrary, the tilt angle (25.8°) between the Rotated Cube {001} < 110 > orientation and its adjacent grains is found to be higher than the twist angle (11.5°). Owing to the smaller twist angle, the crack propagated without

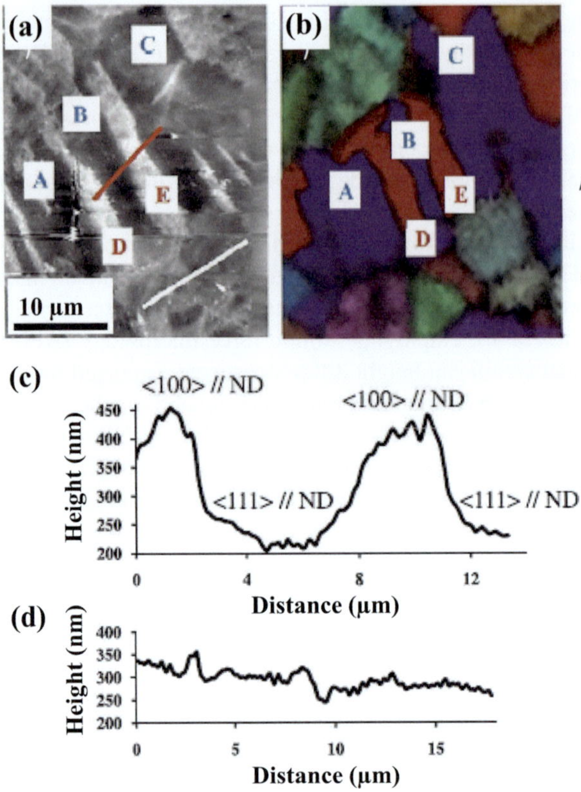

Fig. 1 **a** Topographic map, and **b** the IQ overlaid IPF map of cold rolled and annealed high purity cast copper submerged in 0.1 M NaCl solution for 210 min. The height profiles corresponding to red and white lines in topographic map are shown in **c** and **d**, respectively. Grains A, B and C have near < 111 > orientation while grains D and E have near < 001 > orientation [30]

being deflected in the case of Cube {001} < 100 > and Rotated Cube {001} < 110 > oriented grains. Therefore, it was concluded that {110} orientation grains can easily retard crack propagation and enhance SCC resistance of the material. Similar observations have been made by Li et al. [49] where they have reported an improved SCC resistance in the specimen dominated by high intensity Brass texture {110} < 112 > and a poor SCC resistance in the specimen exhibiting low intensity Brass texture in an aluminum alloy.

2.1.2 BCC

The BCC alloys often exhibit Brass {110} < 112 >, Cube {001} < 100 >, Rotated Cube {001} < 110 >, Goss {110} < 001 >, Rotated brass {111} < 112 > and other texture components such as {111} < 110 >, {112} < 110 > and {110} < 110 >

[11, 53]. In BCC, it is well-established that the {110} planes are the most densely packed, followed by {100} and {111}, respectively [54]. Since the dissolution rate of the densely packed planes are lower than the loosely packed planes, the grains oriented towards the densely packed {110} planes are more resistant to corrosion. Therefore, a higher fraction of {110} // RP oriented grains as well as evolution of Brass {110} < 112 > and Goss {110} < 100 > texture components are beneficial in improving the corrosion resistance in BCC alloys [55, 56].

Similar to FCC, microtexture plays a key role in arresting the cracks during SCC in steels. When intergranular SCC study was conducted on API X-65 steel, it was observed that the cracks tend to propagate along the grain boundaries associated with {001} // RP textured grains, while the cracks get arrested on encountering the grain boundaries of {110} // RP and {111} // RP textured grains [57]. Further, it was reported that the regions just beyond the crack tips (crack arrest regions) exhibit dominant {110} // RP and {111} // RP oriented grains, as shown in Fig. 2a. On the contrary, the cracked regions show {001} // RP oriented grains, as evident from Fig. 2b. Further studies on orientation dependent SCC resistance have revealed that {001} // ND orientations facilitate transgranular crack growth on rolling plane [58, 59].

Studies focusing on hydrogen induced cracking (HIC) in BCC materials have mentioned that the {001} // ND orientations are susceptible to HIC as the crack tends to propagate through {001} // ND oriented grains [59–61]. The higher fraction of {001} // ND oriented grains, associated with high surface energy planes, offer fast dissolution of atoms and reduce the corrosion resistance, as reported by Pourazizi et al. [59]. Furthermore, Hallen et al. [61] have reported that a hot rolled specimen facilitated the propagation of cracks while the warm rolled specimen was resistant to the same owing to the development of {001} // ND fiber in the hot rolled specimen and {111} // ND fiber in the warm rolled specimen, respectively. In another study, Venegas et al. [60] have mentioned that the transgranular and intergranular crack propagation can be significantly reduced by weakening of the {001} // ND fiber and strengthening of the {112} // ND, {111} // ND and {110} // ND fibers.

Further, it was observed that pitting corrosion is strongly dependent on the crystallographic orientations of the material [62, 63]. In this regard, Madrigal-Cano et al.

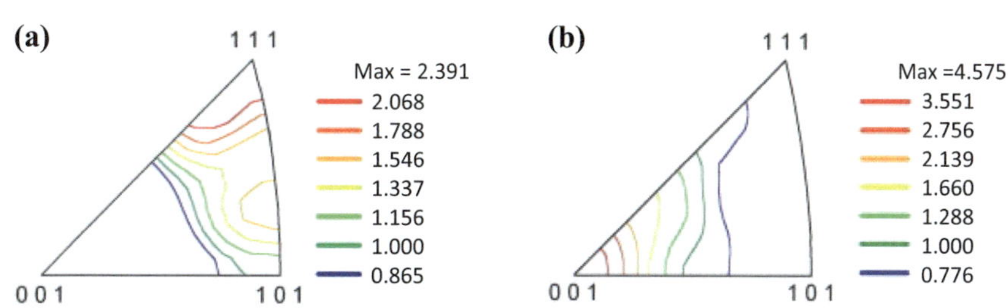

Fig. 2 Inverse pole figure of **a** immediately ahead of crack tip areas and **b** crack tip region after conducting in situ tensile test, exposed to 24 h hydrogen charging [57]

[62] have designed a model using Monte Carlo simulations to determine pits initiation and growth with respect to grain orientation. It was reported that pitting initiation was independent of microtexture. However, frequency of pits growth was found to follow the order of {222} // ND > {112} // ND > {200} // ND > {110} // ND. The effect of microtexture on corrosion resistance of BCC metals and alloy is very limited which necessities further investigations.

2.1.3 HCP

The predominant texture components that can develop in HCP materials are basal texture ({0001} < uvtw >), prismatic texture ({$10\bar{1}0$} < uvtw >), and pyramidal texture ({$11\bar{2}0$} < uvtw >) [22, 37, 64]. Another distinct kind of texture is reportedly observed in some rolled or extruded structures of Mg alloys containing rare earth (RE) elements where a typical RE texture component with < $11\bar{2}1$> directions parallel to the deformation direction is developed [65–68]. Among the textures mentioned above, the basal and prismatic textures are the most commonly found ones in HCP materials since these are the prevailing slip planes [22, 64]. The ideal pole figures of basal and prismatic textures simulated for Mg alloy are shown in Fig. 3a–c [22]. However, based on the c/a ratio, the activity of slip systems varies in HCP materials [69], leading to a change in texture components that develop in them. It has been observed in literatures that the HCP materials having a c/a ratio nearly equal to or above 1.633, predominantly undergo basal slip [22]. Whereas, in case of the HCP materials with a c/a ratio lower than 1.633, the prismatic slip is dominant [22, 64]. Accordingly, the texture components that evolve in HCP materials through deformation vary. These aspects have been discussed thoroughly in many literatures [22, 37, 64, 69]. In the context of this chapter, it is important to focus on the variation in electrochemical response of HCP materials depending upon the type of microtexture that develops in it. The following sections briefly assess the effect of microtexture evolution on different types of corrosion in HCP materials.

In HCP materials, the {0002} basal planes with the highest atomic packing density exhibit slower corrosion rate as compared to loosely packed planes (i.e., {$10\bar{1}0$} and {$11\bar{2}0$} planes) [22, 64, 70]. Fu et al. [7] have carried out a theoretical quantification of surface energies using empirical electron theory (EET) for some important low index crystallographic planes present in Mg based alloys. It was found out that the surface energies associated with {0002}, {$10\bar{1}0$}, and {$11\bar{2}0$} planes are 1.808, 1.868, and 2.156 eV/nm^2, respectively [7, 8]. The different surface energy levels of the distinct crystallographic planes lead to a variation in the electrochemical dissolution rate of the materials as per Eq. 1. The dissolution rates, theoretically calculated using Eq. 1, for the {$10\bar{1}0$} and {$11\bar{2}0$} planes of AZ31 alloy are around 18–20 times higher than that of the basal plane {0001} [8]. Therefore, the surface with strong basal texture exhibits much higher stability in terms of dissolution due to its lower surface energy, whereas, the other two cross sectional surfaces, consisting of {$10\bar{1}0$} and {$11\bar{2}0$} planes, show rapid degradation rate [8].

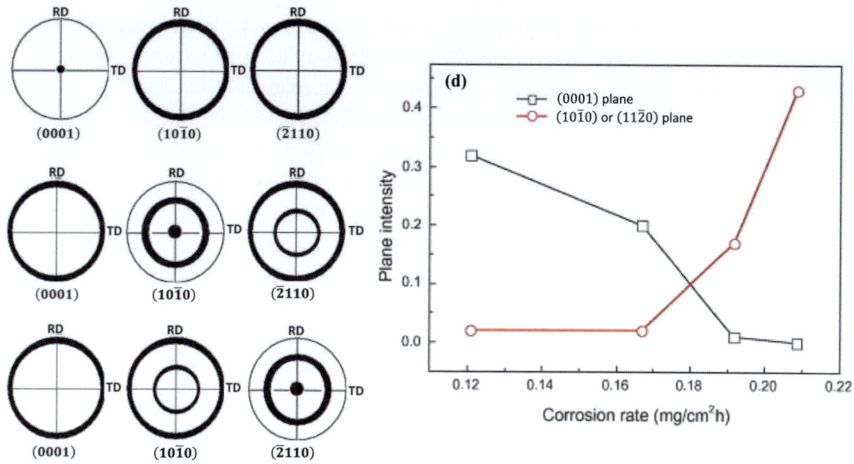

Fig. 3 Simulated ideal pole figures for important texture components obtained for Mg alloy **a** {0001} fiber, **b** {10$\bar{1}$0} fiber, **c** {11$\bar{2}$0} fiber [22]; **d** effect of those texture components on corrosion rate of a Mg alloy (AZ31) in 3.5% NaCl solution [36]

For Mg based HCP materials, it has been reported that the presence of {0001} basal planes in combination with {10$\bar{1}$0} or {11$\bar{2}$0} on a specimen surface can increase the extent of corrosion compared to the specimens that consist of only {10$\bar{1}$0} or {11$\bar{2}$0} or a combination of {10$\bar{1}$0} and {11$\bar{2}$0} type crystallographic planes [8, 29]. This follows from the galvanic effect that arises due to a significant difference in potential existing between the basal and the prismatic planes of the alloy. Xin et al. [36] have experimentally established the relationship between the corrosion rate and texture intensity of these three important crystallographic planes for AZ31 sample, immersed in 3.5% NaCl solution. It can be clearly observed from the Fig. 3d that the presence of higher intensity of basal textures can significantly reduce the corrosion rate, as opposed to the {10$\bar{1}$0} or {11$\bar{2}$0} planes [36]. Evidently, the presence of strong crystallographic texture in a material can introduce "corrosion anisotropy" between the rolled and the cross-sectional surfaces of the specimen, leading to a significant variation in the corrosion behavior of different crystallographic planes [71–73]. Such distinct anisotropy in corrosion properties might not be useful in applications where improvement in overall corrosion resistance of the alloy is important.

It has been observed that activation of twins in HCP materials can appreciably reduce the extent of anisotropy produced due to texture and therefore play a major role in determining the effect of texture on different properties, including corrosion [71, 74–77]. Due to the restricted number of slip systems, twinning is known to be an important secondary deformation mode for HCP materials [69]. It has been found that there exists a critical c/a ratio ($\sim\sqrt{3}=1.732$), below and above which the twins can form in HCP materials upon compression or tension, respectively [22]. Wang et al. [71] have reported that when a rolled AZ31 sample (for Mg, c/a ratio is < 1.732) is subjected to $\sim$ 3% compressive strain along the rolling direction, it activates $\{10\overline{1}2\} < 10\overline{1}0>$ extension twins in the material. The weakening of original basal texture developed in the rolled plate due to activation of extension twins has reportedly eliminated the corrosion anisotropy in the material [71, 75, 77].

Apart from uniform corrosion, texture is also known to have a dominant role in dictating the extent of different types of localized corrosion, where the corrosion rate is governed by the formation of a passive layer. Since the crystallographic orientation of grains present on a surface is reported to influence the development of a passive film and its characteristics [28, 78–80], it is expected that a variation in texture will lead to a difference in corrosion behavior. A study conducted on Ti6Al4V alloy has shown that the crystallographic texture can highly influence the bio-corrosion properties of the alloy by modulating the characteristics of the passive film that forms on the specimen in phosphate buffer solution (PBS) [28]. As observed from the IPF maps shown in Fig. 4a, the grains in the specimen are oriented in a manner such that their basal planes are parallel to the RD-TD surface, whereas, for other two surfaces (i.e., TD-ND and RD-ND), the grains are oriented in some direction between [1120] and [1010] [28]. The potentiodynamic polarization studies conducted on the specimens confirm that the corrosion rate of the RD-TD surface consisting of basal texture is around 3.6×10^{-5} mm/year, which is 5 and 7 times lower than those of TD-ND surface and RD-ND surface, respectively [28]. Such variation in corrosion rate for the three surfaces has been related to the passive film characteristics in the study. The Nyquist plots obtained from electrochemical impedance spectroscopy (EIS) of the specimens in PBS solution are shown in Fig. 4b [28]. The figure clearly demonstrates that the RD-TD surface has much higher corrosion resistance compared to the TD-ND and RD-ND surfaces. Moreover, the values of passive film resistances obtained by fitting the EIS curves using equivalent circuit model are reported to be 2.79, 0.70, and 0.52 $M\Omega.cm^2$ for the RD-TD, TD-ND, and RD-ND surfaces, respectively [28]. The thickness of the passive films calculated using the film capacitance values (obtained from circuit fitting data) is shown in Fig. 4c [28]. The authors of this study have claimed that the formation of a thicker passive layer over the RD-TD sample consisting of basal texture contributes towards its higher corrosion resistance compared to the other two cross sectional surfaces [28, 79]. Having said that, it must be mentioned here that the properties of the surface film also depend on several other factors such as the type of material under consideration, the electrolytic solution, presence of corrosive ions, and other microstructural features of the material [6, 81, 82]. Therefore, the relative importance of crystallographic orientation in comparison to the above factors becomes significant in deciding the role of microtexture on passive film formation and degradation resistance of a material.

2.2 *Effect of Microtexture on Oxidation*

The degradation of a material at high temperature is dependent upon the oxide layers which form on the substrate surface through oxidation process. Substrate microtexture is regarded as an important factor that can regulate the nucleation and growth of

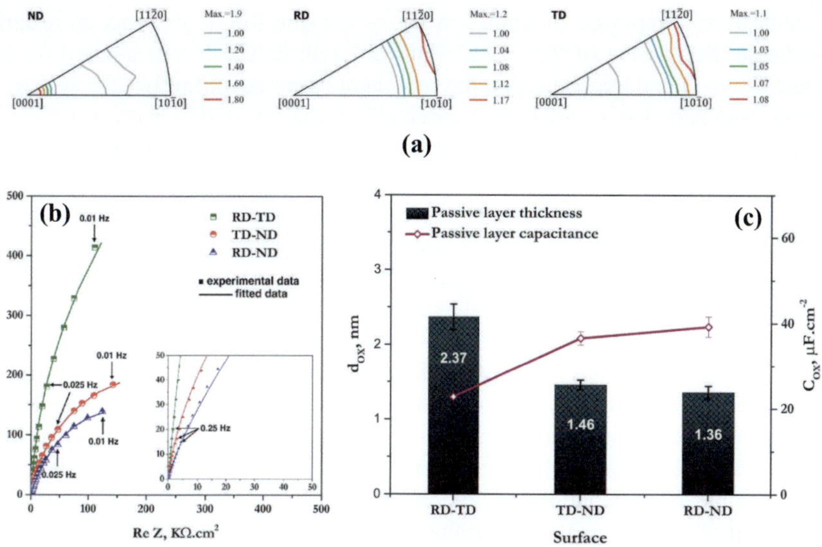

Fig. 4 **a** IPF maps showing the orientation of important planes on different surfaces acquired for heat-treated samples of Ti6Al4V alloy, **b** Nyquist plots obtained for different surfaces in PBS solution, **c** Passive layer thickness calculated from EIS data for the specimens [28]

the oxide layers [83–85]. A few literatures have reported that the texture of the oxide layer develops from substrate microtexture through epitaxial growth, thus resulting in a certain orientation relationship between the two [83, 86, 87]. However, such studies correlating the surface microtexture and oxidation kinetics are quite limited [27, 83, 86, 88, 89]. Moreover, the studies that dealt with establishing a link between the oxide orientation and substrate microtexture show several inconsistencies in result [86, 90, 91].

Among the available studies, a major part is concerned to Fe–Cr alloys, since those alloys are quite convenient options for industrial applications, where medium to high temperature oxidation resistance is desired. It is well known that increasing the amount of Cr in Fe improves the oxidation resistance by forming a dense chromia (Cr_2O_3) layer over the surface. Zhang et al. [92] have reported that oxidation rate of Cr increases with surface orientation in the order of (001) < (110) < (111), whereas, for Ni, the oxidation rate increases in the order of (111) < (110) < (001). Clearly, a less porous and protective passive film can develop with significant strengthening of {111} orientation, as the oxidation of Cr would be promoted, resulting in the instantaneous formation of Cr-rich oxide [93, 94]. Contrarily, the oxide formed on the specimen dominated by {001} orientation would be defective due to the formation of Ni-rich oxide which is less protective in nature [93]. Therefore, the strengthening of {111} orientation is essential in improving the oxidation resistance as it favors the formation of Cr-rich oxide.

Recently, a few studies have revealed a clear correlation between the substrate microtexture and the oxidation kinetics for different steels [27, 83, 86]. The substrate microtexture is reported to control the type of oxide film that forms on it during oxidation. Since each of these oxides has different defect densities based on their structure, the ionic transportation through them varies accordingly [27]. It is worth mentioning here that in case of iron-based alloys, the dry oxidation process typically generates three distinct forms of oxides namely wustite ($Fe_{1-x}O$), magnetite (Fe_3O_4), and hematite (Fe_2O_3), sequentially arranged according to decreasing order of iron content in them, starting from metal-oxide interface towards oxide-atmosphere interface [27, 86, 91]. Among these, the hematite layer is the most stable one as it has the lowest ionic diffusion rate, whereas, the wustite layer is highly defective with the least stability [83, 86]. To understand the effect of substrate microtexture on oxidation process, Stephan-Scherb et al. [86] have exposed Fe–Cr alloys (with Cr content varying between 2–9 wt%) in SO_2 environment at 923 K. It was observed that for Fe-2Cr-0.5Mn alloy, the developed magnetite layer had epitaxial relationship with the ferrite substrate showing $< 110 >_{ferrite} // < 111 >_{magnetite}$ (as shown in Fig. 5a–c). The reason behind the preferential growth of magnetite on (110) planes of ferrite has been explained by the highest atomic packing of those planes. The formation of < 111 > magnetite grains parallel to the < 110 > ferrite grains result in minimization of interfacial energies between the substrate and oxide layer [86]. In the other alloy containing higher amount of chromium (Fe-9Cr-0.5Mn alloy), a mixed magnetite-hematite layer was formed over the ferritic substrate grains. Here the hematite crystals with < 001 > orientation observed to grow favorably on ferrite grains of < 110 > orientation (as described in Fig. 5b–d) [86]. However, practically it is quite unlikely for hematite crystals with < 001 > orientation to grow directly over ferrite grains of < 110 > orientation, since, the process induces a large amount of strain (~4%) at the interface [86]. Therefore, a magnetite layer forms in between the ferrite substrate and hematite layer, leading to an orientation relationship of $< 111 >_{magnetite} // < 100 >_{hematite}$ [86]. The nucleation of < 100 > hematite crystals over the < 111 > oriented crystals of magnetite assist in minimizing the interfacial energy between the two oxide layers, which explains the origin of such orientation relationship between them. This kind of epitaxial relationship influences the compactness of oxide layer, thus affecting the oxidation rate of materials [83, 86, 95]. Such orientational correlation between substrate texture and oxide crystals for other substrate materials has also been reported in other studies [83, 96, 97].

Apart from influencing the orientation of oxide crystals, the substrate microtexture can also affect the thickness of oxide layers, resulting in a variation in oxidation response. In a recent study carried out by Mehtani et al. [83], interstitial free (IF) steel specimens with varying volume fraction of γ-fiber (ND // < 111 >) texture were subjected to oxidation (at 773 K temperature) in order to understand the variation in oxidation behavior with respect to substrate microtexture and oxide layer thickness. It was found that the increase in ND // < 111 > texture intensity in the substrates resulted in an enhanced oxidation rate of the specimens. Further, the strongly textured substrate (volume fraction of ND // < 111 > ~ 0.75) formed a thicker magnetite layer in combination with a very thin outer hematite layer. Whereas, the weakly textured specimen with ND // < 111 > ~ 0.17 showed the presence of almost equal fraction of magnetite and hematite layer. Since magnetite layer consists of higher number

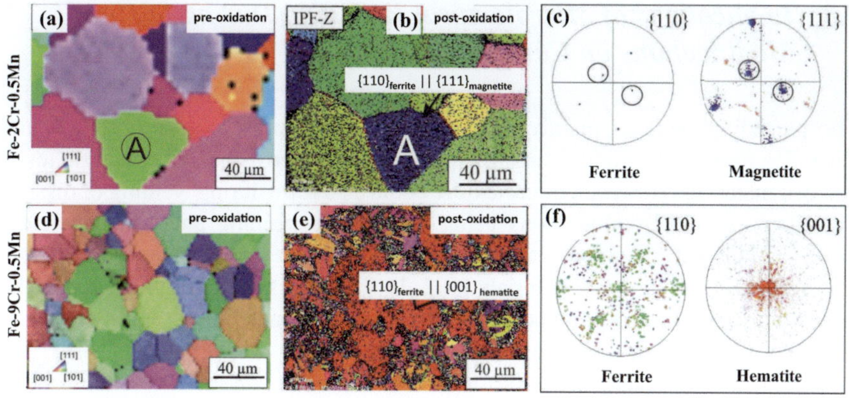

Fig. 5 IPF map of the **a, d** pre-oxidation, and **b, e** post-oxidation specimens in **a–b** Fe-2Cr-0.5Mn **d–e** Fe-9Cr-0.5Mn alloy, respectively. The pole figure indicating orientation relationship between ferrite and magnetite in Fe-2Cr-0.5Mn alloy as well as between ferrite and hematite in Fe-9Cr-0.5Mn alloy is shown in **c** and **f**, respectively [86]

of defects, it leads to enhanced ionic diffusion through magnetite and results in an increased oxidation rate of the specimen, as observed in case of strongly textured specimen. On the other hand, in case of weakly textured specimen, the formation of a protective outer hematite layer resulted in improved oxidation behavior. However, Garner et al. [88] have reported that the oxidation kinetics is independent of the substrate microtexture and the microtexture present on substrate surface does not determine the type of oxide texture that develops on it. In that study, Zr alloy specimens were subjected to oxidation at different temperature ranges and it was concluded that the oxide texture is influenced by the compressive stress generated at the Zr-ZrO_2 interface rather than the substrate texture [88].

Apart from this, several studies have claimed that the surface oxide that formed on the highly packed basal plane of pure Mg is observed to have comparatively lower thickness than on the other planes [71, 83, 98, 99]. This is clearly in contradiction to the above-mentioned fact that a close packed plane leads to the development of a thicker passive film. However, in both the cases the higher compactness of the film contributes towards improved corrosion resistance, since, irrespective of the layer thickness, the integrity of the layer becomes the most crucial factor [100, 101]. It is worth noting here that the investigations focusing on the influence of microtexture on high-temperature oxidation behavior of material is still quite limited and hence requires further investigation.

2.3 *Implications of Microtextural Evolution During Surface Modifications on Corrosion*

The preceding sections mainly dealt with the development of deformation and annealing microtexture and their subsequent effect on corrosion and oxidation resistance of materials. However, microtexture could also form during welding of dissimilar alloys, surface modifications using laser shock peening and coatings (electrodeposition, galvanization, etc.), which may enhance or deteriorate the corrosion resistance of the base materials [102–110]. So, this section will focus on the effect of microtexture, developed during the aforementioned processes, on corrosion resistance of weldments, coatings, etc.

Kangazian and Shamanian [106] have observed that a dissimilar stainless steel/Ni based superalloy weld zone exhibited higher oxide film resistance during the development of random grain orientations, illustrated in Fig. 6a–c. It is quite clear from Nyquist plots shown in Fig. 6c that weld zone with high intensity < 321 > // welding direction (WD) (Fig. 6b) have lower oxidation resistance as compared to weld zone with no preferential orientation (random texture) (Fig. 6a). Moreover, when stainless steel coatings were developed employing friction surfacing (variant of friction stir welding) at low spindle speed, poor pitting resistance of the coatings were reported despite the development of high intensity {111} orientation (associated with densely packed planes) [107]. This was related to the low stacking fault energies of {111} oriented grains which led to the generation of high residual stress during friction surfacing. However, when the friction surfacing was performed at higher spindle speed, the development of randomly oriented grains weakened the (111) orientations which lowered the generation of residual stress and significantly improved the pitting resistance.

It is well known that coatings strongly influence the localized corrosion behavior due to the growth of crystals along preferred orientation. Substantial amount of work has been conducted to alter the crystal growth direction by varying the process parameters such as electrodeposition current density, current mode and temperature [13, 14, 108]. In this regard, β-Sn (body-centered tetragonal crystal structure) coatings were electrodeposited on mild steel which led to development of (100) and (110) growth texture at different current densities [14]. It is worth mentioning here that the (100) are low surface energy planes whereas (110) are high surface energy planes in β-Sn metallic system. In Fig. 7a–j, the inverse pole figure (IPF) maps reveal the dominance of (100) growth texture at lower electrodeposition current densities whereas high surface energy (110) growth texture prevailed at higher current densities. The evolution of (100) growth texture contributed to a greater corrosion resistance of Sn coatings at lower electrodeposition current density, as evident from Fig. 7k. However, the coating fabricated at lowest current density of 5 $mAcm^{-2}$ revealed poor corrosion resistance despite the development of high intensity (100) growth texture (Fig. 7k). This was attributed to the non-uniform stress distribution during electrodeposition [14]. Further, the effect of current modes i.e., direct, pulsed and reverse pulsed on crystal growth and corrosion behavior was studied by electrodepositing

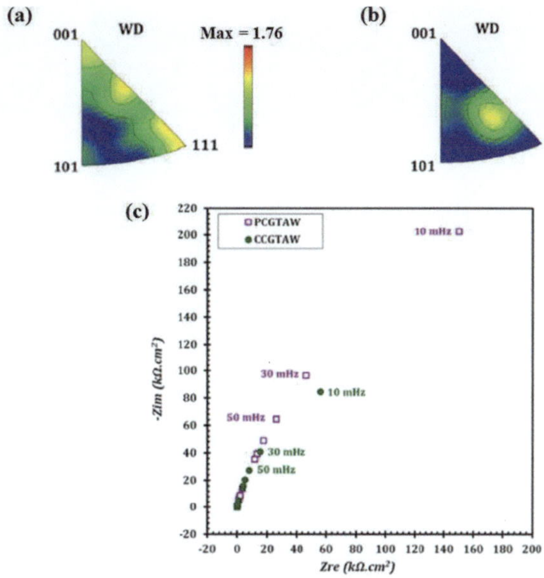

Fig. 6 Welding direction (WD) inverse pole figure map employing **a** pulsed and **b** continuous current modes and **c** the Nyquist plot of the weld zone during gas tungsten arc welding. PCGTAW and CCGTAW in **c** represents pulsed current gas tungsten arc welding and continuous current tungsten arc welding, respectively [106]

nano-crystalline nickel films on pure nickel plate [108]. Corrosion studies revealed that pulsed electrodeposited film exhibited higher corrosion resistance due to the crystallographic growth of {111} orientations. In another study, Zn coatings were electrodeposited at different temperatures which resulted in microtextural variations on the coating surface, as shown in Fig. 8 [13]. The higher corrosion resistance of 275 K deposited coating is related to the growth of grains oriented with basal planes on the coating surface (Fig. 8a). As stated earlier, these basal planes offer higher resistance for the passage of corrosive media, therefore mitigating the propensity of transgranular corrosion. It can be clearly seen from Fig. 8b that the basal planes are inclined to the coating growth direction owing to which the corrosive media interacts with the partial portions of high energy $\left\{10\overline{1}0\right\}$ planes and percolates via the grains. Similarly, the coated surface at 333 K deposition temperature is highly vulnerable to corrosion owing to the interaction of $\left\{10\overline{1}0\right\}$ planes with the corrosive media (Fig. 8c) [13]. Moreover, recent reports have illustrated that microtexture of the Zn coatings can be altered by the use of surfactants or wetting agents [109]. For example, the sodium lauryl sulfate surfactant Zn coating exhibited higher corrosion resistance as compared to coating without surfactant, which was attributed to the evolution of near basal microtexture along coating growth direction and higher fraction of low energy grain boundaries.

In the recent times, nano-sized foreign particles such as carbon nanotubes (CNTs) have been incorporated into conventional metallic coating matrix to alter the initial

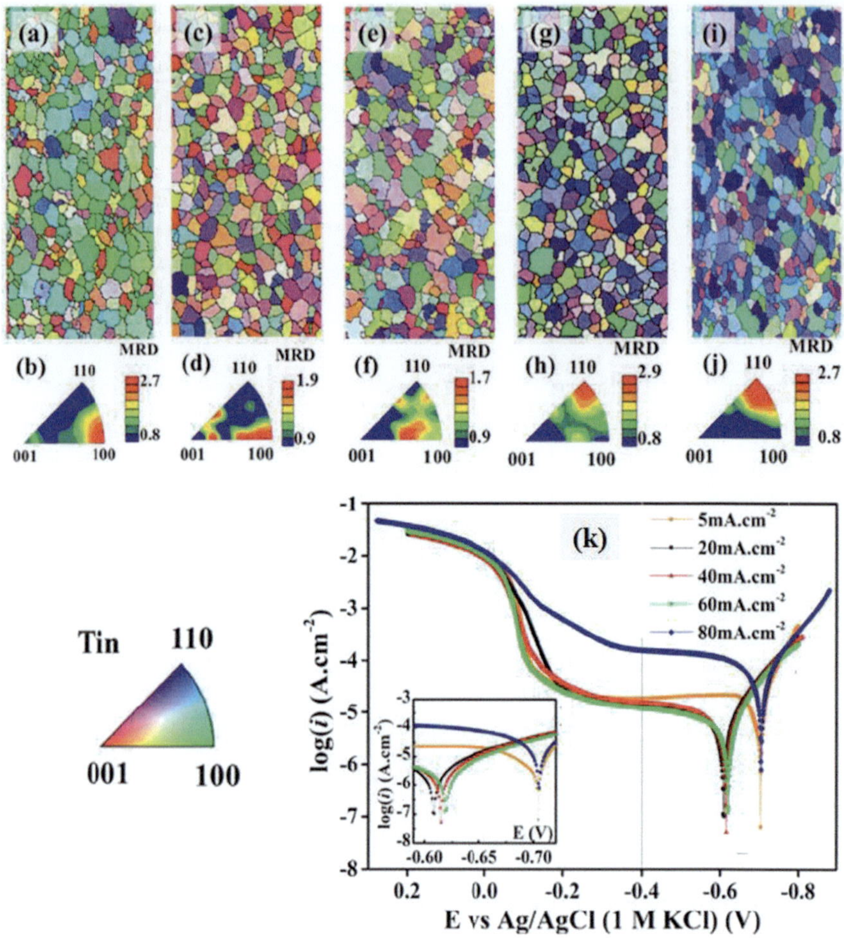

Fig. 7 **a**, **c**, **e**, **g**, **i** IPF maps and corresponding **b**, **d**, **f**, **h**, **j** IPF indicating microtexture intensity of Sn coatings fabricated at electrodeposition current density of **a**, **b** 5 mA cm^{-2}, **c**, **d** 20 mA cm^{-2}, **e**, **f** 40 mA cm^{-2}, **g**, **h** 60 mA cm^{-2}, and **i**, **j** 80 mA cm^{-2}. Potentiodynamic polarization curves for Sn coatings fabricated at different electrodeposition current densities are shown in **k**. The inset shows the enlarged view of the potentiodynamic polarization curves within the potential **e** range of –0.6 to –0.7 V [14]

microtexture and microstructure of metallic coating. Experimental evidences reveal that metal matrix composite coatings such as Cu-CNT, Zn-CNT, Ni-CNT and NiCo-CNT have superior corrosion resistance as compared to conventional metallic coatings [110–113]. In pristine Zn coating, it has been found that the evolution of inert grains with basal texture resisted the transgranular corrosion. The addition of CNTs led to the development of basal texture as well as low energy grain boundaries which played a vital role in restricting transgranular and intergranular corrosion, respectively [111]. Further studies have illustrated that concentration of CNT in the metallic

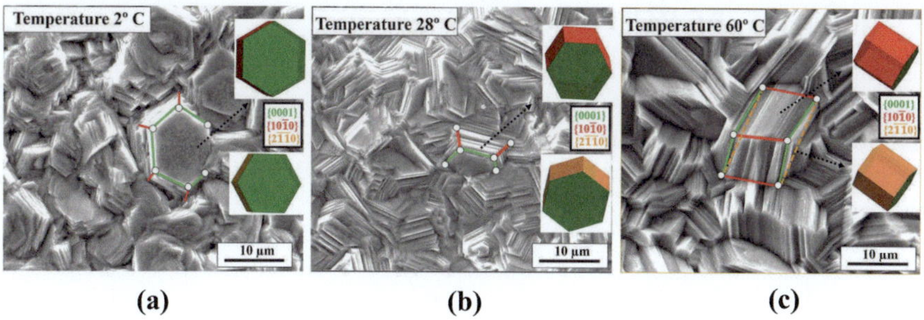

Fig. 8 The SEM micrographs of Zn coatings, electrodeposited at **a** 275 K, **b** 301 K and **c** 333 K. The insets represent the schematic layout of crystallographic planes in Zn coatings [13]

coating matrix significantly influences the corrosion resistance of the coating due to microtextural modification [113]. In this regard, Cu-CNT composite coatings were electrodeposited on mild steel and it was observed that Cu-0.75 wt% CNT coating exhibit lowest corrosion current density and corrosion rate which could be associated with highest {111} orientation and lowest {220} orientation. On similar note, the dominance of (111) densely packed plane orientation in Ni-17 wt % Co coating resulted in superior corrosion resistance [114]. However, in another study of Ni-W alloy coatings, it was reported that Ni-13 wt %W exhibited highest corrosion rate despite the development of high intensity {111} oriented planes along the growth direction [115]. This discrepancy was related to the formation of high fraction of high-angle grain boundaries which allow the corrosive media to seep into the coating surface interface.

Besides coating, laser shock peeing has also been employed to alter near surface microtexture of the material which can subsequently influence the corrosion properties [94]. In this context, laser shock peening was conducted on Al-Li alloy 2060-T8 which significantly strengthened Goss {110} $< 100 >$ and Cube {001} $< 100 >$ texture components during single peening. As discussed earlier, the crystallographic plane {100}, corresponding to Cube texture, has densely packed atoms as compared to {110} plane which belongs to Goss texture {110} $< 001 >$. During multiple peening, the cube texture was weakened with marginal variation in Goss texture owing to which the corrosion resistance of the material was reduced and facilitated pits formation. On the contrary, relatively higher volume fraction of Cube component resulted in dense and uniform α-Al_2O_3 formation which lowered the pitting corrosion during single peening [94].

3 Influence of GBCD on Corrosion

Apart from microtexture, GBCD also plays a dominant role in determining the corrosion resistance of an alloy. Towards this, the influence of GBCD on different types of corrosion like IGC, SCC, pitting corrosion as well as high-temperature oxidation has been mentioned below.

3.1 Influence of GBCD on Sensitization and IGC

IGC is a localized form of corrosion where the grain boundaries or the areas adjacent to them are preferentially attacked on exposure to corrosive environment [1]. IGC is caused by several reasons such as depletion of one of the alloying elements (due to precipitation of second-phase particles at the boundaries), segregation of certain alloying elements or presence of impurities at the grain boundaries [1]. Towards this, it has been widely reported that the sensitization phenomena specifically occurring in the Cr-rich alloys promotes IGC [1, 116–120]. Sensitization proceeds with the predominant precipitation of Cr-rich carbides at grain boundaries, and subsequent formation of Cr-depleted zones along the boundaries [1, 116–121]. The depleted regions are preferentially attacked when exposed to aggressive environments. Various studies have reported that the low Σ ($\Sigma \leq 29$) CSL boundaries, particularly the $\Sigma3^n$ ($n \leq 3$) twin boundaries, are resistant to carbide precipitation as compared to the random high-angle grain boundaries (RHAGBs) owing to their highly ordered structure and lower interfacial energy [122–126] Consequently, the formation of Cr-depleted zones near these "special" boundaries is significantly suppressed in contrast to the RHAGBs [127]. Taking advantage of this fact, several studies have utilized the concept of GBE to enhance the resistance against IGC [116, 118, 128–132]. However, few other studies have argued that the increase in the low Σ ($\Sigma \leq 29$) CSL boundaries may not always lead to superior IGC resistance [133–136]. This could be possibly attributed to the interference from other microstructural features like grain size and retained strain that gets altered during the thermomechanical treatments intended for attaining the GBE microstructure. In this context, the sole influence of GBCD on IGC in stainless steel was recently understood through decoupling the effect of the aforementioned microstructural features [118]. Some of the interesting observations from this study have been shown in Fig. 9. The double loop electrochemical potentiodynamic reactivation (DL-EPR) curves for the GBE-processed ($\Sigma3^n$ ($n \leq 3$) fraction ~0.80) and non-GBE specimens ($\Sigma3^n$ ($n \leq 3$) fraction ~0.46) have been presented in Fig. 9a. The degree of sensitization (DOS) values, as calculated from the DL-EPR curves, were found to be considerably higher for the non-GBE specimen (~17%) as compared to the GBE-processed specimen (~4%). The optical micrographs revealing the surface morphologies of the non-GBE and GBE-processed specimens after the DL-EPR tests are presented in Fig. 9b, c, respectively. As apparent, the grain boundary attack in the non-GBE specimen was observed to be severe, whereas the GBE specimen showed lower grain boundary attack (Fig. 9c). Moreover, the cross-sectional images revealing the percolation depth for both the non-GBE and GBE-processed specimens after the IGC test have been shown in Fig. 9d, e respectively. The presence of higher fraction of $\Sigma 3^n$ ($n \leq 3$) boundaries disrupted the susceptible RHAGBs network and significantly restricted the percolation of electrolyte from the surface in the GBE-processed specimen [118]. These observations confirm that GBCD plays a dominant role in increasing the resistance against sensitization and IGC. On the other hand, few studies have argued that only the $\Sigma 3$ (coherent) boundary is resistant to corrosion whereas some of its higher-order variants ($\Sigma 9$ and $\Sigma 27$) are attacked during corrosion [121, 132, 137, 138]. These discrepancies arise from the fact that the aforementioned studies have only considered the CSL model to predict the relationship between sensitization/IGC and GBCD. Since the CSL model does not take the orientation of the grain boundary plane into account (which plays dominant role in governing the grain boundary-related properties [139, 140]), such discrepancies in the findings are inevitable. In a recent study by An et al. [141], the correlation between the character of the grain boundary (in terms of its misorientation and plane) and its response to IGC has been investigated. It has been observed that the IGC response of the low angle grain boundaries (LAGBs) is chiefly reliant on the misorientation angle while the grain boundary plane (in terms of atom packing density (APD)) played a governing role in case of RHAGBs and the CSL boundaries. Grain boundaries with lower APD are more prone to IGC, while those with higher APD are resistant to corrosion. Evidently, the boundaries with lower APD have higher free volume and consequently higher energy, which provides easy pathways for elemental diffusion as well as favors easier nucleation of Cr-carbides [141].

3.2 *Influence of GBCD on Intergranular SCC*

Intergranular SCC often occur as a result of the nucleation and propagation of cracks along the Cr-depleted zones (formed as a consequence of sensitization), on straining [1, 142, 143]. Several studies have claimed that the low Σ ($\Sigma \leq 29$) CSL boundaries are highly resistant to cracking, as compared to the RHAGBs [144–147]. Towards this, Crawford and Was [145] have suggested that the low Σ ($\Sigma \leq 29$) CSL boundaries promote slip in the abutting grains either by transmitting or by absorbing and re-emitting lattice dislocations. This, in turn, reduces the strain energy accumulated at the grain boundaries as a consequence of deformation. The decrease in strain energy facilitates further deformation and consequently, alleviates the propensity for the initiation of cracks [145]. Since the cracking susceptibility is highly governed by the grain boundary character, several researchers have adopted the GBE approach to enhance resistance against the intergranular SCC in various materials [147–153]. During GBE, large grain clusters with tree-ring structures and intricate morphologies are formed as a result of multiple twinning [151]. As these grain clusters consist of crack-resistant $\Sigma 3^n$ ($n < 3$) boundaries, the crack follows a zigzag path along the outer boundaries of these clusters, which, in turn, reduces the SCC susceptibility

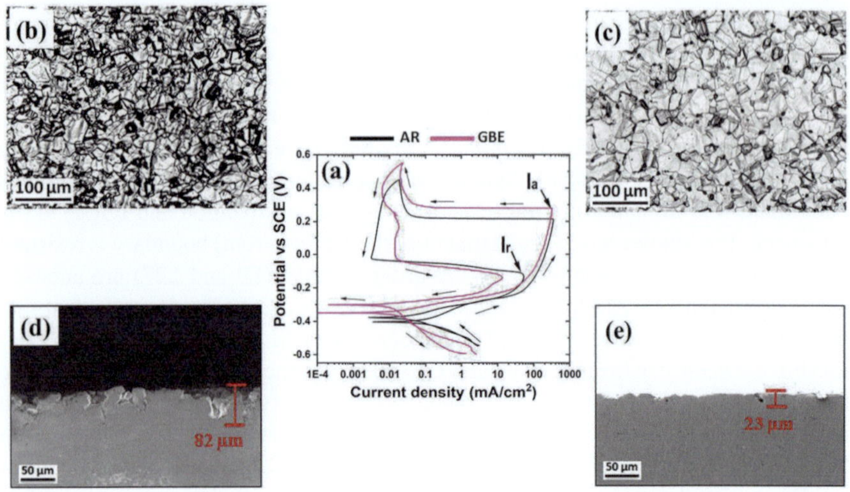

Fig. 9 **a** The DL-EPR curve of the non-GBE and GBE-processed 304L specimens. The optical micrograph of the non-GBE and GBE-processed specimens post DL-EPR test is shown in **b** and **c**, respectively. The cross-sectional SEM image revealing the percolation depth in **d** non-GBE, and **e** GBE specimens post aqueous corrosion test for 24 h [118]

[151]. However, it has been reported by few researchers that the beneficial effects of GBE are pronounced only at lower extents of strain [154, 155]. On the application of a higher amount of strain, grain rotation takes place which transforms the special boundaries to RHAGBs just before the fracture [154, 155].

3.3 Influence of GBCD on Passive Film Behavior and Pitting Corrosion

Pitting corrosion resistance primarily depend upon the stability as well as the compactness of passive oxide layer [156]. In the past, few investigations have claimed that passive film breakdown, even at the nano level, causes initiation of pits [157–159]. The stability of the passive layer formed at the boundaries and junctions is reported to be poor and thereby, the layer breaks down on exposure to aggressive ions resulting in the initiation of the pits [160]. Moreover, the passivation efficiency of the boundaries largely varies depending upon their character [161]. These findings pave the pathway towards the use of GBE in enhancing the passive film properties. Recently, Pradhan et al. [162] investigated the influence of GBE (in terms of GBCD and triple junction distribution) on the passive layer stability with the help of EIS and Mott-Schottky analyses. The polarization resistance, determined from the diameter of the capacitive loop in the Nyquist plot in Fig. 10a, is considerably greater in the GBE-processed specimen than the non-GBE specimen. Moreover, the values

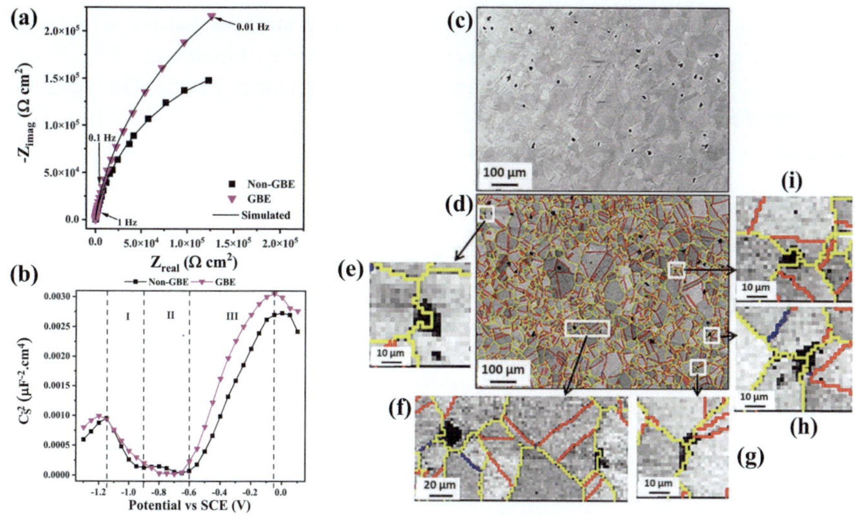

Fig. 10 **a** Nyquist plots of non-GBE and GBE-processed specimens tested in 3.5 wt. % NaCl electrolyte [162]; **b** Mott-Schottky plots of non-GBE and GBE-processed specimens tested in 3.5 wt. % NaCl electrolyte at 1000 Hz [162]; **c** SEM image; and **d** IQ superimposed grain boundary maps (Color code: Σ3—red, Σ9—blue, Σ27—green, and RHAGBs—yellow) of the non-GBE specimen after potentiodynamic polarization test in 3.5 wt. % NaCl [165]. The marked sections in **d** have been enlarged and shown in (**e**–**i**) to depicts the formation of pits at triple junctions

of charge transfer (R_{ct}) and passive film (R_p) resistance, calculated by fitting the Nyquist plot with an equivalent electrical circuit model, were found to be higher in the GBE-processed specimen ($R_p \sim 17 \times 10^4$ Ω.cm^2, and $R_{ct} \sim 5.8 \times 10^5$ Ω. cm^2) as compared to its non-GBE counterpart ($R_p \sim 6.4 \times 10^4$ Ω.cm^2, and $R_{ct} \sim 3.3 \times 10^5$ Ω. cm^2). This suggests that the passive layer is highly resistant in the GBE-processed specimen. The defect concentration in the passive film, as assessed from the Mott-Schottky plots presented in Fig. 10b, was found to be considerably lower in the GBE specimen (1.42 x 10^{21} cm^{-3}) than the non-GBE specimen (2.35 x 10^{21} cm^{-3}). The lower defect concentration in the GBE-processed specimen has been attributed to the higher frequency of $\Sigma 3^n$ ($n \leq 3$) boundaries (~0.78) along with an enhanced fraction of J_2 and J_3 types of triple junctions which inhibit the formation of defects owing to their lower energy. As a result, the passive film developed over the GBE specimen is more compact and protective. Regarding the influence of GBCD on pitting corrosion, several studies have claimed that the CSL ($\Sigma \leq 29$) boundaries have greater resistance to the formation of pits [160, 163, 164]. Recently, a comprehensive investigation focusing on the implication of GBCD on pitting corrosion in 304 austenitic stainless steel has been carried out by Pradhan et al. [165] with the help of potentiostatic and potentiodynamic tests. The SEM micrograph of the non-GBE condition post potentiodynamic polarization test, presented in Fig. 10c, reveals the formation of pits throughout the microstructure. The corresponding image quality overlaid color-coded grain boundary map (Fig. 10d) along with the enlarged views

(Fig. 10e–i)of the regions marked in Fig. 10d clearly indicate that majority of the pits have nucleated at the RHAGBs whereas the low energy boundaries (Σ3, Σ9, and Σ27), as well as the J_2- and J_3-type junctions are resistant to pitting. Owing to the presence of higher fraction of $\Sigma3^n$ ($n \leq 3$) boundaries and J_2- and J_3-type junctions, the GBE specimen showed higher resistance to pitting.

3.4 Influence of GBCD on Oxidation and High-Temperature Hot Corrosion

Studies focusing on the intergranular oxidation phenomena have reported that the RHAGBs are highly susceptible to oxidation whereas the low-Σ CSL boundaries are resistant to oxidation owing to their ordered structure and lower interfacial energy [95, 166, 167]. Since oxidation is a diffusion-controlled phenomenon [168, 169] and grain boundaries are effective diffusion paths for the reactive elements to percolate inside the alloys [170, 171], multiple studies have utilized the concept of tailoring the GBCD by increasing the fraction of the low-Σ CSL boundaries to control the extent of oxidation [148, 172–177]. The higher frequency of low-Σ CSL boundaries disrupts the RHAGB networks that preferentially act as easy diffusion paths for the entry of oxygen [148, 175–177]. Consequently, this results in a lower internal oxidation attack and thereby, improved oxidation performance [148, 175]. Additionally, the disrupted networks also hinder the percolation of harmful species like chlorine or sulphur into the alloy, thereby increasing the resistance against high-temperature hot corrosion, which is known to be an accelerated form of oxidation [172]. Recently, Vikram et al. [178] adopted fractal analysis to investigate the oxidation response of a GBE-processed high entropy alloy and developed an empirical relationship linking the oxide layer thickness with the fractal number (parameter used to evaluate the connectivity of the RHAGBs). The oxide layer thickness was observed to vary linearly with the fractal number i.e., it is lowest for the specimen having the highest fraction of low-Σ CSL boundaries (lowest fractal number), and vice-versa. The introduction of a higher fraction of low-Σ CSL boundaries decreases the overall diffusion occurring through the grain boundaries. As a result, the grain boundary diffusion becomes equivalent to the bulk diffusion in the alloy. Since the bulk diffusion is uniform in nature, it leads to the evolution of a thin, yet dense and continuous oxide layer, which imparts superior corrosion resistance to the alloy [174, 176].

4 Summary and Concluding Remark

The implications of microtexture and grain boundary misorientation on electrochemical degradation behavior of materials have been systematically reviewed in this chapter. The first part of the chapter clearly outlined the predominant microtextures that develop in different materials (through deformation and surface modification processes) and their effect on various forms of corrosion, both at ambient and high temperature. Evidently, the variation in corrosion response with change in microtexture is primarily driven by the extent of thermodynamic energy associated with that textured surface. Thereby, a microtexture containing a large fraction of densely packed crystallographic planes (possessing lower surface energy) parallel to surface can essentially lead to a substantial reduction in material dissolution. Accordingly, a micro-textured surface consisting of {111}, {110}, and {0001} set of planes are found to be the most resistant to environmental degradation for FCC, BCC, and HCP materials, respectively. However, the presence of a strongly textured surface is often found to induce severe anisotropy in corrosion with respect to the other two cross-sectional surfaces and therefore "texture weakening" has been considered as a suitable method to improve the overall corrosion resistance of the specimen. Moreover, it has been observed that the orientation statistics of individual grains alone are not always sufficient to explain the correlation between microtexture and corrosion. The orientation relationship existing between the neighboring grains can also play a predominant role in determining the corrosion response. For instance, the presence of two adjacent planes having significant difference in electrochemical potential are highly probable to induce severe micro-galvanic corrosion in material. Further, some important parameters related to different localized corrosions, like the pit development and their characteristics, the propagation and deflection of cracks, and passive film behavior, have shown a correlation to microtexture of material. Even at elevated temperatures, the corrosion resistance of a material is found to be a function of substrate microtexture, since there exists an epitaxial relationship between the oxide layer and the substrate texture. Moreover, the growth of crystals along preferred direction during surface modifications using coatings, laser shock peening etc., are also observed to significantly influence the corrosion properties of the material. Such correlations existing between microtexture and corrosion definitely indicate that the corrosion response can be significantly tuned through regulating the grain orientation aspects suitably. However, substantial investigation is still necessary in this field for in-depth understanding of the topic and also to explain the contrasting results that still exist in this field. Moreover, the utilization of artificial intelligence (AI)-aided microtextural designing is a highly promising approach which can be extensively explored in future to suitably tune the surface functionality and subsequently impart a control over corrosion, as suggested by Lantada et al. [179].

The second part of the chapter mainly pointed out the influence of GBCD on different types of corrosion in alloys. A microstructure having higher proportion of $\Sigma3^n$ ($n = 1, 2$, and 3) boundaries is found to be resistant to different types of corrosion owing to its lower interfacial energy in comparison to the RHAGBs. As apparent from the discussions till now, the researchers have mainly focused on explaining the corrosion response as a function of $\Sigma3^n$ fraction and its spatial arrangements, solely considering the CSL model. However, grain boundary plane is also a crucial factor in regulating the corrosion behavior and hence needs to be considered to get a broader view concerning the effect of grain boundary character on environmental degradation behavior of materials.

References

1. Fontana MG (1910) Corrosion Engineering, 3rd edn. Tata McGraw-Hill, New York
2. Bairagi D, Mandal S (2022) A comprehensive review on biocompatible Mg-based alloys as temporary orthopaedic implants: Current status, challenges, and future prospects. J Mag Alloy 10:627–669
3. Frankel GS, Vienna JD, Lian J, Scully JR, Gin S, Ryan JV, Wang J, Kim SH, Windl W, Du J (2018) A comparative review of the aqueous corrosion of glasses, crystalline ceramics, and metals. Npj Mat Degrad 2 15
4. Song GL (2012) The effect of texture on the corrosion behavior of AZ31 Mg Alloy. JOM 64:671–679
5. Venegas V, Caleyo F, Baudin T, Hallen JM, Penelle R (2009) Role of microtexture in the interaction and coalescence of hydrogen-induced cracks. Corros Sci 51:1140–1145
6. Jyotheender KS, Kumar MKP, Srivastava C (2021) Influence of surfactant polarity on the evolution of micro-texture, grain boundary constitution and corrosion behavior of electrodeposited Zn coatings. Surf Coat Tech 423:127594
7. Fu BQ, Liu W, Li ZL (2009) Calculation of the surface energy of hcp-metals with the empirical electron theory. App Sur Sci 255:9348–9357
8. Song GL, Mishra R, Xu Z (2010) Crystallographic orientation and electrochemical activity of AZ31 Mg alloy. Electrochem Commun 12:1009–1012
9. Song GL, Xu Z (2010) The surface, microstructure and corrosion of magnesium alloy AZ31 sheet. Electrochim Acta 55:4148–4161
10. Wang SG, Tian EK, Lung CW (2000) Surface energy of arbitrary crystal plane of bcc and fcc metals. J Phy Chem Sol 61:1295–1300
11. Suwas S, Ray RK (2014) Crystallographic texture of materials. Springer, London, London
12. Wahab JA, Ghazali MJ, Derman MN (2020) Corrosion resistance of micro-textured surface modified alumina-titania coating. IOP Conf Ser Mat Sci Eng 743:012004
13. Jyotheender KS, Kumar MKP, Srivastava C (2021) Low temperature electrogalvanization: texture and corrosion behavior. App Sur Sci 559:149953
14. Gupta A, Srivastava C (2022) Electrodeposition current density induced texture and grain boundary engineering in Sn coatings for enhanced corrosion resistance. Corr Sci 194:109945
15. Watanabe T (1984) An approach to grain boundary design for strong and ductile polycrystals. Res Mech Int J Struc Mech Mater Sci 11:47–84
16. Engler O, Randle V (2008) Introduction to Texture Analysis: Macrotexture, Microtexture, and Orientation Mapping, 2nd edn. CRC Press, New York
17. Watanabe T (2011) Grain boundary engineering: historical perspective and future prospects. J Mater Sci 46:4095–4115
18. Randle V (2010) Grain boundary engineering: an overview after 25 years. Mater Sci Tech 26:253–261
19. Randle V (2004) Twinning-related grain boundary engineering. Acta Mat 52:4067–4081
20. Prithiv TS, Bhuyan P, Pradhan SK, Sarma VS, Mandal S (2018) A critical evaluation on efficacy of recrystallization vs. strain induced boundary migration in achieving grain boundary engineered microstructure in a Ni-base superalloy. Acta Mat 146:187–201
21. Randle V (1999) Mechanism of twinning-induced grain boundary engineering in low stacking-fault energy materials. Acta Mat 47:4187–4196
22. Wang Y, Huang J (2003) Texture analysis in hexagonal materials. Mat Chem Phy 81:11–26
23. Wei HL, Mazumder J, DebRoy T (2015) Evolution of solidification texture during additive manufacturing. Sci Rep 5:16446
24. Sanyal S, Kanodia S, Saha R, Bandyopadhyay TK, Mandal S (2019) Influence of hard plate hot forging temperature on the microstructure, texture and mechanical properties in a lean Mg–Zn–Al alloy. J Alloy Comp 800:343–354
25. Verma KK, Kumar S, Suwas S (2021) Evolution of microstructure and texture during hot rolling and subsequent annealing of the TZ73 magnesium alloy and its influence on tensile properties. Mater Sci Eng A 821:141480

26. Sridharan N, Wolcott P, Dapino M, Babu SS (2016) Microstructure and texture evolution in aluminum and commercially pure titanium dissimilar welds fabricated using ultrasonic additive manufacturing. Scr Mater 117:1–5
27. Mehtani HK, Khan MI, Jaya BN, Parida S, Prasad MJNV, Samajdar I (2021) The oxidation behavior of iron-chromium alloys: The defining role of substrate chemistry on kinetics, microstructure and mechanical properties of the oxide scale. J Alloy Comp 871:159583
28. Amirnejad M, Rajabi M, Jamaati R (2021) The effect of crystallographic texture as a distinct effective parameter on the biocorrosion performance of Ti6Al4V alloy in PBS solution. Corro Sci 179:109100
29. Gerashi E, Alizadeh R, Langdon TG (2022) Effect of crystallographic texture and twinning on the corrosion behavior of Mg alloys: a review. J Mag Alloy 10:313–325
30. Lapeire L, Lombardia EM, Verbeken K, Graeve ID, Kestens LAI, Terryn H (2013) Effect of neighboring grains on the microscopic corrosion behavior of a grain in polycrystalline copper. Corro Sci 67:179–183
31. Fang H, Liu H, Yan Y, Luo X, Xu X, Chu X, Lu Y, Yu K, Wang D (2021) Evolution of texture, microstructure, tensile strength and corrosion properties of annealed Al–Mg–Sc–Zr alloys. Mater Sci Eng A 804:140682
32. Dong S, Chen X, Plante ECL, Gussev M, Leonard K, Sant G (2020) Elucidating the grain-orientation dependent corrosion rates of austenitic stainless steels. Mater Des 191:108583
33. Lüth H (2010) Solid Surfaces, Interfaces and Thin Films, 6th edn. Springer, Berlin Heidelberg, Berlin, Heidelberg
34. Groß A (2009) Theoretical Surface Science: A Microscopic Perspective, 2nd edn. Springer, Berlin Heidelberg, Berlin, Heidelberg
35. Parvate S, Dixit P, Chattopadhyay S (2020) Superhydrophobic surfaces: insights from theory and experiment. J Phy Che B 124:1323–1360
36. Xin R, Li B, Li L, Liu Q (2011) Influence of texture on corrosion rate of AZ31 Mg alloy in 3.5wt.% NaCl. Mater Des 32:4548–4552
37. Chapuis A, Liu Q (2015) Simulations of texture evolution for HCP metals: Influence of the main slip systems. Comp Mater Sci 97:121–126
38. Oldani C, Silvetti S (2000) Microstructure and texture evolution during the annealing of a lamination steel. Scr Mater 43:129–134
39. Rakshit R, Sarkar A, Panda SK, Mandal S (2022) Influence of out-of-plane stretch forming induced different strain paths on micro-texture evolution, slip system activity and Taylor factor distribution in Al–Li alloy. Mater Sci Eng A 830:142267
40. Babu KA, Athreya CN, Mozumder YH, Sarma VS, Mandal S (2020) A comprehensive study on texture development and twin-related domain evolution following hot compression in a super austenitic stainless steel. Metall Mater Trans A 51:3535–3551
41. Liao L, Li J, Xu F, Zhang W (2021) Cold deformation for ameliorating the corrosion resistance of 654SMO in a high-temperature simulated seawater environment: the combined effect of texture and twinning boundaries. J Mater Sci 56:17505–17526
42. Wen YN, Zhang JM (2007) Surface energy calculation of the fcc metals by using the MAEAM. Sol Stat Comm 144:163–167
43. Fu B, Liu W, Li Z (2010) Calculation of the surface energy of fcc-metals with the empirical electron surface model. App Surf Sci 256:6899–6907
44. Cai G, Pang Y, Huang Y, Misra RDK (2019) Roles of inclusion, texture and grain boundary in corrosion resistance of low-nickel austenite stainless steel containing Ce. ISIJ Int 59:2302–2310
45. Dwivedi D, Lepková K, Becker T (2017) Carbon steel corrosion: a review of key surface properties and characterization methods. RSC Adv 7:4580–4610
46. Magnin T, Chieragatti R, Oltra R (1990) Mechanism of brittle fracture in a ductile 316 alloy during stress corrosion. Acta Metall Mater 38:1313–1319
47. Meletis EI, Hochman RF (1984) The crystallography of stress corrosion cracking in face centered cubic single crystals. Corro Sci 24:843–862

48. Zhu L, Yan Y, Li J, Qiao L, Li Z, Volinsky AA (2015) Stress corrosion cracking at low loads: surface slip and crystallographic analysis. Corro Sci 100:619–626
49. Li Z, Yi D, Tan C, Wang B (2020) Investigation of the stress corrosion cracking behavior in annealed 5083 aluminum alloy sheets with different texture types. J Alloy Comp 817:152690
50. Rao ACU, Vasu V, Govindaraju M, Srinadh KVS (2016) Stress corrosion cracking behaviour of 7xxx aluminum alloys: a literature review. Trans Nonfer Met Soc China 26:1447–1471
51. Bonab MAM (2019) Effects of different parameters on initiation and propagation of stress corrosion cracks in pipeline steels: a review. Metals (Basel) 9:1–18
52. Wu W, Liu Z, Bai S, Li F, Liu M, Wang A (2017) Anisotropy in fatigue crack propagation behavior of Al-Cu-Li alloy thick plate. Mater Char 131:440–449
53. Sarkar A, Sanyal S, Bandyopadhyay TK, Mandal S (2019) Implications of microstructure, Taylor factor distribution and texture on tensile properties in a Ti-added Fe-Mn-Al-Si-C steel. Mater Sci Eng A 767:138402
54. Callister WD (1980) Materials Science and Engineering: An Introduction, 7th ed. John Wiley & Sons, Inc. USA
55. Park SA, Kim JG, He YS, Shin KS, Yoon JB (2014) Comparative study on the corrosion behavior of the cold rolled and hot rolled low-alloy steels containing copper and antimony in flue gas desulfurization environment. Phy Met Metall 115:1285–1294
56. Cano MM, Maya LH, Hallen JM, Luna MC, Estrada EMA, Manh TL (2018) Model for the correlation between anodic dissolution resistance and crystallographic texture in pipeline steels. Mater 11:1432
57. Arafin MA, Szpunar JA (2009) A new understanding of intergranular stress corrosion cracking resistance of pipeline steel through grain boundary character and crystallographic texture studies. Corro Sci 51:119–128
58. Blanco A, Hallen JM, Nguyen TS, Manh TL (2021) Influence of crystallographic texture on susceptibility to stress corrosion cracking mechanism of API 5L X52 steel for sour service. Eng Fail Anal 119:105002
59. Pourazizi R, Bonab MAM, Szpunar JA (2020) Role of texture and inclusions on the failure of an API X70 pipeline steel at different service environments. Mater Char 164:110330
60. Venegas V, Caleyo F, Hallen JM, Baudin T, Penelle R (2007) Role of crystallographic texture in hydrogen-induced cracking of low carbon steels for sour service piping. Metall Mater Trans A 38:1022–1031
61. Hallen JM, Venegas V, Caleyo F (2013) Crystallographic texture helps reduce HIC cracking in pipeline steels. In: 13th International Conference Fracture 2013, ICF vol 1, pp 768–774
62. Cano MM, Hallen JM, Estrada EMA, Manh TL (2019) Effect of crystallographic texture and microstructure on pitting corrosion behavior of low carbon steels: A Monte Carlo model. Comp Mater Sci 161:394–402
63. Zhang LN, Szpunar JA, Dong JX, Ojo OA, Wang X (2018) Dependence of crystallographic orientation on pitting corrosion behavior of Ni-Fe-Cr alloy 028. Metall Mater Trans B 49:919–925
64. Partridge PG (1967) The crystallography and deformation modes of hexagonal close-packed metals. Metall Rev 12:169–194
65. Stanford N (2010) Micro-alloying Mg with Y, Ce, Gd and La for texture modification—A comparative study. Mater Sci Eng A 527:2669–2677
66. Bohlen J, Yi S, Letzig D, Kainer KU (2010) Effect of rare earth elements on the microstructure and texture development in magnesium–manganese alloys during extrusion. Mater Sci Eng A 527:7092–7098
67. Jiang MG, Xu C, Nakata T, Yan H, Chen RS, Kamado S (2016) Rare earth texture and improved ductility in a Mg-Zn-Gd alloy after high-speed extrusion. Mater Sci Eng A 667:233–239
68. Stanford N, Barnett MR (2008) The origin of "rare earth" texture development in extruded Mg-based alloys and its effect on tensile ductility. Mater Sci Eng A 496:399–408
69. Griffiths D (2014) Understanding texture weakening in magnesium rare earth alloys. The University of Manchester (United Kingdom)

70. Duley P, Bairagi D, Bairi LR, Bandyopadhyay TK, Mandal S (2021) Effect of microstructural evolution and texture change on the in-vitro bio-corrosion behaviour of hard-plate hot forged Mg-4Zn-0.5Ca-0.16Mn (wt%) alloy. Corro Sci 192:109860
71. Wang BJ, Xu DK, Xin YC, Sheng LY, Han EH (2017) High corrosion resistance and weak corrosion anisotropy of an as-rolled Mg-3Al-1Zn (in wt.%) alloy with strong crystallographic texture. Sci Rep 7:1–14
72. Li CQ, Xu DK, Yu S, Sheng LY, Han EH (2017) Effect of icosahedral phase on crystallographic texture and mechanical anisotropy of Mg–4%Li based alloys. J Mater Sci Tech 33:475–480
73. Bohlen J, Nürnberg MR, Senn JW, Letzig D, Agnew SR (2007) The texture and anisotropy of magnesium-zinc-rare earth alloy sheets. Acta Mater 55:2101–2112
74. Sabbaghian M, Mahmudi R, Shin KS (2019) Effect of texture and twinning on mechanical properties and corrosion behavior of an extruded biodegradable Mg–4Zn alloy. J Mag Alloy 7:707–716
75. Wang BJ, Xu DK, Dong JH, Ke W (2014) Effect of the crystallographic orientation and twinning on the corrosion resistance of an as-extruded Mg–3Al–1Zn (wt.%) bar. Scr Mater 88:5–8
76. Yu H, Xin Y, Chapuis A, Huang X, Xin R, Liu Q (2016) The different effects of twin boundary and grain boundary on reducing tension-compression yield asymmetry of Mg alloys. Sci Rep 6:29283
77. Zou G, Peng Q, Wang Y, Liu B (2015) The effect of extension twinning on the electrochemical corrosion properties of Mg–Y alloys. J Alloy Comp 618:44–48
78. Hoseini M, Shahryari A, Omanovic S, Szpunar JA (2009) Comparative effect of grain size and texture on the corrosion behaviour of commercially pure titanium processed by equal channel angular pressing. Corro Sci 51:3064–3067
79. Davepon B, Schultze JW, König U, Rosenkranz C (2003) Crystallographic orientation of single grains of polycrystalline titanium and their influence on electrochemical processes. Sur Coat Tech 169–170:85–90
80. Martin E, Azzi M, Salishchev GA, Szpunar J (2010) Influence of microstructure and texture on the corrosion and tribocorrosion behavior of Ti-6Al-4V. Trib Int 43:918–924
81. Park K, Ahn S, Kwon H (2011) Effects of solution temperature on the kinetic nature of passive film on Ni. Electro Acta 56:1662–1669
82. Yi G, Liu X, Zheng C, Zhang H, Xu C, Cui YW, Liu S (2021) Characteristics of passive films formed on as-cast Ti–6Al-4V in hank's solution before and after transpassivation. Fron Mater 7:1–11
83. Mehtani HK, Khan MI, Durgaprasad A, Deb SK, Parida S, Prasad MJNV, Fullwood D, Doherty RD, Samajdar I (2020) Oxidation behavior of interstitial free steel: the defining role of substrate crystallographic texture. Acta Mater 190:43–57
84. Kim BK, Szpunar JA (2001) Orientation imaging microscopy for the study on high temperature oxidation. Scr Mater 44:2605–2610
85. Akhiani H, Nezakat M, Penttilä S, Szpunar J (2015) The oxidation resistance of thermo-mechanically processed Incoloy 800HT in supercritical water. J Super Fluid 101:150–160
86. Scherb CS, Menneken M, Weber K, Jácome LA, Nolze G (2020) Elucidation of orientation relations between Fe-Cr alloys and corrosion products after high temperature SO2 corrosion Corro. Sci 174:108809
87. Cranstoun GKL, Lynch JT (1980) Field ion microscope studies of the oxidation of iron. App Sur Sci 5:161–179
88. Garner A, Frankel P, Partezana J, Preuss M (2017) The effect of substrate texture and oxidation temperature on oxide texture development in zirconium alloys. J Nucl Mater 484:347–356
89. Luborsky FE, Koch EF, Morelock CR (1963) Crystallographic orientation and oxidation of submicron whiskers of iron, iron - Cobalt, and cobalt. J App Phy 34:2905–2909
90. Subbaraman R, Deshmukh SA, Sankaranarayanan SKRS (2013) Atomistic insights into early stage oxidation and nanoscale oxide growth on Fe(100), Fe(111) and Fe(110) surfaces. J Phy Chem C 117:5195–5207

91. Oh SJ, Cook DC, Townsend HE (1998) Characterization of iron oxides commonly formed as corrosion products on steel. Hyper Inter 112:59–66
92. Zhang LN, Ojo OA (2020) Corrosion behavior of wire arc additive manufactured Inconel 718 superalloy. J Alloy Comp 829:154455. https://doi.org/10.1016/j.jallcom.2020.154455
93. Zhang Y, Cheng F, Wu S (2021) Improvement of pitting corrosion resistance of wire arc additive manufactured duplex stainless steel through post-manufacturing heat-treatment. Mater Char 171:110743
94. Karthik D, Jiang J, Hu Y, Yao Z (2021) Effect of multiple laser shock peening on microstructure, crystallographic texture and pitting corrosion of Aluminum-Lithium alloy 2060–T8. Sur Coat Tech 421:127354
95. Phaniraj MP, Kim DI, Cho YW (2011) Effect of grain boundary characteristics on the oxidation behavior of ferritic stainless steel. Corro Sci 53:4124–4130
96. Brito P, Pinto H, Genzel C, Klaus M, Pyzalla AK (2012) Epitaxial stress and texture in thin oxide layers grown on Fe-Al alloys. Acta Mater 60:1230–1237
97. Garner A, Gholinia A, Frankel P, Gass M, Maclaren I, Preuss M (2014) The microstructure and microtexture of zirconium oxide films studied by transmission electron backscatter diffraction and automated crystal orientation mapping with transmission electron microscopy. Acta Mater 80:159–171
98. Travessa DN, Guedes GVB, Oliveira ACD, Cardoso KR, Roche V, Jorge AM (2021) The effect of surface laser texturing on the corrosion performance of the biocompatible β-Ti12Mo6Zr2Fe alloy. Sur Coat Tech 405:126628
99. Wallinder D, Pan J, Leygraf C, Bauer AD (1998) EIS and XPS study of surface modification of 316LVM stainless steel after passivation. Corro Sci 41:275–289
100. Bakradze G, Jeurgens LPH, Mittemeijer EJ (2010) Oxide-film growth kinetics on Zr(0001) and Zr(1010) single-crystal surfaces. Sur Inter Anal 42:588–591
101. Yu X, Jiang Z, Zhao J, Wei D, Zhou C, Huang Q (2015) Microstructure and microtexture evolutions of deformed oxide layers on a hot-rolled microalloyed steel. Corro Sci 90:140–152
102. Li P, Wang S, Dong H, Wen G, Yu F, Ma Y, Lei Z (2021) Effect of inhomogeneous microstructure evolution on the mechanical properties and corrosion behavior of rotary friction welded AA2024 joints. Mater Char 178:111306
103. Jinlong L, Hongyun L (2013) Effect of surface burnishing on texture and corrosion behavior of 2024 aluminum alloy. Sur Coat Tech 235:513–520
104. Tsutsumi Y, Ishimoto T, Oishi T, Manaka T, Chen P, Ashida M, Doi K, Katayama H, Hanawa T, Nakano T (2021) Crystallographic texture- and grain boundary density-independent improvement of corrosion resistance in austenitic 316L stainless steel fabricated via laser powder bed fusion. Addi Manu 45:102066
105. Asgari H, Toroghinejad MR, Golozar MA (2008) The role of texture and microstructure in optimizing the corrosion behaviour of zinc hot-dip coated steel sheets. ISIJ Int 48:628–633
106. Kangazian J, Shamanian M (2019) Micro-texture and corrosion behavior of dissimilar joints of UNS S32750 stainless steel/UNS N08825 Ni-based superalloy. Mater Char 155:109802
107. Guo D, Kwok CT, Chan SLI (2019) Spindle speed in friction surfacing of 316L stainless steel—How it affects the microstructure, hardness and pitting corrosion resistance. Sur Coat Tech 361:324–341
108. Nasirpouri F, Sanaeian MR, Samardak AS, Sukovatitsina EV, Ognev AV, Chebotkevich LA, Hosseini MG, Abdolmaleki M (2014) An investigation on the effect of surface morphology and crystalline texture on corrosion behavior, structural and magnetic properties of electrodeposited nanocrystalline nickel films. App Sur Sci 292:795–805
109. Jyotheender KS, Kumar MKP, Srivastava C (2021) Influence of surfactant polarity on the evolution of micro-texture, grain boundary constitution and corrosion behavior of electrodeposited Zn coatings. Sur Coat Tech 423:127594
110. Jyotheender KS, Arora S, Srivastava C (2021) Solving the corrosion behaviour and cobalt content correlation anomaly in electrodeposited Ni–Co coatings by analysis of coating micro-texture, strain and grain boundary constitution. Phil Mag 101:1823–1835

111. Jyotheender K, Srivastava C (2021) Engineering the micro-texture of Zn coating by carbon nanotube incorporation for enhanced corrosion resistance behavior. Micro Microana 27:2498–2499
112. Kamboj A, Raghupathy Y, Rekha MY, Srivastava C (2017) Morphology, texture and corrosion behavior of nanocrystalline copper–graphene composite coatings. JOM 69:1149–1154
113. Aliyu A, Srivastava C (2021) Correlation between growth texture, crystallite size, lattice strain and corrosion behavior of copper-carbon nanotube composite coatings. Sur Coat Tech 405:126596
114. Bakhit B, Akbari A (2013) Nanocrystalline Ni-Co alloy coatings, Electrodeposition using horizontal electrodes and corrosion resistance. J Coat Tech Res 10:285–295
115. Aliyu A, Jyotheender KS, Srivastava C (2021) Texture and grain boundary engineering in nickel coating with tungsten addition and its effect on the coating corrosion behavior. Sur Coat Tech 412:127079
116. Kaithwas CK, Bhuyan P, Pradhan SK, Mandal S (2018) Materials Characterization Microstructure evolution during low-strain thermo-mechanical processing and its repercussion on intergranular corrosion in alloy 600H. Mater Char 145:582–593
117. Pradhan SK, Prithiv TS, Mandal S (2017) Through-thickness microstructural evolution during grain boundary engineering type thermomechanical processing and its implication on sensitization behavior in austenitic stainless steel. Mater Char 134:134–142
118. Pradhan SK, Bhuyan P, Mandal S (2018) Individual and synergistic influences of microstructural features on intergranular corrosion behavior in extra-low carbon type 304L austenitic stainless steel. Corro Sci 139:319–332
119. Pradhan SK, Bhuyan P, Kaithwas C, Mandal S (2018) Strain-annealing based grain boundary engineering to evaluate its sole implication on intergranular corrosion in extra-low carbon type 304L austenitic stainless steel. Metall Mater Trans A 49:2817–2831
120. Kumar SB, Prasad SB, Kain V, Reddy J (2013) Methods for making alloy 600 resistant to sensitization and intergranular corrosion. Corro Sci 70:55–61
121. Jones R, Randle V (2010) Sensitisation behaviour of grain boundary engineered austenitic stainless steel. Mater Sci Eng A 527:4275–4280
122. Kokawa H, Shimada M, Sato SY (2000) Grain-boundary structure and precipitation in sensitized austenitic stainless steel. JOM 52:34–37
123. Trillo EA, Murr LE (1998) A TEM investigation of $M_{23}C_6$ carbide precipitation behaviour on varying grain boundary misorientations in 304 stainless steels. J Mater Sci 33:1263–1271
124. Trillo EA, Murr LE (1999) Effects of carbon content, deformation and interfacial energetics on carbide precipitation and corrosion sensitization in 304 stainless steel. Acta Metall 47:235–245
125. Kokawa H (2005) Weld decay-resistant austenitic stainless steel. J Mater Sci 40:927–932
126. Zhou Y, Aust KT, Erb U, Palumbo G (2001) Effects of grain boundary structure on carbide precipitation in 304L stainless steel. Scr Mater 45:49–54
127. Bi HY, Kokawa H, Wang ZJ, Shimada M, Sato SY (2003) Suppression of chromium depletion by grain boundary structural change during twin-induced grain boundary engineering of 304 stainless steel. Scr Mater 49:219–223
128. Shimada M, Kokawa H, Wang ZJ, Sato SY, Karibe I (2002) Optimization of grain boundary character distribution for intergranular corrosion resistant 304 stainless steel by twin-induced grain boundary engineering. Acta Mater 50:2331–2341
129. Lehockey EM, Palumbo G, Lin P, Brennenstuhl AM (1997) On the relationship between grain boundary character distribution and intergranular corrosion. Scr Mater 36:1211–1218
130. Lin P, Palumbo G, Erb U, Aust KT (1995) Influence of grain boundary character distribution on sensitization and intergranular corrosion of alloy 600. Scr Metall Mater 33:1387–1392
131. Michiuchi M, Kokawa H, Wang ZJ (2006) Twin-induced grain boundary engineering for 316 austenitic stainless steel. Acta Mater 54:5179–5184
132. Shi F, Gao RH, Guan XJ (2020) Application of grain boundary engineering to improve intergranular corrosion resistance in a Fe–Cr–Mn–Mo–N high-nitrogen and nickel-free austenitic stainless steel, Acta Metall. Sin 33:789–798

133. Ahmedabadi PM, Kain V, Muralidhar KV, Samajdar I (2013) On the role of residual strain in controlling sensitisation of twin-boundary engineered type 304 stainless steel. J Nucl Mater 432:243–251
134. Chowdhury SG, Singh R (2008) The influence of recrystallized structure and texture on the sensitization behaviour of a stable austenitic stainless steel (AISI 316L). Scr Mater 58:1102–1105
135. Singh R, Chowdhury SG, Chattoraj I (2008) Modification of sensitization resistance of AISI 304L stainless steel through changes in grain size and grain boundary character distributions. Metall Mater Trans A 39:2504–2512
136. Singh R, Chowdhury SG, Kumar BR, Das SK, De PK, Chattoraj I (2007) The importance of grain size relative to grain boundary character on the sensitization of metastable austenitic stainless steel. Scr Mater 57:185–188
137. Hu C, Xia S, Li H, Liu T, Zhou B, Chen W, Wang N (2011) Improving the intergranular corrosion resistance of 304 stainless steel by grain boundary network control. Corro Sci 53:1880–1886
138. Shi F, Tian PC, Jia N, Ye ZH, Qi Y, Liu CM, Li WX (2016) Improving intergranular corrosion resistance in a nickel-free and manganese-bearing high-nitrogen austenitic stainless steel through grain boundary character distribution optimization. Corro Sci 107:49–59
139. Randle V (2006) Special boundaries and grain boundary plane engineering. Scr Mater 54:1011–1015
140. Mandal S, Pradeep KG, Zaefferer S, Raabe D (2014) A novel approach to measure grain boundary segregation in bulk polycrystalline materials in dependence of the boundaries' five rotational degrees of freedom. Scr Mater 81:16–19
141. An D, Griffiths TA, Konijnenberg P, Mandal S, Wang Z, Zaefferer S (2018) Correlating the five parameter grain boundary character distribution and the intergranular corrosion behaviour of a stainless steel using 3D orientation microscopy based on mechanical polishing serial sectioning. Acta Mater 156:297–309
142. Babout L, Marrow TJ, Engelberg D, Withers PJ (2006) X-ray microtomographic observation of intergranular stress corrosion cracking in sensitised austenitic stainless steel. Mater Sci Tech 22:1068–1075
143. Stratulat A, Duff JA, Marrow TJ (2014) Grain boundary structure and intergranular stress corrosion crack initiation in high temperature water of a thermally sensitised austenitic stainless steel, observed in situ. Corro Sci 85:428–435
144. Pan Y, Adams BL, Olson T, Panayotou N (1996) Grain-boundary structure effects on intergranular stress corrosion cracking of Alloy X-750. Acta Mater 44:4685–4695
145. Crawford DC, Was GS (1992) The role of grain boundary misorientation in intergranular cracking of Ni-16Cr-9Fe in 360 °C argon and high-Purity water. Metall Trans A 23:1195–1206
146. Telang A, Gill AS, Kumar M, Teysseyre S, Qian D, Mannava RS, Vasudevan KV (2016) Iterative thermomechanical processing of alloy 600 for improved resistance to corrosion and stress corrosion cracking. Acta Mater 113:180–193
147. Ishibashi R, Horiuchi T, Kuniya J, Yamamoto M, Tsurekawa S, Kokawa H, Watanabe T, Shoji T (2005) Effect of grain boundary character distribution on stress corrosion cracking behavior in austenitic stainless steels. Mater Sci For 475–479:3863–3866
148. Krupp U (2008) Improving the resistance to intergranular cracking and corrosion at elevated temperatures by grain-boundary-engineering-type processing. J Mater Sci 43:3908–3916
149. Rahimi S, Engelberg DL, Duff JA, Marrow TJ (2009) In situ observation of intergranular crack nucleation in a grain boundary controlled austenitic stainless steel. J Micro 233:423–431
150. Telang A, Gill SA, Tammana D, Wen X, Kumar M, Teysseyre S, Mannava RS, Qian D, Vasudevan KV (2015) Surface grain boundary engineering of Alloy 600 for improved resistance to stress corrosion cracking. Mater Sci Eng A 648:280–288
151. Liu T, Xia S, Bai Q, Zhou B, Zhang L, Lu Y, Shoji T (2018) Three-dimensional study of grain boundary engineering effects on intergranular stress corrosion cracking of 316 stainless steel in high temperature water. J Nucl Mater 498:290–299

152. Tingguang L, Xia S, Bai Q, Zhou B, Lu Y, Shoji T (2019) Evaluation of grain boundary network and improvement of intergranular cracking resistance in 316L stainless steel after grain boundary engineering. Mater 12:242–259
153. Tingguang L, Bai Q, Ru X, Xia S, Lu Y, Shoji T (2019) Grain boundary engineering for improving stress corrosion cracking of 304 stainless steel. Mater Sci Tech 35:477–487
154. West EA, Was GS (2009) IGSCC of grain boundary engineered 316L and 690 in supercritical water. J Nuc Mater 392:264–271
155. Alexandreanu B, Capell B, Was GS (2001) Combined effect of special grain boundaries and grain boundary carbides on IGSCC of Ni-16Cr-9Fe-xC alloys. Mater Sci Eng A 300:94–104
156. Parangusan H, Bhadra J, Thani NA (2021) A review of passivity breakdown on metal surfaces : influence of chloride- and sulfide-ion concentrations, temperature, and pH. Emer Mater 4:1187–1203
157. Marcus P, Maurice V, Strehblow H (2008) Localized corrosion (pitting): a model of passivity breakdown including the role of the oxide layer nanostructure. Corro Sci 50:2698–2704
158. Amri J, Souier T, Malki B, Baroux B (2008) Effect of the final annealing of cold rolled stainless steels sheets on the electronic properties and pit nucleation resistance of passive films. Corro Sci 50:431–435
159. Smialowska SZ (2002) Mechanism of pit nucleation by electrical breakdown of the passive film. Corro Sci 44:1143–1149
160. Palumbo G, Aust KT (1989) Triple-line corrosion in high purity nickel. Mater Sci Eng A 113:139–147
161. Bettayeb M, Maurice V, Klein LH, Lapeire L, Verbeken K, Marcus P (2019) Combined in situ microstructural study of the relationships between local grain boundary structure and passivation on microcrystalline copper, Electro. Acta 305:240–246
162. Pradhan SK, Bhuyan P, Bairi LR, Mandal S (2021) Comprehending the role of individual microstructural features on electrochemical response and passive film behaviour in type 304 austenitic stainless steel. Corro Sci 180:109187
163. Meng G, Li Y, Shao Y, Zhang T, Wang Y, Wang F, Cheng X, Dong C, Li X (2016) Effect of microstructures on corrosion behavior of nickel coatings : (II) competitive effect of grain size and twins density on corrosion behavior. J Mater Sci Tech 32:465–469
164. Qi J, Huang B, Wang Z, Ding H, Xi J, Fu W (2017) Dependence of corrosion resistance on grain boundary characteristics in a high nitrogen CrMn austenitic stainless steel. J Mater Sci Tech 33:1621–1628
165. Pradhan SK, Bhuyan P, Mandal S (2019) Influence of the individual microstructural features on pitting corrosion in type 304 austenitic stainless steel. Corro Sci 158:108091
166. Yamaura S, Igarashi Y, Tsurekawa S, Watanabe T (2000) Grain boundary engineering for the control of oxidation embrittlement. Prop Comp Inorg Solid 2:27–37
167. Yamaura S, Igarashi Y, Tsurekawa S, Watanabe T (1999) Structure-dependent intergranular oxidation in Ni - Fe polycrystalline alloy. Acta Mater 47:1163–1174
168. Rabbani F, Ward LP, Strafford KN (2000) A comparison of the growth kinetics and scale morphology for three superalloys at 930 °C in air and low PO_2 Environments. Corro Sci 54:139–153
169. Hussain N, Shahid KA, Khan IH, Rahman S (1995) Oxidation of high-temperature alloys (superalloys) at elevated temperatures in air.I. Oxid Met 43:363–378
170. Deepak K, Mandal S, Athreya CN, Kim D, Boer B, Sarma VS (2016) Implication of grain boundary engineering on high temperature hot corrosion of alloy 617. Corro Sci 106:293–297
171. Bhuyan P, Pradhan SK, Mitra R, Mandal S (2019) Evaluating the efficiency of grain boundary serrations in attenuating high- temperature hot corrosion degradation in Alloy 617. Corro Sci 149:164–177
172. Stölken JS, Kinney JH, Yang BS, Krupp U (2005) The relationship between grain boundary character and the intergranular oxide distribution in IN718 superalloy. Adv Eng Mater 7:2003–2006
173. Tan L, Rakotojaona L, Allen TR, Nanstad KR, Busby TJ (2011) Microstructure optimization of austenitic Alloy 800H (Fe – 21Cr – 32Ni). Mater Sci Eng A 528:2755–2761

174. Athreya CN, Deepak K, Kim D, Boer B, Mandal S, Sarma VS (2019) Role of grain boundary engineered microstructure on high temperature steam oxidation behaviour of Ni based superalloy alloy. J Alloy Comp 778:224–233
175. Trindade VB, Krupp U, Wagenhuber PhEG, Virkar MY, Christ JC (2005) Studying the role of the alloy-grain-boundary character during oxidation of Ni-base alloys by means of the electron back-scattered diffraction technique. Mater High Temp 22:207–212
176. Tan L, Ren X, Sridharan K, Allen TR (2008) Corrosion behavior of Ni-base alloys for advanced high temperature water-cooled nuclear plants. Corro Sci 50:3056–3062
177. Tan L, Allen TR, Busby JT (2013) Grain boundary engineering for structure materials of nuclear reactors. J Nuc Mater 441:661–666
178. Vikram RJ, Gaddam S, Kalsar R, Acharya S, Suwas S (2019) A fractal approach to predict the oxidation and corrosion behavior of a grain boundary engineered low SFE high entropy alloy. Materia 7:100398
179. Lantada AD, Martínez FF, Hengsbach S, Rupp F, Thelen R, Bade K (2020) Artificial intelligence aided design of microtextured surfaces: application to controlling wettability. Nanomater 10:2287

Chapter 13
SEM-EBSD Studies of Tectonically Deformed Rocks and Kinematic Analysis

Manish A. Mamtani

1 Introduction

Scanning electron microscope based electron backscatter diffraction (SEM-EBSD) studies have gained considerable importance in the past few decades in textural studies of various materials. There is considerable published literature available on the basic principles involved in EBSD data acquisition as well as the history of development of the field [1]. In geosciences, SEM-EBSD has been in vogue since the latter half of the last century [2, 3]. One of the prominent areas within geosciences, where EBSD has been extensively used is "Structural Geology". This deals with the study of deformed rocks. Texture analysis, deciphering deformation mechanisms of minerals, fabric quantification and evaluating kinematics, form an important component of various structural geological investigations; in all of these, SEM-EBSD data have been playing an important role since the past few decades. The reader is referred to Prior et al. [3, 4] for a summary of applications of EBSD in the textural analysis of rocks. The objective of this write-up is to particularly review and highlight the use of EBSD in the kinematic analysis of "tectonically" deformed rocks. Since the kinematics recorded in a sample must often be related to regional deformation/tectonics, the importance of reference frame of study is also elaborated here. In addition, rocks are mostly polymineralic and preparing a well-polished surface for SEM-EBSD analysis is often a challenge. Thus, the recent developments in preparing polished geological samples are discussed before presenting examples that highlight the use of EBSD studies in kinematic analysis of deformed rocks.

M. A. Mamtani (✉)
Department of Geology and Geophysics, Indian Institute of Technology, Kharagpur, India
e-mail: mamtani@gg.iitkgp.ac.in

S. Suwas and D. P. Field (eds.), *Advances in Texture, Microtexture, and Allied Techniques*, Materials Horizons: From Nature to Nanomaterials,
https://doi.org/10.1007/978-981-96-5346-1_13

2 Reference Frame for Kinematic Studies of Deformed Rocks

When rocks in nature are subjected to tectonic stresses, they accommodate strain, which is visualized by a strain ellipsoid with three principal axes ($X > Y > Z$); these are oriented at right angles to one another. Several processes may occur during deformation (and straining), viz. rotation of minerals in the rock, (re)crystallization, grain boundary sliding, intracrystalline deformation, solution transfer among others. This leads to the development of preferred orientation of minerals in rocks. In tectonically deformed rocks, this preferred orientation of minerals results in the development of planar fabric (foliation), and/or linear fabric (lineation), and the rocks are referred to as "tectonites" [5–7]. It is well-established that foliation (e.g., schistosity in a schist or gneissosity in quartzo-feldspathic gneiss) are parallel to the XY plane of the strain ellipsoid [8, 9]. The direction of extension in a deformed rock defines the X-direction of the strain ellipsoid; Sander [10] referred to this as the direction of tectonic transport. The X-direction is identified in rocks by the presence of a "stretching" lineation marked by preferentially oriented minerals lying on the XY (foliation) plane. Figure 1a schematically highlights this relationship between these structures that develop in a rock and the principal axes of the strain ellipsoid. For kinematic analysis, it is the norm to investigate structures in a section parallel to the stretching lineation and perpendicular to the foliation; this is the XZ section and is the kinematic reference frame. Figure 1b,c show some of the commonly developed shear structures in tectonically deformed rocks that are used to decipher the sense of shear (kinematics). The reader is referred to various published works for a comprehensive account of shear sense indicators [6, 11].

Since rocks in nature undergo deformation in a regional stress field, the structures that develop are commonly related to regional tectonics and hence regional kinematics. This necessitates that the field structures such as foliations/lineations mentioned above, must also be well-defined in the geographic reference frame. It is therefore the norm to collect oriented samples of deformed rocks in the field. This oriented sample collection is a very critical exercise in structural geology, particularly if the objective is to work out kinematics. An example of this is presented in Fig. 2, which shows field photographs of a mylonite from Kushtagi region of Dharwar Craton (southern India). The photographs showing the foliation (XY) and stretching lineation (X) are presented in the geographic reference frame (see caption of Fig. 2 for details). An oriented rock slice from the XZ section (Fig. 2c) of this field outcrop is shown in Fig. 3a. The geographic orientation (half-arrow pointing to the right marked 330°) in Fig. 3a helps relate the rock slice with the field outcrop in Fig. 2c. Several quartz clasts are marked (Q1 to Q5) on this rock slice and from the inclination of the quartz clast Q1 (top left of Fig. 3a), the sense of shear can be inferred to be top-towards-330°. In order to investigate the shear structures in detail, as well as to perform textural studies of various mineral phases in the rock, it is necessary to prepare oriented thin section from such rock slices. Care has to be taken by a researcher (scientist) as well as the sample-preparator to ensure that during the rock

Fig. 1 **a** Schematic block diagram highlighting the relation between foliation plane, stretching lineation and the three principal axes of the strain ellipsoid (X > Y > Z). XY represents the foliation plane. Stretching lineation that marks the X-direction of the strain ellipsoid lies on the foliation plane. The section parallel to the stretching lineation and perpendicular to the foliation is the XZ plane, which is the kinematic reference frame. All shear structures used by structural geologists to decipher the sense of shear and related kinematic analysis are investigated in this XZ section. Some commonly developed shear structures in tectonically deformed rocks are shear bands (**b**) and rotated porphyroclasts (**c**) (location: north of Angul, eastern India). The sense of shear is dextral and sinistral, respectively, in (**b**) and (**c**)

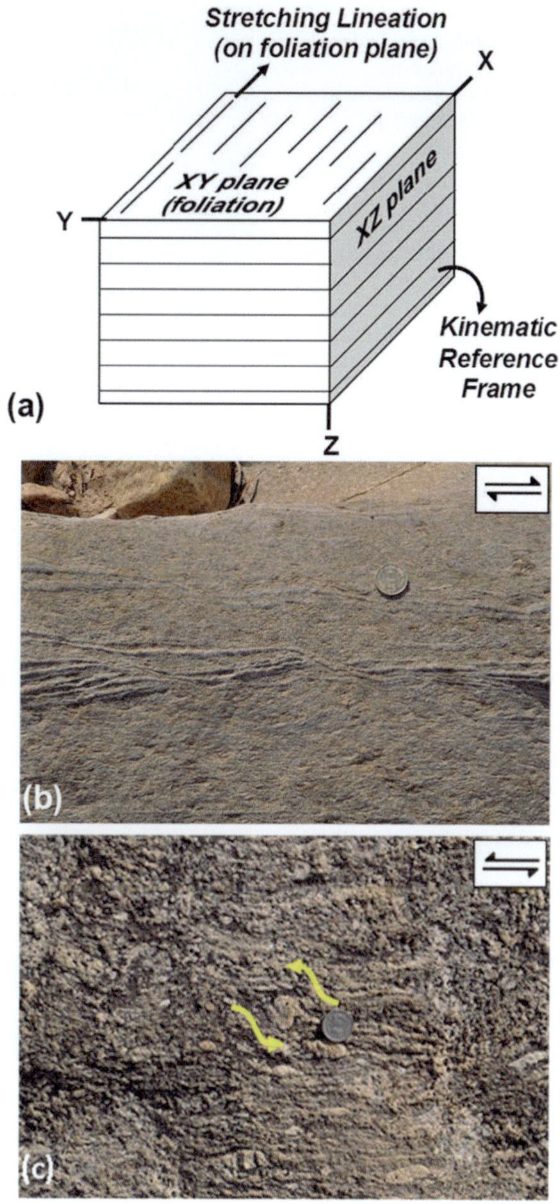

thin section preparation the geographic reference frame is well-marked on the slide. This allows investigation of the microstructures in the correct geographic orientation within the kinematic reference frame. This aspect is highlighted in Fig. 3b, which shows rock thin sections of the large rock slice in Fig. 3a. It maybe noted that the thin sections were prepared by first polishing and mounting the surface visible on

Fig. 2 Photographs documenting an example of identifying kinematic reference frame in the field and the importance of viewing it in the geographical reference frame. See text for discussion and Fig. 3 for further elaboration with regards to microstructures and kinematics in this rock. The field photographs are from a mylonite from the vicinity of Kushtagi region (Dharwar Craton, southern India). XY is the foliation plane that contains stretching lineation (thick horizontal white arrows in a,b). The oblique thin white arrows in **a** point to trace of the foliation on the XZ plane. The half arrow (black) in (**b**) and (**c**) points to the exact direction (i.e., 330°). Plan view of the field outcrop in (**c**) is the plane parallel to the stretching lineation and perpendicular to the foliation (i.e., XZ section), which is the kinematic reference frame. Oriented rock slice from (**c**) and its oriented rock thin section are shown in Fig. 3

rock slice (in Fig. 3a) onto the glass slide. As a consequence, the orientation of the sample gets inverted and the half-arrow marked 330° points to the left in Fig. 3b. The positions of quartz clasts (Q1 to Q5) can be compared between Fig. 3a,b, and it is clear that in this inverted geographic reference frame, the sense of shear in Fig. 3b is also top-towards-330° (similar to that in Fig. 3a).

Fig. 3 Documentation of the importance of maintaining full knowledge of the geographic reference frame along with the kinematic reference frame during the preparation of oriented rock thin sections for kinematic studies. **a** Oriented rock slice from the mylonite field outcrop in Fig. 2c. The rock slice is cut parallel to the XZ plane of the strain ellipsoid, which is the kinematic reference frame. The half-arrow marked 330° (pointing to the right) defines the geographic reference frame, which helps relate the rock slice with the field outcrop in Fig. 2c. Five quartz clasts (Q1 to Q5) are marked. The sense of shear in this rock slice is top-towards-330°, which is deciphered from the inclination of the quartz clast (Q1). **b** Oriented rock thin sections prepared from the oriented rock slice in (**a**). Since the latter was very large, three thin sections had to be made to cover the structure of the entire slice. The top surface in (**a**) was polished and mounted on the glass slide as a consequence of which the geographic reference frame is inverted. Hence the half-arrow marked 330° points to the left in the oriented rock thin section. Compare the positions of clasts Q1 to Q5 in (**b**) with those in (**a**). The inclination of the quartz clast Q1 in (**b**) indicates that the sense of shear in the thin section is top-towards-330°, which is same as in the rock slice (**a**)

The above description gives a detailed account of the importance of identifying the kinematic reference frame and relating it to geographic reference frame in kinematic analysis. This forms the foundation of any further studies on such thin sections using either the light microscope (for fabric quantification using petrography) or SEM-EBSD analysis (for determining deformation mechanisms and slip systems in minerals, kinematic and vorticity analyses, etc.). However, all deformed rocks do not always develop a visible foliation and/or lineation. Often, foliated rocks such as gneiss contain a well-defined foliation, but the foliation does not have a clearly developed stretching lineation (Fig. 4a). Deformed quartzites that lack any significant amount of other mineral phase (like micas) do not necessarily develop a field foliation and/or lineation despite being extensively strained (Fig. 4b). There are several other rocks such as granite that emplace because of regional deformation but do not develop visible foliation and/or lineation (Fig. 4c). In such rocks, it is challenging to

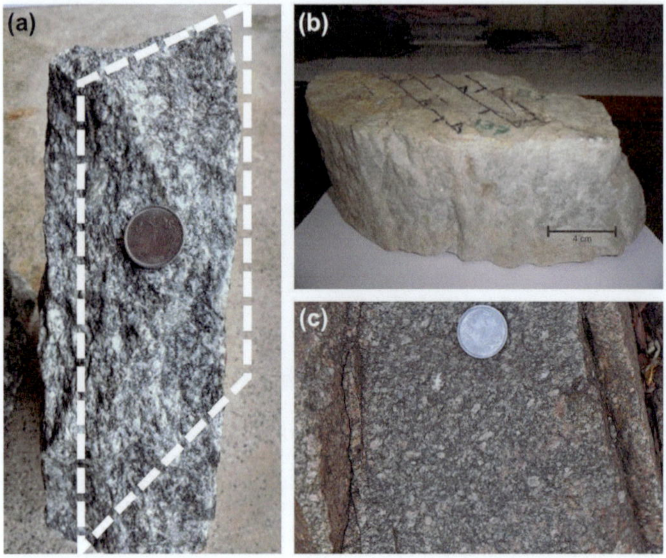

Fig. 4 Examples of deformed rocks that do not develop foliation and/or lineation. **a** Gneiss (Dharwar Craton, southern India) having a foliation (white dashed box), but no stretching lineation. **b** Quartzite (Angul region, eastern India) with no foliation and lineation. **c** Granite (Dharwar Craton, southern India) showing coarse phenocrysts of feldspar with no mesoscopically (visibly) decipherable foliation and lineation

determine the reference frame for kinematic analysis, and the geologist must rely on alternative methods to identify the same. The measurement of anisotropy of magnetic susceptibility (AMS) has been used for this purpose quite extensively in the past in structural geology [12–16]. AMS involves placing a sample (cylindrical or cubic) in an external magnetic field in different orientations and measuring the induced magnetization in each orientation. The magnetic induction takes place along the "*easy*" axis (crystallographic or shape) of the different minerals present in the sample [17]. Therefore, rocks with randomly oriented minerals are magnetically isotropic since they have the same magnetic susceptibility in different orientations. However, deformed rocks have a preferred orientation of minerals and hence possess an AMS. This helps identify the fabric in a rock, even if visible foliation and/or lineation is not well-developed. AMS is visualized as a magnetic susceptibility ellipsoid, a second-order tensor with three principal axes, $K_1 > K_2 > K_3$. The K_1K_2 plane is the magnetic foliation, K_3 is the pole to the magnetic foliation, and K_1 is the magnetic lineation; the reader is referred to Tarling et al. [17] for details about AMS and formulae of various parameters.

For the present study, it is important to note that the orientations of principal axes of the AMS ellipsoid ($K_1 > K_2 > K_3$) have been equated with those of the strain ellipsoid ($X > Y > Z$) as reported in several past studies [17–20]. In rocks that have visible foliation and stretching lineation (e.g., Fig. 5a–i), the magnetic foliation (K_1K_2 plane) is parallel to the visible foliation (XY), and the magnetic lineation (K_1) is parallel to the stretching lineation, thus defining the X-direction of the strain ellipsoid (Fig. 5a–ii). On the other hand, if a rock has foliation, but no stretching lineation (e.g., Fig. 5b–i), then the magnetic foliation has been known to have similar orientation as the visible field foliation (Fig. 5b–ii). Based on several similar studies, it is well-established that in tectonically deformed rocks, the orientations of the three principal axes of the strain ellipsoid ($X > Y > Z$) are, respectively, correlatable with the orientations of the three principal axes of the AMS ellipsoid ($K_1 > K_2 > K_3$). Thus, the K_1K_3 section of the AMS ellipsoid can be equated with XZ of the strain ellipsoid, which makes AMS a very useful tool in the identification of the kinematic reference frame for petrofabric studies even if the rock contains no visible foliation and/or stretching lineation [13–16, 21, 22]. As shown in Fig. 5c, the above forms the basis of defining the kinematic reference frame for EBSD studies. If the rock has a visible foliation and stretching lineations, then a rock thin section parallel to the XZ section of the strain ellipsoid can be prepared directly from field orientation data. In this situation, all EBSD data obtained in the XZ reference frame would be plotted on a lower hemisphere equal area projection (i.e., pole figure), where the horizontal diameter (dashed horizontal line in Fig. 5c-i) represents the field foliation (XY) trace. The stretching lineation (X) is located on this trace and lies on the circle (black dots in Fig. 5c-i). On the other hand, if a deformed rock does not have well-defined visible foliation and/or stretching lineation, then the thin section is prepared parallel to the K_1K_3 plane of the AMS ellipsoid. In this situation, all EBSD data obtained in the K_1K_3 reference frame would be plotted on a lower hemisphere equal area projection where the horizontal line (dashed in Fig. 5c–ii) represents the magnetic foliation (K_1K_2) trace. The magnetic lineation (K_1) is located on this trace and lies on the circle (squares in Fig. 5c–ii). In this write-up, the author will present examples from tectonically deformed rocks, where kinematic analysis has been done using EBSD data acquisition in thin section prepared parallel to the XZ or the K_1K_3 reference frames depending upon whether foliation and/stretching lineations were present or absent. However, before going to the examples, it is important for the reader to be aware of the challenges that lie in final polishing stage of sample preparation for EBSD study of rocks. This is described in the next section.

3 Sample Preparation for EBSD Analysis of Rocks

Since rocks are polymineralic and polycrystalline, they comprise minerals (phases) of different rheologies and strengths. For e.g., a micaceous quartzite contains quartz, which is a very hard mineral, as compared to mica, which is soft. Most EBSD studies of rocks are done on thin sections. Procedure for preparing a rock thin section for

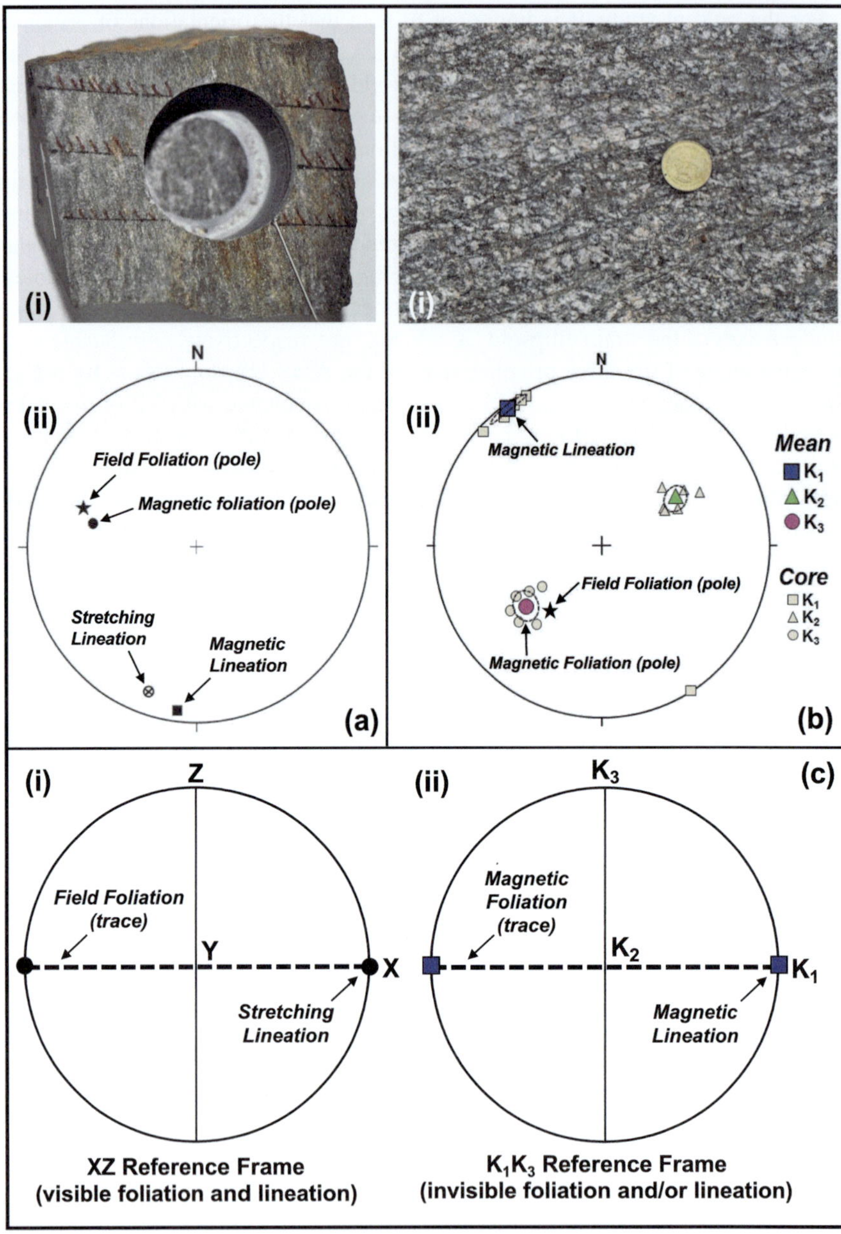
(i)
N
(ii)
Field Foliation (pole)
Magnetic foliation (pole)
Stretching Lineation
Magnetic Lineation
(a)
(i)
N
(ii)
Magnetic Lineation
Mean
K1
K2
K3
Field Foliation (pole)
Core
K1
K2
K3
Magnetic Foliation (pole)
(b)
(i)
Z
Field Foliation (trace)
Y
X
Stretching Lineation
XZ Reference Frame
(visible foliation and lineation)
(ii)
K3
(c)
Magnetic Foliation (trace)
K2
K1
Magnetic Lineation
K1K3 Reference Frame
(invisible foliation and/or lineation)

◀**Fig. 5** Diagrams to highlight importance of anisotropy of magnetic susceptibility (AMS) in establishing the kinematic reference frame for microstructural and SEM-EBSD analysis. **a** Block sample of quartzite (Naxos, Greece) containing foliation and stretching lineation (i), and lower hemisphere equal area projection of field and magnetic orientation data (ii). Note the similarity in the orientations of field and magnetic foliation, as well as stretching and magnetic lineation (after Renjith et al. [15]; Renjith [50]). **b** Field photograph of gneiss (Dharwar Craton) containing foliation but no stretching lineation (i) and lower hemisphere equal area projection of field and magnetic orientation data (ii) (after Mamtani et al., 2017). Note the similarity in the field and magnetic foliation. See text for detailed description. **c** Lower hemisphere equal area projections designating the kinematic reference frame based on field and/or AMS analysis. Fig. **c**-i shows XZ reference frame for rocks that contain visible field foliation as well as stretching lineation. The horizontal dashed line is the field foliation (XY) trace and the black dots on it lying on the circle represent the stretching lineation (X-direction). Fig. **c**-ii is the K_1K_3 reference frame for rocks that lack visible field foliation and/or stretching lineation, where the horizontal dashed line is the magnetic foliation (K_1K_2) trace. The squares lying on this trace lying on the circle represent the magnetic lineation (K_1)

routine petrography and SEM/EPMA studies is well-established. It involves mechanical polishing of rock slices mounted on a glass slide with various grades of different abrasives, with the final polishing being done with diamond paste. However, for SEM-EBSD analysis, further step of polishing is required. This is because the sample (thin section) surface must be damage free to yield very good EBSD patterns that can be well-indexed. This super-polished damage free surface can be obtained by further mechanical polishing of the rock thin section with colloidal silica solution. Depending upon the mineral phase to be indexed, the timing of this polishing stage may vary. For e.g., in the authors experience, for obtaining good EBSD patterns of quartz, 6 h of polishing with colloidal silica is desirable, while for a magnetite the time is as high as 10 h [23].

The high polishing time mentioned above, also implies that the softer mineral phases tend to abrade. As a consequence, it is challenging to obtain good EBSD data from minerals such as micas and amphiboles, which are soft. This challenge has been circumvented by the application of Broad Ion Beam Polishing (BIBP) [24]. Recently, the author and his research group have also been involved in establishing a *modus operandi* referred to as Mamtani's Fabric Analysis Lab (MFAL) protocol for BIBP of rock thin sections using the Leica EM TIC 3X triple broad ion beam milling system [25]. Accordingly, the final stage of mechanical polishing with colloidal silica is reduced to 30 min, and this is followed by a total of 35 min of BIBP. For BIBP, the tilt angle of the thin section (degree) with reference to ion beam (Argon), acceleration voltage (kV) and source current (mA) can be controlled along with lateral and rotary movements. The MFAL protocol involves two steps—in step-1, cleaning of the rock thin section surface is done for 5 min (1.5°/4 kV/1.5 mA). In step-2, polishing is done for 30 min (1.5°/4 kV/3 mA). In the authors experience, these conditions provide very good EBSD patterns for soft as well as hard mineral phases in rocks [25, 26].

4 SEM-EBSD Data from Tectonically Deformed Rocks and Kinematic Analysis—Some Examples

With regards to kinematic analysis in structural geology using crystallographic data, quartz is the most common mineral that has been studied by geoscientists. This is because (a) the slip systems in quartz and their activation with temperature are very well established and documented (b) quartz is one of the most commonly present minerals in rocks and (c) microstructures and recrystallization mechanisms in quartz and their relation with temperature are also well-calibrated [6, 27, 28]. Even prior to the advent of fully automated SEM-EBSD data acquisition, geoscientists commonly used the Universal stage (U-stage) to measure the quartz *c*-axis orientations [29]. It is well-established that if a rock undergoes non-coaxial progressive deformation, then different slip systems in quartz get activated. The most common, with increasing temperature, are basal < a >, rhomb < a >, prism < a > and prism[c] slip. Figure 6a is a schematic pole figure (lower hemisphere equal area projection) representing the areas where quartz *c*-axes generally plot when the different slip systems are activated. In tectonically deformed rocks that have undergone shearing by non-coaxial flow, the distribution of quartz *c*-axes on the pole figures is asymmetric (Fig. 6b), which has been classically used to decipher the sense of shear [6]. It maybe noted that the reference frame of the pole figures in Fig. 6 is XZ of the strain ellipsoid. Hence, Y-axis maxima are typical of medium temperature deformation (prism < a > slip), while X-axis maxima are typical of high-temperature deformation (prism[c] slip). The quartz pole figures in Fig. 6b form the basis of most interpretations that are done using quartz crystallographic preferred orientation (CPO) data acquired from SEM-EBSD analysis. The author would now like to present a few examples here where SEM-EBSD data have been acquired in the XZ reference frame (for rocks containing visible foliation and stretching lineation) or the K_1K_3 reference frame (for rocks lacking visible foliation and/or stretching lineation). Data from each example can be compared with the classical quartz pole figures (Fig. 6b) to decipher kinematics as well as slip systems. In addition, examples are also discussed where the microscale kinematics provides information about the regional tectonics and kinematics. It may be noted that all the SEM-EBSD data presented here have been acquired using the Carl Zeiss Auriga Compact FEG-SEM fitted with NordlysMax2 EBSD detector (Oxford Instruments, UK) housed in the Central Research Facility (CRF, IIT Kharagpur, India), where the data acquisition was done using Aztec software (Oxford Instruments, UK). The post-acquisition data processing involving preparation of pole figures (lower hemisphere equal area) and inverse pole figure (IPF) maps was done with HKL Channel 5 software (Oxford Instruments, UK) in the authors' Fabric Analysis Lab (FAL) housed in the Department of Geology and Geophysics (IIT Kharagpur, India).

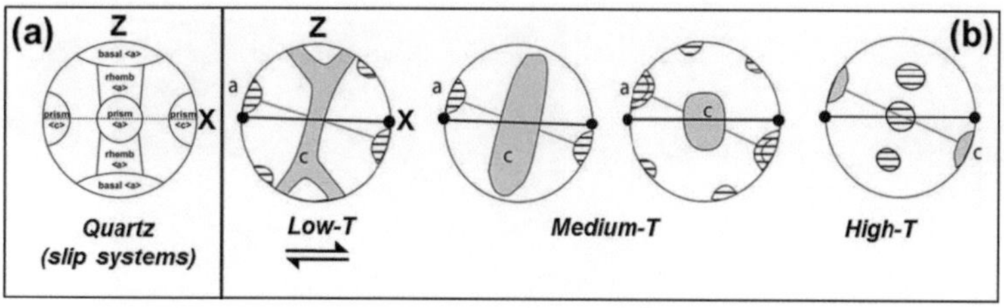

Fig. 6 **a** Schematic pole figure (lower hemisphere equal area projection) representing common slip systems in quartz (after Schmid and Casey [27]) in XZ reference frame. **b** Schematic contoured pole figures of quartz crystallographic preferred orientation that commonly develop in deformed metamorphosed rocks with increase in temperature under dextral non-coaxial progressive deformation; the *c*-axes are in gray, while the a-axes are striped; the representation is in the XZ reference frame (modified after Passchier and Trouw [6]). The distribution of quartz *c*- and *a*-axes in pole figures obtained from SEM-EBSD data form the basis of deciphering kinematics in tectonically deformed rocks. See text for details

The graphite bearing quartzite shown in Fig. 5a has visible foliation and stretching lineation, as a consequence of which the XZ reference frame is decipherable from these visible structural elements. The texture of the rock in a rock thin section prepared parallel to the XZ section and its EBSD data are presented in Fig. 7. The thin section shows quartz grains that are elongated and define a shape preferred orientation (SPO) (Fig. 7a); this is also obvious in the IPF map of the sample (Fig. 7b). The SPO is parallel to the trace of the foliation in the sample, which is marked by the white line in the pole figures of quartz *c*- and *a*- axes (Fig. 7c). The quartz *c*-axes form an asymmetrical crossed girdle that is indicative of dominantly basal < a > slip and typical of dextral sense of shear (compare with Fig. 6). In contrast to the graphite bearing quartzite (Fig. 5a), the quartzite in Fig. 4b lacks any visible foliation and lineation. As discussed above, the slip systems of quartz and the sense of shear using quartz CPO must be investigated in the XZ reference frame, which in rocks such as the one in Fig. 4b needs to be deciphered from AMS analysis; the K_1K_3 plane defines kinematic reference frame in such rocks, and the quartz CPO must be investigated in thin sections prepared parallel to it. Such a case is presented in Fig. 8, where the quartzite rock thin section was prepared parallel to the K_1K_3 plane. The photomicrograph of the quartzite (Fig. 8a) shows dynamically recrystallized quartz grains that are elongated and clearly indicative of the fact that the rock is deformed. However, in the hand specimen this rock did not show any foliation/lineation (like Fig. 4b). Since the SEM-EBSD data acquisition of quartz was done in the K_1K_3 reference frame (Fig. 8b,c), the horizontal white line in the pole figures (Fig. 8c) represents the trace of the magnetic foliation (compare with Fig. 5c-ii). The quartz *c*-axes form a maxima around K_1 (Fig. 8c), which is equated with X of the strain ellipsoid (as discussed above). Hence, in the K_1K_3 reference frame, which is equivalent of XZ section, the pole figures of quartz *c*- and *a*-axes are typical of dextral sense of shear by dominantly prism[c] slip (compare with Fig. 6).

In structural geology, the kinematics that is worked out at the microscale must often be related to the geographic reference frame and further be related to larger scale

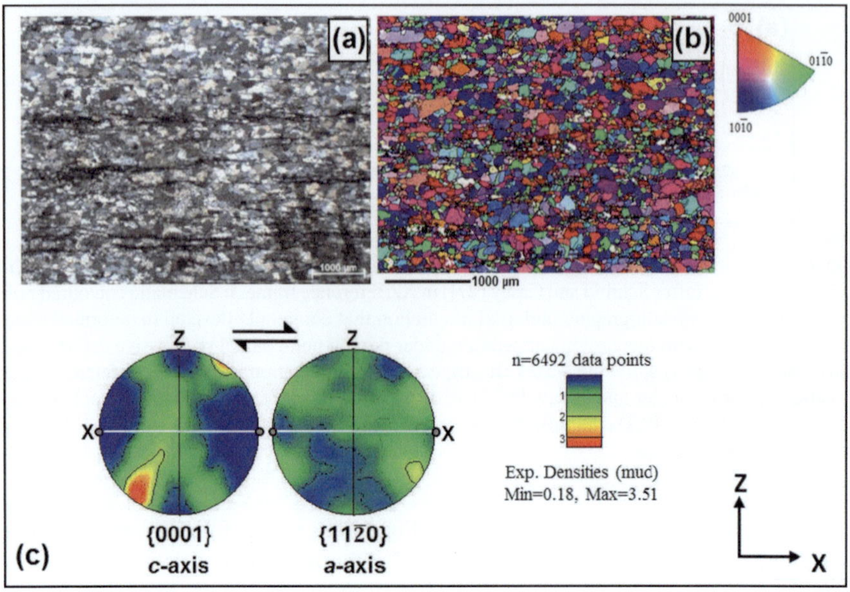

Fig. 7 Kinematic analysis from EBSD data in graphite bearing quartzite from Naxos (Greece) (modified after Renjith, 2018). The sample had visible foliation and a stretching lineation (Fig. 5a) and thus the rock thin section for microstructural investigation is XZ section. **a** Photomicrograph of the region of interest (ROI) subjected to quartz EBSD data acquisition. **b** IPF map of quartz in the ROI. **c** Pole figures (lower hemisphere equal area) of quartz *c*- and a- axes in the XZ reference frame. The white horizontal line is the trace of the XY (foliation) plane and the stretching lineation (X) lies on the circle (shown by the gray dot). The quartz *c*-axis form a typical asymmetrical crossed girdle that indicates a dextral sense of shear, and the slip system is dominantly basal < a > slip (compare with Fig. 6b)

deformation, kinematics and/or tectonics. One such example is presented from the tonalite trondhjemite granodiorite (TTG) gneiss (Dharwar Craton, southern India; Fig. 9). This gneiss has a well-defined foliation (gneissosity) that is visible and measurable in the field, but it did not have a stretching lineation (Fig. 4a and 5b show this gneiss). Hence, in order to identify the XZ equivalent reference frame for kinematic studies, AMS analysis was done and the K_1K_3 section was identified. Subsequently, a thin section parallel to the K_1K_3 plane was prepared and the microstructural as well as EBSD analysis were done in this reference frame (Fig. 9a). The geographic reference frame of the thin section was well-defined and the top left of the photomicrograph (Fig. 9a-i) points towards 325°. The distribution of quartz *c*-axis in this K_1K_3 reference frame is asymmetrical that is indicative of sinistral shearing. In the geographic reference frame, the sense of shear is top-towards-325°. Mondal et al. [30] have discussed the paleostress and kinematics of this part of the Dharwar Craton in detail, and it has been envisaged that the rocks in this region underwent shearing as shown in Fig. 9b. The gneiss sample whose data have been presented here (Fig. 9a) comes from the vicinity of a shear zone to the south of

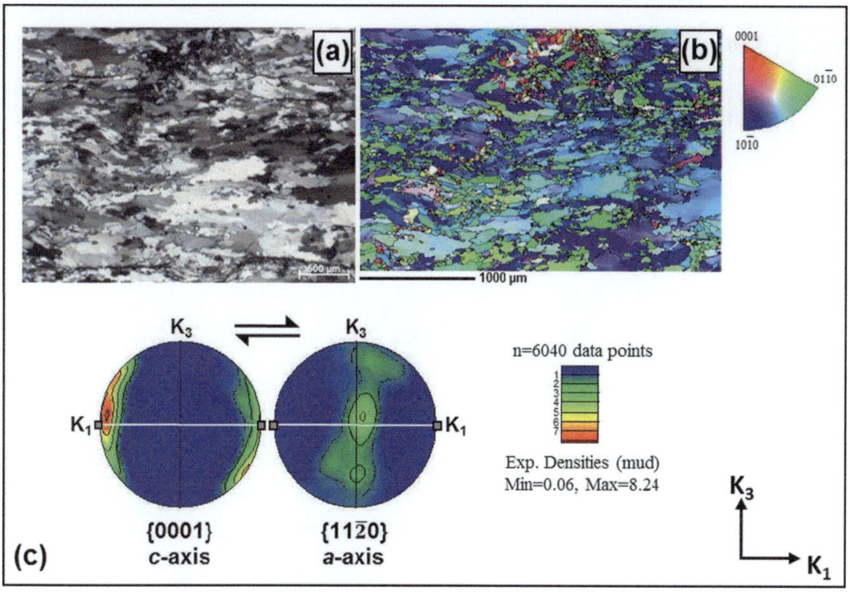

Fig. 8 Kinematic analysis from EBSD data in quartzite from Angul region (eastern India) (modified after Renjith [50]). The sample had no visible foliation and lineation (like in Fig. 4b). The rock thin section for microstructural investigation was prepared based on AMS data and is a K_1K_3 section. **a** Photomicrograph of the region of interest (ROI) subjected to quartz EBSD data acquisition. **b** IPF map of quartz in the ROI. **c** Pole figures (lower hemisphere equal area) of quartz *c*- and *a*- axes in the K_1K_3 reference frame. The white horizontal line in is the trace of the K_1K_2 (magnetic foliation) plane and the magnetic lineation (K_1) lies on the circle (shown by the gray square). The quartz *c*-axis form maxima at K_1, which is equivalent to X of the strain ellipsoid (see text for discussion). The sense of shear is dextral and the slip system is inferred to be dominantly prism[c] slip (compare with Fig. 6)

Mulgund town (Fig. 9b), and the microscale kinematics deciphered from EBSD data of quartz complements the regional tectonics related to D3 (third) deformation in this part of the Dharwar Craton.

Another example where kinematics deciphered from EBSD can provide unique information about the larger scale deformation comes from an oriented quartz vein sample collected from Hutti region of Dharwar Craton (southern India). This is discussed through Figs. 10and 11. Quartz veins are a common occurrence in naturally deformed rocks. Structures such as foliations, fractures, and joints developed during tectonic deformation provide pathways through which mineral bearing hydrothermal fluids can percolate and eventually some of these structures undergo dilation to allow precipitation of minerals. Fluid pressure fluctuation and related processes are involved during this phenomenon that also play a critical role in deposition of economically important minerals (ores). Hence, the kinematics of vein formation and its relation to regional deformation is important in structural geology [16, 31–34]. The Hutti region in Dharwar Craton (southern India) is a known province of

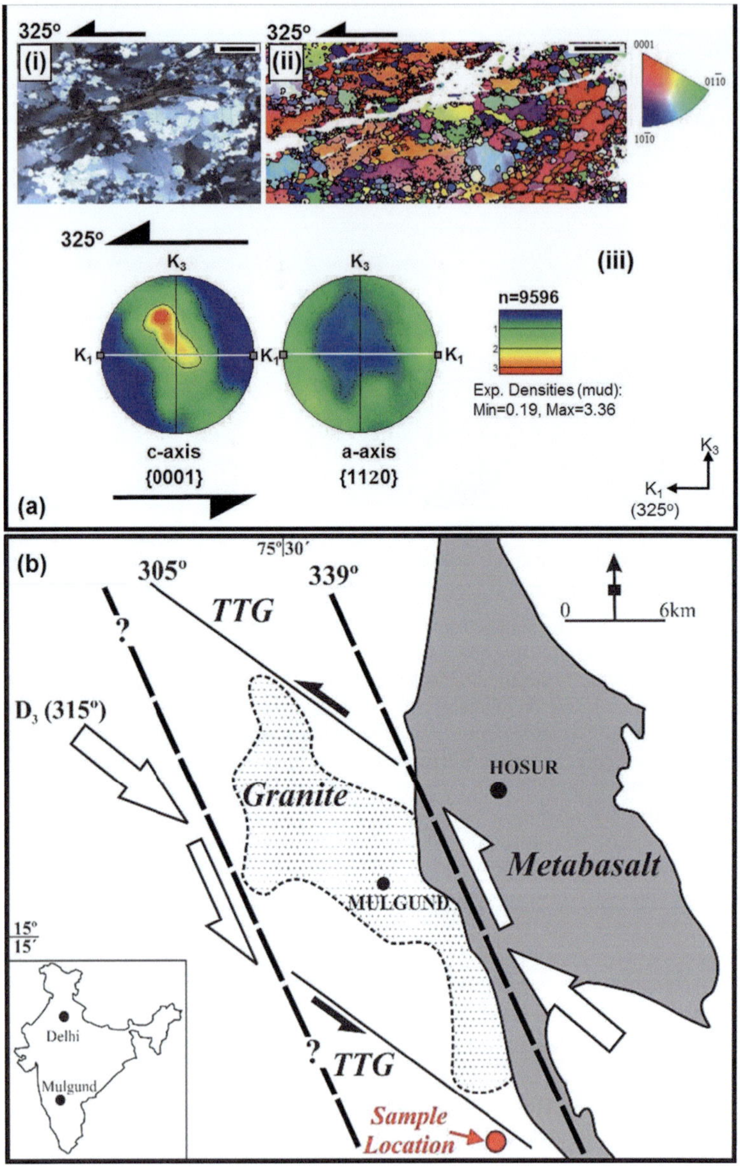

325°
(i)
325°
(ii)
0001
01$\bar{1}$0
10$\bar{1}$0
325°
K3
K3
(iii)
n=9596
K1
K1
K1
K1
Exp. Densities (mud):
Min=0.19, Max=3.36
c-axis
{0001}
a-axis
{1120}
K3
K1
(325°)
(a)
(b)
75°|30′
305°
339°
TTG
0
6km
D3 (315°)
HOSUR
Granite
Metabasalt
MULGUND
15°
15′
Delhi
Mulgund
TTG
Sample
Location

◂**Fig. 9** Kinematic analysis from EBSD data in gneiss from Dharwar Craton (southern India) (modified after Mamtani, et al. [14]). The sample had a gneissosity (visible foliation) but no stretching lineation (Figs. 4a and 5b–i). The rock thin section for microstructural investigation was prepared based on AMS data and is a K_1K_3 section; the orientation of the sample in geographic reference frame is also shown. **a**-i Photomicrograph of the region of interest (ROI) subjected to quartz EBSD data acquisition. (**a**-ii) IPF map of quartz in the ROI. Scale bar in (i) and (ii) is 500 μm. (**a**-iii) Pole figures (lower hemisphere equal area) of quartz *c*- and *a*- axes in the K_1K_3 reference frame. The white horizontal line is the trace of the K_1K_2 (magnetic foliation) plane and the magnetic lineation (K_1) lies on the circle (shown by the gray square). The slip system is dominantly a combination of rhomb < a > and prism < a > slip (compare with Fig. 6a), and the sense of shear in the geographical reference frame is top-towards 325°. The sample location of this gneiss, which is geologically referred to as tonalite trondjemite granodiorite gneiss (TTG) is shown on a generalized geological map in **b** (modified after Mondal and Mamtani [30]). The sense of shear deciphered from quartz CPO data (based on EBSD data) complements the top-to-NW regional kinematics in the Mulgund region of the Dharwar Craton (southern India) that is related to regional third (D3) deformation

gold mineralization, which occurs in quartz veins. One such occurrence of quartz veins (along with calcite) in host-rock (metavolcanic) exposed on an underground mine face is shown in Fig. 10a along with its orientation in the geographical reference frame. Foliation and/or lineation are not well-decipherable from the exposed surface. Goswami et al. [16] carried extensive AMS investigations on the host-rock metavolcanics and established that the exposed face is sub-parallel to the K_1K_3 section of the AMS ellipsoid; this reference frame is marked on Fig. 10a. An oriented quartz-calcite vein sample (Fig. 10b) was taken from the exposed face (from the boxed area in Fig. 10a) and its thin section was prepared for microstructural studies. Due process described earlier (Fig. 3) was followed to ensure that the geographical reference frame is well-recognizable in the thin section in which the orientations get inverted (see caption of Fig. 10b,c). Under the petrographic microscope, the quartz veins occur either as lensoidal pods (marked extensional quartz vein in Fig. 10c) or as laminated veins; the latter contain elongated dynamically recrystallized quartz grains that mark the trace of the magnetic foliation (K_1K_2) plane. Quartz CPO from EBSD analysis of the lensoidal extensional quartz vein (Fig. 10c and 11a) in the K_1K_3 reference frame is presented in Fig. 11b. It maybe noted that the K_1K_2 trace in the present case is vertical shown by white line on the pole figures (Fig. 11b) and the magnetic lineation lies on the circle (gray boxes marked K_1 in Fig. 11b). The quartz *c*-axis form maxima around K_1, which is equivalent to X-direction of the strain ellipsoid. This implies quartz deformed dominantly by prism[c] slip and the sense of shear is dextral. In the kinematic evolution of extensional veins under simple shear, it is well known that the maximum instantaneous axis (ISA_{max}) is oriented 45° to the flow plane and the extensional veins form perpendicular to ISA_{max} [6, 35]. Accordingly, the kinematics of formation of extensional vein in Fig. 11a can be inferred to have formed perpendicular to ISA_{max} under dextral simple shear in the thin section (Fig. 11c). It was mentioned earlier (Fig. 10) that from collecting the oriented vein sample from the underground mine face to the preparation of rock thin section, the orientation gets inverted. Hence, to relate the kinematics from the rock thin section to the underground mine face, the orientation must be reverted back to the original. This is documented in Fig. 11d and it is established that on the underground mine

Fig. 10 **a** Quartz-calcite vein in underground mine in Hutti. The orientation of the exposed face in the geographic reference frame is given at the bottom left; the exposed face is sub-parallel to K_1K_3 plane of AMS ellipsoid (Goswami et al. [16]). The dashed box shows the area from where the oriented vein sample was collected, which is shown in **b**. Arrows O and P are marked to keep a track of the orientation in geographical reference frame during thin section preparation. Oriented thin section from a part of this vein was prepared for microstructural and EBSD studies in K_1K_3 reference frame. **c** Photomicrograph of the oriented thin section of the vein. Note the inverted position of arrows O and P, and also the geographic orientation, as compared to **b**. An asymmetrical lensoidal shaped extensional quartz vein as well as laminated quartz vein are marked in **c**. Figure is modified after Goswami et al. [16]. EBSD data of the extensional quartz vein and its kinematics are presented in Fig. 11

face that contains the vein, the sense of shear was sinistral down-dip. This fits well with the geometry of the quartz veins on the face of the wall that are expected to form when the vein opening develops perpendicular to ISA_{max} (Fig. 11e).

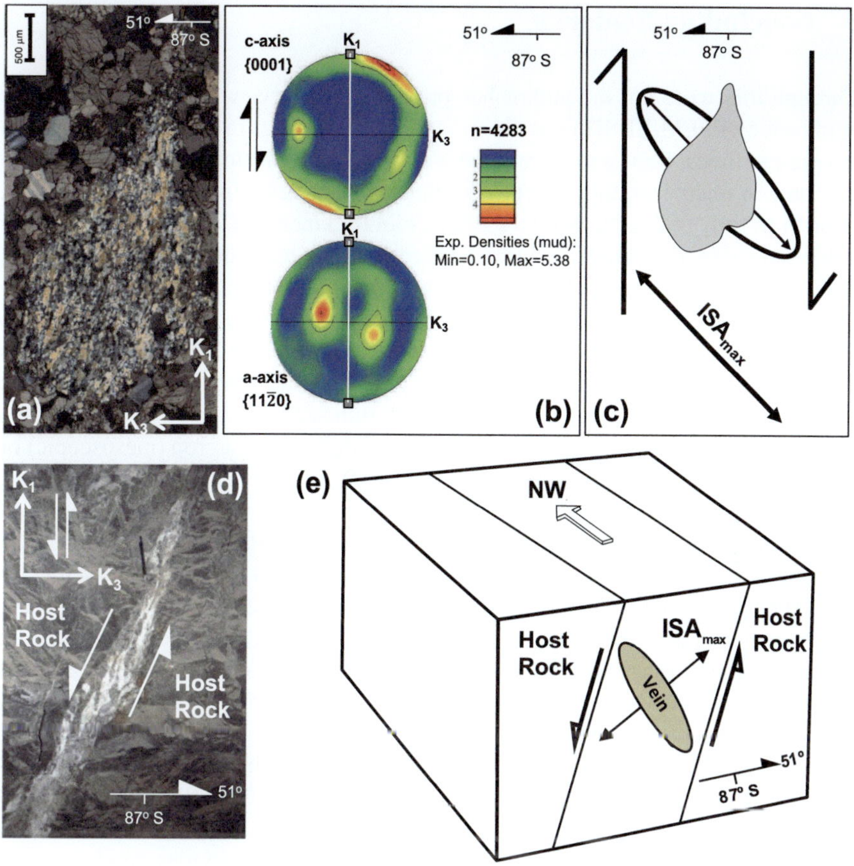

Fig. 11 **a** Photomicrograph of the lensoidal shaped extensional quartz vein in thin section (Fig. 10c) prepared parallel to the K_1K_3 reference frame of AMS ellipsoid. Quartz EBSD data were acquired from this vein. **b** Pole figures of quartz *c*- and *a*- axes from the vein in K_1K_3 reference frame. The white vertical line represents the trace of the magnetic foliation (K_1K_2) plane. Magnetic lineation, K_1 on this trace lies on the circle (gray squares). The quartz CPO is typical of prism[c] slip and indicates dextral sense of shear. **c** Schematic diagram documenting the kinematics involved in development of the extensional vein in the thin section under the dextral simple shear. The space to allow formation of the vein develops perpendicular to the maximum instantaneous stretching axis (ISA_{max}). Kinematic interpretation of shearing and formation of vein on the actual wall-face in underground mine is highlighted in (**d** and **e**). Note that the geographic orientation on the wall-face gets inverted back (as it was in Fig. 10a), and the sense of shear becomes sinistral. The shear sense fits with down-dip movement along the contact between vein and host-rock (metavolcanic). Figure is modified after Goswami et al. [16]

5 Concluding Remarks

Through this write-up, the author has provided an overview to the reader on the application of SEM-EBSD in performing kinematic studies in structural geology. In the present times, SEM's that are fitted with an EBSD detector are quite commonly available in many institutes and/or geoscience departments in the world. However, in order to fully utilize the potential of this tool for kinematic studies of tectonically deformed rocks, it is critical to have detailed information of the sample orientation, both, with respect to the geographic north, as well as the strain and/or AMS ellipsoid. EBSD data must be acquired in an oriented thin section prepared parallel to the kinematic reference frame, and in the present article, the author has shared his experience in this regard. It is only with very careful understanding of the reference frame, can EBSD data from an oriented rock thin section be used to decipher the sense of shear (kinematics) and then extrapolate this to the outcrop (mesoscopic) or regional scale. The examples discussed in Figs. 9, 10 and 11 document the importance of the above statement. Apart from deciphering the sense of shear (discussed in this article), quartz CPO can also be used to quantify vorticity [36]. Moreover, the opening angle of quartz *c*-axes (on pole figures) has been used to gauge temperature. The opening angle increases with increasing temperature (metamorphic grade), which corresponds to transition from dominant basal < a >, rhomb < a > and prism < a > slip to prism[c] slip; this also, respectively, corresponds to transition from crossed girdle *c*-axis fabrics to point maximum fabrics that are oriented at low angle to X (stretching lineation) as shown in Fig. 6b [37, 38].

SEM-EBSD studies have also been extremely useful in deciphering the deformation processes and mechanisms of minerals such as calcite [39], K-feldspar [40], garnet [41], gypsum [42], magnetite [23, 25], mica [43], olivine [44], amphibole [45, 46], and omphacite [26]. IPF map of a particular mineral provides good evidence to infer intracrystalline deformation in a mineral. In addition IPF maps also help identify region of interest (ROI) for preparing misorientation profiles in order to identify and/or classify subgrain boundaries as low-angle grain boundaries or high-angle grain boundaries; this has been done extensively using IPF map of magnetite by Mamtani et al. [25]. As shown in Fig. 12a (modified after Mamtani et al. [25]), the magnetite IPF map has been placed in 3D perspective of the kinematic (XZ) reference frame; the geographic orientation is also shown. This magnetite was hosted in a mylonite schist sample (Slipsiken region, Sweden) that has foliation as well as stretching lineation, and the rock thin section was prepared parallel to the XZ plane; this allows evaluation of different grain boundaries and microstructure in the kinematic reference frame using the IPF map. An important development in recent years has been the use of IPF maps in identifying ROI for excavating thin films (lamellae) using focused ion beam (FIB) for the investigation of nanostructures under TEM. It is the norm to extract lamellae normal to the plane of view of the sample using FIB [47]. Important to note here is that even if the rock thin section is parallel to the kinematic reference frame (such as in Fig. 12a), the thin film (lamella) extracted would be "*Plane Normal*" (PN), and hence not parallel to the kinematic reference frame. Such "*Plane Normal*" (PN) FIB lamellae were extracted by Mamtani et al. [25] from the magnetite grain (Fig. 12a) and nanostructures observed under TEM were documented. In order to better relate nanostructures with the strain ellipsoid and put them into a larger scale perspective, it is necessary to extract "*In-Plane*" (IP) FIB lamella from the kinematic reference frame, i.e., extract the thin film parallel to the plane of view (XZ section). Recently, Mamtani et al. [48] have carried out IP-FIB lamella extraction from magnetite hosted in a rock thin section that was prepared parallel to the kinematic reference frame; the method described by Li et al. [49] was used for this IP lamella extraction. For easy visualization of the difference between PN and IP lamella extraction, the author has used the analogy of slicing cheese with knife; in Figs. 12b,c, this is referred to as "*Cheese & Knife Example*", where a hypothetical reference frame (X > Y > Z of strain ellipsoid) is marked. PN slicing of the cheese block with a knife is done in Fig. 12b, which is analogous to extracting a "*Plane Normal*" (PN) FIB lamella from XZ reference frame of a thin section in Fig. 12d. On the other hand, IP slicing of the cheese block with a knife is done in Fig. 12c, which is analogous to extracting an "*In-Plane*" (IP) FIB lamella from XZ reference frame of a thin section in Fig. 12e. Mamtani et al. [48] have indicated that TEM studies from such IP lamella extracted parallel to the kinematic reference frame help in understanding nanotectonics and nanokinematics (also see [51]).

Grain size, shape and orientation of long axis of the grain shape are important in several structural geological investigations, e.g., for identifying grain size and shape orientation variations between samples in a geological terrain and their relation to strain. Collecting these statistical data is challenging from light microscopy and routine petrography because several grains of a particular mineral (e.g., quartz or micas) often form clusters and grain boundaries cannot be distinguished. However, EBSD analysis uses crystallographic data to map grain boundaries, and helps generate grain boundary map of a particular mineral phase, along with automatic calculation of grain size, shape (aspect ratio) and orientation of long grain shape axis. This provides an extremely useful tool for fabric quantification to structural geologists [50].

The author concludes by stating that in the coming decades, SEM-EBSD analysis of tectonically deformed rocks will be even more in vogue than it is today, and will be increasingly used in large number of structural geological investigations. In addition, an integration of EBSD data with those from other techniques viz. AMS and TEM will become more prevalent in the near future. And the author hopes that this write-up will provide a thrust to geoscientists interested in applying SEM-EBSD to their research work.

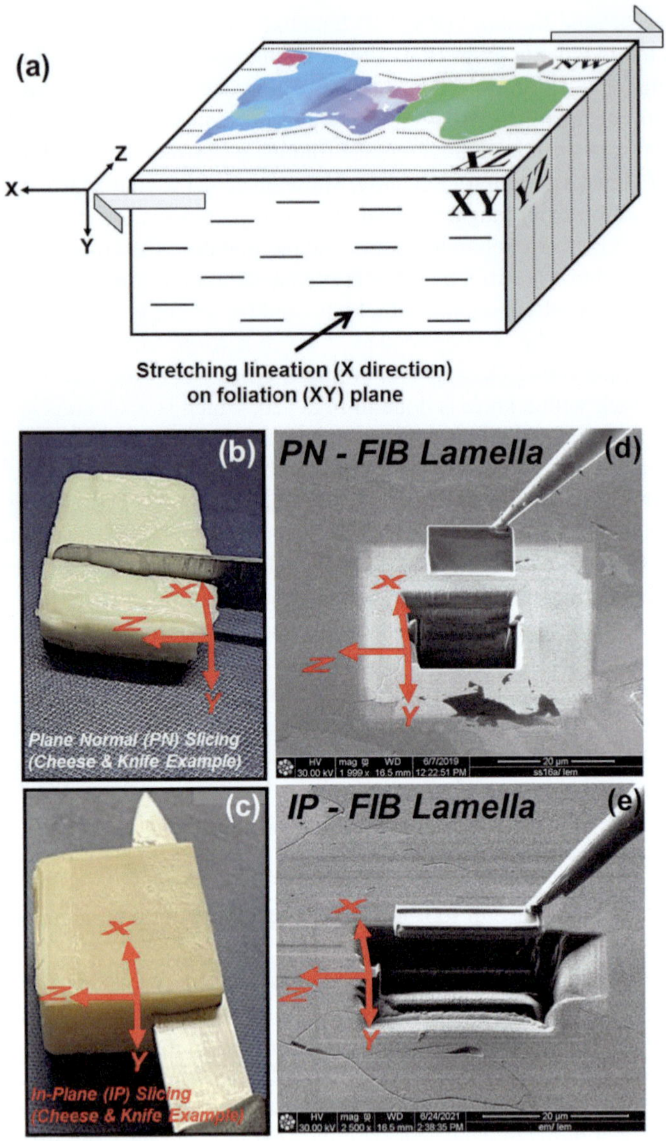
(a)
NW
XZ
XY
YZ
X
Y
Z
Stretching lineation (X direction)
on foliation (XY) plane
(b)
Plane Normal (PN) Slicing
(Cheese & Knife Example)
(c)
In-Plane (IP) Slicing
(Cheese & Knife Example)
(d)
PN - FIB Lamella
(e)
IP - FIB Lamella

◂**Fig. 12** **a** IPF map of magnetite prepared from EBSD data collected in XZ section of strain ellipsoid (mylonitized schist, Slipsiken region, Sweden). To get a 3D perspective of the IPF map in the kinematic and geographical reference frames, it has been placed in a schematic block diagram (modified after Mamtani et al. [25]). Such IPF maps guide in identification of region of interest for extraction of thin film/lamella for TEM studies. See text for discussion. **b–d** Documentation of the philosophy involved in the extraction of "*Plane Normal*" (PN) and "*In-Plane*" (IP) lamella (thin film) from the kinematic reference frame for TEM studies using the "*Cheese & Knife Example*". X, Y, Z are analogous to the three principal axes of the strain ellipsoid (X > Y > Z) in (**b**) and (**c**); the plan view represents the XZ plane (kinematic reference frame). Slicing of cheese with knife is PN in (**b**) and IP in (**c**). TEM sample extraction using Focused Ion Beam (FIB) is PN in (**d**), and IP in (e). The process involved in excavation of lamella in (**d**) and (**e**) is analogous to the slicing documented in (**b**) and (**c**), respectively, through *"Cheese & Knife Example"*

Acknowledgements The author is grateful to Satyam Suwas for inviting and motivating him to contribute this article. Thanks are due to Indian Institute of Technology (IIT) Kharagpur (India) for providing all logistic facilities and work environment to setup the Fabric Analysis Lab (FAL) in the Department of Geology and Geophysics. EBSD data presented here have been exclusively generated using the SEM housed in Central Research Facility (IIT Kharagpur), and the author is grateful to Niloy Bhowmik for his help with data acquisition. Special thanks to A.R. Renjith and Shalini Goswami, who extensively applied EBSD in their doctoral research while working in the FAL; their work has helped enhance the knowledgebase shared in this article. The Ministry of Earth Sciences (MoES) is thanked for providing funds to establish the Broad Ion Beam Polishing facility for EBSD sample preparation in FAL (IIT Kharagpur) through project MoES/P.O.(Geo)/169/2017. SEM images of magnetite presented in this write-up were acquired at the Laboratory for Electron Microscopy (Karlsruhe Institute of Technology, Germany) during the authors research stays in Karlsruhe that were funded by the DAAD (in 2019) and the Alexander von Humboldt (AvH) Foundation (in 2021); Erich Müller is thanked for the imaging. Discussions during these visits with Boris Reznik, Olivia Wenzel, Agnes Kontny, Christoph Hilgers, Erich Müller, and Heike Störmer on nanostructures are acknowledged, although the author takes full responsibility of the interpretations and terminologies used here in this regard. Finally, the author remains ever grateful to Sandra Piazolo, who introduced him to the wonderful world of SEM-EBSD (in the year 2009) through a collaboration made possible by funding from the AvH foundation, and the motivation, as well as freedom to explore new avenues, provided by R.O. Greiling.

References

1. Prior DJ, Mariani E, Wheeler J (2009) EBSD in the earth sciences: applications, common practice, and challenges. In Schwartz AJ Kumar M, Adams BL, Field DP (eds) Electron backscatter diffraction in materials science, Springer, pp 345–360
2. Lloyd GE, Hall MG (1981) Application of scanning electron-microscopy to the study of deformed rocks. Tectonophy 78:687–698
3. Prior DJ, Boyle AP, Brenker F, Cheadle MC, Day A, Lopez G, Peruzzo L, Potts GJ, Reddy S, Spiess R, Timms NE, Trimby P, Wheeler J, Zetterström L (1999) The application of electron backscatter diffraction and orientation contrast imaging in the SEM to textural problems in rocks. Amer Miner 84:1741–1759
4. Prior DJ, Mariani E, Wheeler J (2009) EBSD in the earth sciences: applications, common practice, and challenges, In Schwartz AJ, Kumar M, Adams BL, Field DP (eds) Electron backscatter diffraction in materials science, Springer, pp 345–360
5. Hobbs BE, Means WD, Williams PF (1976) An Outline of Structural Geology. John Wiley, New York

6. Passchier CW, Trouw RAJ (2005) Microtectonics. Springer, Berlin
7. Bouchez JL, Nicolas A (2021) Principles of rock deformation and tectonics. Oxford University Press, UK
8. Ghosh SK (1982) The problem of shearing along axial plane foliations. J Struc Geol 4:63–67
9. Passchier CW, Myers JS, Kröner A (1990) Field geology of high-grade gneiss terrains. Springer-Verlag, Berlin
10. Sander B (1970) An introduction to the study of fabrics of geological bodies, Pergamon Press, p 641
11. Hanmer S, Passchier CW (1991) Shear sense indicators: a review. Geolo Surv Canada 90:71
12. Mamtani MA, Pal T, Greiling RO (2013) Kinematic analysis using AMS data from a deformed granitoid. J Struc Geol 50:119–132
13. Mamtani MA (2014) Magnetic fabric as a vorticity gauge in syntectonically deformed granitic rocks. Tectonophy 629:189–196
14. Mamtani MA, Abhijith V, Lahiri S, Rana V, Bhatt S, Goswami S, Renjith AR (2017) Determining the reference frame for kinematic analysis in S-tectonites using AMS. J Geolo Soc Ind 90:5–8
15. Renjith AR, Mamtani MA, Urai JL (2016) Fabric analysis of quartzites with negative magnetic susceptibility–does AMS provide information of SPO or CPO of quartz? J Struc Geo 82:48–59
16. Goswami S, Mamtani MA, Virendra R (2018) Quartz CPO and kinematic analysis in deformed rocks devoid of visible stretching lineations: an integrated AMS and EBSD investigation. J Struc Geo 115:270–283
17. Tarling DH, Hrouda F (1993) The Magnetic Anisotropy of Rocks. Chapman and Hall, London, p 217p
18. Borradaile GJ, Henry B (1997) Tectonic applications of magnetic susceptibility and its anisotropy. Ear Sci Rev 42:49–93
19. Sen K, Mamtani MA (2006) Magnetic fabric, shape preferred orientation and regional strain in granitic rocks. J Strucl Geo 28:1870–1882
20. Borradaile GJ, Jackson MJ (2010) Structural geology, petrofabrics and magnetic fabrics (AMS, AARM, AIRM). J Struc Geo 32:1519–1551
21. Archanjo C, Launeau P, Bouchez JL (1995) Magnetic fabrics vs. magnetite and biotite shape fabrics of the magnetite-bearing granite pluton of Gameleiras (Northeast Brazil). Phy Ear Planet Inter 89:63–75
22. Nké BEB, Njanko T, Mamtani MA, Njonfang E, Rochette P (2018) Kinematic evolution of the Mbakop Pan-African granitoids (Western Cameroon-Domain): an integrated AMS and EBSD approach. J Struc Geo 111:42–63
23. Mamtani MA, Piazolo S, Kontny A, Greiling RO, Hrouda F (2011) Process of magnetite fabric development during granite deformation. Ear Planet Sci Lett 308:77–89
24. Halfpenny A, Hough RM, Verrall M (2013) Preparation of samples with both hard and soft phases for electron backscatter diffraction: examples from gold mineralization. Microsc Microan 19:1007–1018
25. Mamtani MA, Reznik B, Kontny A (2020) Intracrystalline deformation and nanotectonic processes in magnetite from a naturally deformed rock. J Struc Geo 135:104045
26. Dey A, Sen K, Mamtani MA (2022) Electron Back Scatter Diffraction study of ultra-high pressure Tso Morari eclogites (trans-Himalayan collisional zone):implications for strain regime transition from constrictional to plane strain during exhumation. Lithosphere 2022:7256746
27. Schmid SM, Casey M (1986) Complete fabric analysis of some commonly observed quartz [c]-axis patterns. In: Hobbs BE, Heard HC (eds) Mineral and rock deformation: laboratory studies. Geo Monogr 36:263–286
28. Stipp M, Stünitz H, Heilbronner R, Schmid SM (2002) The eastern Tonalite fault zone: a natural laboratory for crystal plastic deformation of quartz over a temperature range from 250–700°C. J Struc Geo 24:1861–1884
29. Kile DE (2009) The Universal Stage: The past, present, and future of a mineralogical research instrument, Geochel News, 140

30. Mondal TK, Mamtani MA (2016) Palaeostress analysis of normal faults in granite; implications for interpreting riedel shearing related to regional deformation. J Geolo Soc London 173:216–227
31. Jolly RJH, Sanderson DJ (1997) A Mohr circle reconstruction for the opening of a pre-existing fracture. J Struc Geo 19:887–892
32. Mondal TK, Mamtani MA (2013) 3-D Mohr circle construction using vein orientation data from Gadag (southern India)—implications to recognize fluid pressure fluctuation. J Struc Geo 56:45–56
33. Lahiri S, Mamtani MA (2016) Scaling the 3-D Mohr circle and quantification of paleostress during fluid pressure fluctuation—application to understand gold mineralization in quartz veins of Gadag (southern India). J Struc Geo 88:63–72
34. Lahiri S, Rana V, Bhatt S, Mamtani MA (2020) Paleostress and statistical analysis using quartz veins from mineralized and non-mineralized zones: application for exploration targeting. J Struc Geo 133:104006
35. Ramsay JG, Huber MI (1983) The techniques of modern structural geology, 1: Strain analysis. Academic Press, London
36. Xypolias P (2010) Vorticity analysis in shear zones: a review of methods and applications. J Struc Geo 32:2072–2092
37. Kruhl JH (1998) Reply: prism- and basal-plane parallel subgrain boundaries inquartz: a microstructural geothermobarometer. J Metamor Geo 16:142–146
38. Law RD (2014) Deformation thermometry based on quartz *c*-axis fabrics and recrystallization microstructures: a review. J Struc Geo 66:129–161
39. Bestmann M, Prior DJ, Grasemann B (2006) Characterisation of deformation and flow mechanics around porphyroclasts in a calcite marble ultramylonite by means of EBSD analysis. Tectonophy 413:185–200
40. McLaren S, Redd SM (2008) Automated mapping of K-feldspar by electron backscatter diffraction and application to 40Ar/39Ar dating. J Struct Geol 30:1229–1241
41. Bestmann M, Habler G, Heidelbach F, Thöni M (2008) Dynamic recrystallization of garnet and related diffusion processes. J Struc Geo 30:777–790
42. Hildyard RC, Prior DJ, Mariani E, Faulkner DR (2009) Crystallographic preferred orientation (CPO) of gypsum measured by electron backscatter diffraction (EBSD). J Micro 236:159–164
43. Dempsey ED, Prior DJ, Mariani E, Toy VG, Tatham DJ (2011) Mica-controlled anisotropy within mid-to-upper crustal mylonites: an EBSD study of mica fabrics in the Alpine Fault Zone, New Zealand. Geo Soc London, Special Publications, 360.33–47
44. Wallis D, Hansen LN, Britton TB, Wilkinson AJ (2017) Dislocation interactions in olivine revealed by HR-EBSD. J Geophy Res Sol Ear 122:7659–7678
45. Aspiroz MD, Lloyd GE, Fernández C (2007) Development of lattice preferred orientation in clinoamphiboles deformed under low-pressure metamorphic conditions, a SEM/EBSD study of metabasites from the Aracena metamorphic belt (SW Spain). J Struc Geo 29:629–645
46. Elyaszadeh R, Prior DJ, Sarkarinejad K, Mansouri H (2018) Different slip systems controlling crystallographic preferred orientation and intracrystalline deformation of amphibole in mylonites from the Neyriz mantle diapir, Iran. J Struc Geo 107:38–52
47. Giannuzzi LA, Stevie FA (2005) Introduction to focused ion beams: instrumentation, theory, techniques, and practice. Springer, New York
48. Mamtani MA, Kontny A, Hilgers C, Reznik B, Wenzel O, Müller E, Störmer H, Renjith AR (2022) Nanostructures in oriented thin films—From Microtectonics to nanotectonics and nanokinematics in deformed rocks. Abstract: deformation mechanisms, rheology and tectonics (DRT) conference, Catania (Italy). Società Geologica Italiana, Roma 2022, p 18. https://doi.org/10.3301/ABSGI.2022.01%20
49. Li C, Habler G, Baldwin LC, Abart R (2018) An improved FIB sample preparation technique for site-specific plan-view specimens: a new cutting geometry. Ultramicro 184:310–317
50. Renjith AR (2018) Fabric analysis of quartzites with low magnetic susceptibility-evaluating the relationship between magnetic susceptibility, shape preferred orientation and crystallographic preferred orientation (unpublished PhD thesis, IIT Kharagpur)

51. Mamtani MA, Wenzel O, Kontny A, Hilgers C, Müller E, Renjith AR, Llorens M-G, Gomez-Rivas E (2023) "In-plane" site-specific FIB lamella extraction from deformed magnetite and the investigation of low angle grain boundaries under TEM. J Struct Geol 174:104937. https://doi.org/10.1016/j.jsg.2023.104937

MIX
Papier aus verantwortungsvollen Quellen
Paper from responsible sources
FSC® C105338

If you have any concerns about our products,
you can contact us on
ProductSafety@springernature.com

In case Publisher is established outside the EU,
the EU authorized representative is:
Springer Nature Customer Service Center GmbH
Europaplatz 3, 69115 Heidelberg, Germany

Printed by Libri Plureos GmbH
in Hamburg, Germany